RETURN TO FLIGHT

Inside NASA's Space Shuttle Missions in
the Wake of the Columbia Disaster

Dr. James F. Peters

Copyright © 2015 Dr. James F. Peters
All rights reserved.

ISBN: 1503227332
ISBN 13: 9781503227330

DEDICATION

This book is dedicated to the men and women lost in the agency's space exploration program. We honor them by celebrating their lives, their bravery and advancements in human spaceflight.

CONTENTS

Preface vii
List of Abbreviations ix
Chapter 1: The *Columbia* Is Lost 1
Chapter 2: Picking Up the Pieces 14
Chapter 3: *Columbia* Accident Investigation Board .. 34
Chapter 4: The Amazing Shuttle Program 57
Chapter 5: External Tank Overview 78
Chapter 6: Orbiter: The Debris Target 88
Chapter 7: Hooked on the Space Program 97
Chapter 8: Prelude to Disaster 112
Chapter 9: Debris Analysis 101 124
Chapter 10: Debris Risk Assessments 136
Chapter 11: Failure to Meet Debris Requirements ... 155
Chapter 12: Dream Job 172
Chapter 13: STS-114 Flight Preparations 193
Chapter 14: STS-114 Go for Launch 210
Chapter 15: NASA Dodges a Bullet 220
Chapter 16: Saving the Shuttle Program 241
Chapter 17: John Muratore Is Replaced 258
Chapter 18: Return to Flight Part II 272
Chapter 19: STS-121 Flight Readiness Review 285
Chapter 20: Debris Integration Group 302

Chapter 21: Night Launch Debris Surprise 313
Chapter 22: 2007—Rough Start for NASA 333
Chapter 23: Loss of Space Station Attitude Control.. 347
Chapter 24: Prepare to Launch the Rescue Mission.. 362
Chapter 25: "Tankenstein" 377
Chapter 26: We Could Be Wrong.................... 389
Chapter 27: Debris Summit........................ 405
Chapter 28: The Perfect Debris Storm 422
Chapter 29: End of an Era for Human Spaceflight ...442

PREFACE

Return to Flight is about the passion and dedication of the men and women who supported this nation's human spaceflight program and strove to make the next shuttle mission the safest ever flown in the wake of the space shuttle *Columbia* disaster. I chaired the Debris Integration Group responsible for correcting the debris problems that led to the *Columbia* accident on STS-107 and returning the shuttle fleet to operational flight status. After the first Return to Flight mission (STS-114), I led a large team of NASA engineers and contractors dedicated to identifying and solving debris issues. No one wanted the job, but I came to love it. The job delved into the abstract world of probabilistic risk assessment and advanced analytic modeling to determine the "safety of flight" and extent to which the well-being of the astronaut crew and space shuttle itself would be jeopardized. The very existence of the shuttle program hung in the balance. The debris team performed this difficult task and when asked did the impossible. After the *Columbia* accident in February 2003, NASA identified debris as the highest risk to the Space Shuttle Program. It remained the number one safety issue for the next eleven missions. The debris team performed risk assessments on more than five hundred potential debris sources, identified the highest debris risk areas, and implemented corrective measures on the orbiter, external tank, solid rocket booster, and launchpad to minimize or eliminate those risks. Each mission posed unique debris surprises, and the only thing the team could count on was the unexpected. Despite the numerous debris challenges—not to mention intense pressure and criticism from both inside and outside the agency—the team remained focused and steadfast in its efforts to work through the issues and optimize the shuttle program's safety.

This book tells the story of the debris challenges during NASA's Return to Flight (RTF) missions. My vantage point might differ from others who have reported the story because many versions of the events exist. I relied primarily on the words of those who lived it, audio recordings, written logs, and other documents in an effort to present the most accurate content and fullest historical context. I condensed some descriptions for clarity and focused on the key characters for brevity. Though I often had to choose between conflicting viewpoints and sources, I tried to balance the story by interviewing those involved. In some cases I modified their exact words in order to provide the reader context, but I did not knowingly violate the story line. The interpretation of events and actions of the key characters as well as any errors that might be present are my own.

Dr. James F. Peters

List of Abbreviations

AEDC	Arnold Engineering Development Center
AFRSI	advanced flexible reusable surface insulation
APU	auxiliary power unit
ASTT	aerodynamically sensitive transport time
BSM	booster separation motor
BTA	booster trowelable ablative
CAIB	*Columbia* Accident Investigation Board
CEIT	crew element integration testing
CFD	computational fluid dynamics
CMG	control moment gyroscope
CoFR	certification of flight readiness
CSCS	contingency shuttle crew support
DAT	damage assessment team
DTA	debris transport analysis
DVR	debris verification review
ECLSS	Environment Control and Life-Support Systems
EI	entry interface
ELVIS	Enhanced Launch Vehicle Imaging System
EVA	extravehicular activity
FIT	final inspection team
FMEA	failure modes and effects analysis
FOD	foreign object debris

FOLD	Foam Observed Loss Database
FRCI	fibrous refractory composite insulation
GNC	guidance, navigation, and control
HAC	heading alignment cylinder
HRSI	high-temperature reusable surface insulation
HST	Hubble Space Telescope
IDBR	integrated debris hazard report
IFA	in-flight anomaly
IFR	ice/frost ramp
IIFA	integrated in-flight anomaly
IPR	integrated problem report
ISERP	Integrated Safety and Engineering Review Panel
JSC	Johnson Space Center
KSC	Kennedy Space Center
LCC	launch commit criteria
LEFM	linear elastic fracture mechanics
LH2	liquid hydrogen
LON	launch on need
LO2	liquid oxygen
LRSI	low-temperature reusable surface insulation
MAF	Michoud Assembly Facility
MCC	Mission Control Center
MECO	main engine cutoff
MER	mission evaluation room
MET	mission elapsed time
MMT	Mission Management Team
MPS	main propulsion system
MSFC	Marshall Space Flight Center
NACA	National Advisory Committee for Aeronautics
NDE	nondestructive evaluation
NDR	NASA debris radar
NESC	NASA Engineering Safety Center
NIRD	NASA imagery reporting database
NUBs	nonuseful bodies
OBSS	orbiter boom sensor system

LIST OF ABBREVIATIONS

OMS	orbital maneuvering system
OPF	orbiter processing facility
OSB-II	Operations Support Building II
PAL	protuberance air load
PMA	portable mating adapter
PRA	probabilistic risk assessment
PRACA	problem reporting and corrective action
PRCB	Program Requirements Control Board
PSEI	propulsion systems engineering and integration
PWT	progressive wave tube
RCC	reinforced carbon-carbon
RCS	reaction control system
RMS	remote manipulator system
RPM	rendezvous pitch maneuver
RSRM	reusable solid rocket motor
RTF	Return to Flight
SEI	systems engineering and integration
SFOC	Space Flight Operations Contract
SICB	Systems Integration Control Board
SI CDH	Science Instrument Command and Data Handling System
SILC	shuttle ice liberation coating
SLA	super lightweight ablator
SLS	space and life sciences
SOMD	space operations mission director
SSME	space shuttle main engine
TAEM	terminal area energy management
TCS	thermal control systems
TPS	thermal protection system
USA	United Space Alliance
VAB	Vehicle Assembly Building
WLEIDS	Wing Leading Edge Impact Detection System

CHAPTER 1

The *Columbia* Is Lost

Mission Control
Johnson Space Center, Houston, Texas

"Lock the doors," Leroy Cain ordered calmly. The young flight director in mission control at the Johnson Space Center (JSC) was in charge of *Columbia*'s reentry after its sixteen-day science mission. The order was given shortly after the space shuttle was scheduled to land at Kennedy Space Center on February 1, 2003, at 8:16 a.m. (CST). This signaled the start of a NASA contingency action plan in response to a Space Shuttle Program accident. Several minutes had passed since communication and telemetry from *Columbia* were lost and mission control had issued the contingency action plan order. Rodney Rocha, the lead engineer for the orbiter, was worried. After observing a massive piece of foam strike the orbiter during ascent the week before, he had begged for additional imagery data but was denied. Now the tire pressure alerts and the flashing left main landing gear indicator, which had transitioned to an unknown state, stumped him. If the foam strike had damaged the triple seals around the landing gear doors, hot gases would have entered the tire well and caused the tire to deflate and damage the landing gear. His greatest concern prior to the emergency declaration was attempting a landing with defective landing gear and a flat tire.

Moments before Leroy's order was given, Joyce Grush, one of Rodney's colleagues, approached Rodney with tears in her eyes. She wept as she informed Rodney communications with *Columbia* had been lost. Leroy's order stunned Rodney and most of the flight controllers, who now realized *Columbia* and its crew were in serious jeopardy. It slowly dawned on Rodney that the repercussions of the debris strike to *Columbia*'s left wing might be far worse than a damaged tire and landing gear. Rodney knew it would be impossible for the orbiter's structure to withstand the reentry heat if the debris strike had compromised the vehicle's thermal protection. It was a real possibility that the debris strike observed during ascent was somehow responsible for the loss of communication with *Columbia* and its crew. Amid the sounds of whirling computer fans and quiet sobbing in the control room, the flight controllers quietly executed the JSC contingency plan. The plan had been established after the *Challenger* accident in 1986, but no one ever expected to use it. Once Leroy declared a contingency situation had occurred, the flight control team refocused their attention from the *Columbia* landing to the necessary operations to determine the magnitude of the problem, carry out search and rescue, and maintain public safety. Leroy's job was to notify his chain of command up to the NASA administrator, who would then notify the president of the United States. Local law enforcement would also need to be alerted in the area of lost communications, and military search and rescue assets would need to be mobilized. Other NASA employees and the general public were quickly becoming aware of the dire situation by viewing the Associated Press video footage of *Columbia*'s breakup on all the local and major news channels. The flight control team activated physical and information-related security measures and continued to collect, preserve, and impound all data and information concerning the flight.

By now the flight control team had seen the fiery streaks of *Columbia* breaking apart. They transitioned to the duties of a mishap-response team and shifted their attention to recovering the crew. The NASA administrator, Sean O'Keefe, activated the International Space Station and Space Shuttle Mishap Interagency Investigation Board about an hour after the contingency action plan was initiated, and O'Keefe named Admiral

Harold W. Gehman Jr. (US Navy, retired) as its chair. Senior leaders at NASA headquarters in Washington, DC, met as part of the Headquarters Contingency Action Team, and they quickly notified astronauts' families, the president, and members of Congress. President Bush telephoned Israeli Prime Minister Ariel Sharon to inform him of the loss of *Columbia*'s crew, which included Israeli astronaut Ilan Ramon. Several hours later President Bush addressed the nation with a short and somber declaration: "The *Columbia* is lost. There are no survivors."

Entry flight director Leroy Cain held his hand over his face in disbelief as workers in mission control waited for a signal that never came from the *Columbia* crew minutes before the scheduled landing on February 1, 2003. (Courtesy of NASA.)

Preparing for Reentry—Aboard *Columbia*

After successfully completing a sixteen-day Spacehab microgravity research mission, the seven crewmembers of *Columbia* were anxious to return home. The Spacehab was the pressurized structure mounted in the shuttle cargo bay that provided the space needed for the crew to conduct their microgravity experiments. The STS-107 crew included Commander Rick Husband, Pilot William "Willie" McCool, Payload Commander Michael Anderson, Mission Specialists Kalpana Chawla, David Brown, and Laurel Clark, and Payload Specialist Ilan Ramon—the first Israeli astronaut. At 1:30 a.m. (CST) on February 1, 2003, the entry flight control team began duty in the mission control center (MCC). The team was not working on any issues or problems related to the planned deorbit and reentry of *Columbia*. Although the crew had been informed about the debris impact to the left wing during ascent, the team indicated no concerns and treated the reentry like any other. The team worked through the deorbit preparation checklist and reentry checklist procedures. Weather forecasters, with the help of pilots in the shuttle training aircraft, evaluated landing site weather conditions at the Kennedy Space Center (KSC). At the time of the deorbit decision (about twenty minutes before the initiation of the deorbit burn), all weather observations and forecasts were within the guidelines set by the established flight rules. All systems were normal.

Entry is divided into three major phases based on the unique aerodynamic, thermal dynamic, and software requirements needed to maintain vehicle control. The initial entry phase begins at the deorbit burn and ends approximately five minutes later at the entry interface (EI) point, which occurs at an altitude of 75 miles. When the deorbit burn is initiated, the orbiter is still in low Earth orbit at an altitude of about 100 miles. It is traveling at 17,500 miles per hour and is approximately 5,000 miles from the landing site. During this phase the primary objective is to dissipate the tremendous amount of energy the orbiter possesses when it enters the atmosphere. This prevents it from burning up (from too steep an entry angle) or skipping out of the atmosphere (from too shallow an entry angle) and helps it stay within its structural limits. The orbiter uses small jets as part of the reaction control system to maneuver because the aerodynamic forces the vehicle generates are too small to be effective at such a high altitude. The second phase extends from EI to the terminal area energy management (TAEM) interface. The altitude of the vehicle is approximately 83,000 feet. Its velocity is 1,700 miles per hour, and it's at a distance of fifty-two miles from the runway. During TAEM the goal is to manage the orbiter's energy while traveling along the imaginary heading alignment cylinder (HAC), which lines up the vehicle on the runway centerline. As the vehicle descends into the atmosphere during this phase, the aerosurfaces become active. When the orbiter is in atmospheric flight, it is flown by varying the aerodynamic forces it generates while moving through the atmosphere—as with any other plane. TAEM extends to the approach and landing capture zone, which is the point when the orbiter is on a glide slope and heading toward the runway centerline with the proper airspeed. This is typically below 10,000 feet in altitude. The third and final entry phase is the approach and landing phase. This ends when the wheels come to a full stop on the runway.

Go for Reentry

Shortly after 7:00 a.m. (CST), Leroy polled the mission control room for a go-no-go decision of the deorbit burn, and at 7:10 a.m., the crew was notified they were go for deorbit burn. As the orbiter flew upside down and tailfirst over the Indian Ocean at an altitude of 175 miles, Commander Rick Husband and Pilot Willie McCool executed the deorbit burn using *Columbia*'s two orbital maneuvering system engines. The deorbit burn slowed the shuttle from its orbital velocity of 17,500 miles per hour to begin its reentry into the atmosphere. During the deorbit burn, the crew felt about 10 percent of the effects of gravity, and no problems were reported during the burn. Everything was going as planned. At 7:44:09 a.m. (CST), *Columbia* reached discernible atmosphere and EI over the Pacific Ocean. As *Columbia* descended from space into the atmosphere, the heat produced by air molecules colliding with the wing's leading edge caused temperatures to rise steadily. They reached an estimated 2,500°F over the next six minutes. Those on the flight deck—Commander Husband, Pilot McCool, and

Mission Specialists Clark and Chawla—were now starting to see bright flashes envelop the orbiter as superheated air molecules discharged light (a normal phenomenon). At 7:49:32 a.m. *Columbia* was traveling at approximately Mach 24.5.[1] *Columbia* executed a roll to the right and began a preplanned banking turn to manage lift and limit the orbiter's rate of descent and heating. At 7:50:53 a.m. they were traveling at Mach 24.1, and the shuttle entered a ten-minute period of peak heating. During this time the thermal stresses were at their maximum. By 7:52:00 a.m., nearly eight minutes after entering the atmosphere and some three hundred miles west of the California coastline, the wing leading edge temperatures reached 2,650°F. While traveling at Mach 23 one minute later, the orbiter's wing leading edge was heated to above 2,800°F. Seconds after *Columbia* crossed from California into Nevada airspace, it traveled at Mach 22.5 and an altitude of 43 miles. Observers on the ground witnessed a sudden brightening that caused a noticeable streak in the orbiter's luminescent trail. Unbeknownst to the crew eighteen similar events occurred during the next four minutes as *Columbia* streaked over Utah, Arizona, New Mexico, and Texas. At 7:58:20 a.m. they traveled at 39 miles and Mach 19.5. *Columbia* crossed from New Mexico into Texas, and about this time it shed a thermal protection system tile. This was the westernmost piece of debris to eventually be recovered.

Columbia's crew received the first indication of a problem at 7:58:39 a.m. (CST). This was almost fifteen minutes after reentering the atmosphere. The first of four fault messages flashed on the onboard backup flight software monitor. An audible alarm accompanied this. The fault message indicated a loss of pressure on the left main landing gear tire. Rodney and the flight control team in mission control also observed these indications. Rick and Willie immediately called up the fault page to review the information. One of the failure scenarios they practiced during training was a circuit breaker trip that resulted in half the tire pressure sensors being disabled. A circuit breaker trip would

---

[1] Mach is a dimensionless speed ratio whereby the velocity of an object is divided by the speed of sound. At Mach twenty-five an object is traveling at twenty-five times the speed of sound or over fifteen times faster than the speed of a rifle bullet.

have disabled some sensors for all the tires (left main gear, right main gear, and nose gear), but this failure signature was different. Only the tire pressure sensors on the left main gear were affected. At 7:58:48 Rick began a call to mission control, but the call broke off and was not repeated. Brief communication interruptions often occur due to the tracking and data relay satellite antenna angles changing relative to the orbiter's transceivers, so this specific communication dropout was not particularly surprising. At 7:59:06, ten seconds after the fourth of four tire pressure fault messages, telemetry indicated the left main gear down lock sensor transferred to on. According to this sensor, the landing gear had deployed. However, other sensors indicated the landing gear door was still closed and the landing gear was locked in the up and stowed position.

Rick and Willie were trying to resolve the conflicting signals, but the left landing gear position indicator displayed a "barber pole"—an indeterminate position. Post-accident data analysis and recovered debris indicated the left landing gear was indeed locked in the up position and the landing gear door was closed. The signal indicating the gear was down was a false signal likely triggered by damage to the sensor system. Rick and Willie continued their attempts to diagnose the situation. The problem involved the same landing gear as the tire pressure messages, and this indicated a potential landing gear deployment problem. Twenty seconds later the crew was beginning to feel a slight turn to the left as the orbiter yaw and roll rates exceeded the ability of the aileron trim to compensate for the changing drag of the now-deformed left wing. One second later the reaction control system jets activated to try to compensate for changing drag and inability of the aerosurfaces to control the vehicle. Although reaction control system jets typically pulse throughout entry to adjust the orbiter's flight path, the jets were now firing continuously as the orbiter attempted to counteract the increased left wing drag and resulting yaw motion. A small light on a panel in front of the commander was illuminated as the jets continued to fire. At 7:59:32 a.m., mission control informed the crew pressure readings had been lost on both left main landing gear tires. They told the crew they were evaluating the indications, and they added that the flight control team did not understand the crew's last transmission. Rick acknowledged the call from mission control but was interrupted midsentence. ("Roger, uh...") It was the last communication from the *Columbia* crew. This last telemetry signal received in mission control marked the cessation of all audio and real-time data to the MCC from the shuttle. Videos made by observers on the ground at 8:00:18 a.m. revealed the orbiter was disintegrating. A communication dropout lasting a few seconds was expected as the orbiter switched from one satellite to another, but MCC personnel had begun to recognize problems were occurring with *Columbia*. The telemetry signatures, however, were such that they were unable to complete analysis of the wide-ranging and seemingly unrelated problems before contact was lost. There were no indications either to the crew or mission control that the loss of audio communications and real-time data were more than a brief condition. As far as those in mission control could tell, *Columbia* only had a potential issue with landing gear deployment. This was a nontrivial event, but the crew had time to troubleshoot the problem. Changing drag on the left wing was just beginning to develop into a potentially recognizable problem. By this time the ill-fated crew probably already recognized the severity of the situation.

Columbia Out of Control

Mission Specialist Michael Anderson was responsible for completing post deorbit burn tasks. Since Michael was last seated, the rapid maneuvers of the orbiter as it began to tumble out of control likely caught him off guard. Rick and Willie probably tried in vain to regain control of the vehicle. Everyone on the flight deck could probably see the lighting and horizon changes through the windows and the changes on the vehicle attitude displays. The forces the crew experienced were increasing dramatically and began to differ from the nominal, expected accelerations. *Columbia* was now in a highly oscillatory thirty to forty-degree-per-second flat spin. The orbiter's belly generally faced into the velocity vector, which was still nearly parallel to the ground. The vehicle was moving along its trajectory in excess of Mach fifteen. Varying g-forces (peaking at three g's) were tossing the crewmembers around and reaching a level that likely was inducing nausea, dizziness, and disorientation in some. It is highly likely Michael was desperately wrestling with his seat restraints and had not yet donned his helmet when the vehicle started to spin. As alarms began ringing on the flight deck, Rick and Willie braced themselves, and they attempted to diagnose and correct the orbiter systems. Forty-three seconds after the communications and telemetry were lost and the vehicle began spinning out of control and shedding debris, Willie managed to return the orbiter to autopilot mode in a last-ditch effort to regain control. He and Rick must have noticed all three hydraulic systems had zero pressure and zero quantities in the reservoirs because Willie tried to restart the pumps. Without hydraulic pressure and with the vehicle out of control, Willie tried to restart two of the three auxiliary power units (APUs) and restore hydraulic pressure. His years of pilot experience and training enabled him to perform the necessary steps to initiate a restart of the APUs and activate switches for two of the three hydraulic circulation pumps. Turning on the hydraulic circulation pump was not on the emergency checklist. Nonetheless it provided some limited hydraulic pressure and revealed good systems knowledge by Willie and Rick as they worked to attempt to restore orbiter control.

Columbia continued to shed debris, and ground-based video from 8:00:09 to 8:00:18 (CST) shows a thin, relatively consistent trail. This suggested the conditions remained steady for a short period of time. *Columbia*'s crew was experiencing increasing g-forces generated from the increasing aerodynamic forces and growing stresses on the orbiter. At 8:00:18 a.m. (forty-six seconds after the loss of communication with mission control) ground video shows a significant and catastrophic event occurred to the orbiter. The GPS miniaturized airborne global receiver experiment, which was located in the middeck and powered by a fuel cell in the payload bay, lost power when the orbiter began to break up. Less than a second later, the modular auxiliary data system recorder, which was also located in the crew module and powered from the payload bay, experienced a total power loss. The aerodynamic loads and thermal degradation of the orbiter structure caused the forward and mid-body orbiter segments to separate and breach the pressurized cabin. That event was the start of a period of several seconds where the orbiter underwent a major structural breakup, and there were visual indications the orbiter was in multiple pieces. *Columbia*'s cabin lights and displays suddenly went dark. Once the forward fuselage began to break away, the exposed crew module rapidly disintegrated from the combined effects of the high g-loads, aerodynamic forces, and thermal loads. The crew module instantly depressurized, rendered the crew unconscious, and exposed them to temperatures hotter than a blast furnace.

The effects of cabin depressurization on the crew would have depended on the rate of depressurization. Human beings are able to live and function within the "physiological zone," which is from zero to ten thousand feet in altitude. From ten thousand to fifty thousand feet (the "physiologically deficient zone"), oxygen is required to protect humans against hypoxia. Above fifty thousand feet (the "space equivalent zone"), protecting an individual from the lack of atmospheric pressure requires a sealed cabin or pressure suit. Additional physiological problems occur within the space equivalent zone above 63,500 feet. This includes boiling of body fluids (Armstrong's limit) in an unprotected individual. Pressure suits allow astronauts and military pilots to operate in the space equivalent zone, and several historically relevant cases have proven the need for suits. There have been significant events of lethal potential where a suit alone or in combination with other survival methods or vehicle design prevented or might have prevented injury or loss of life. This includes test pilot Bill Weaver's survival of the breakup of his SR-71 at 78,000 feet at Mach 3 on January 25, 1966. He credited his survival to the pressurized suit because it provided oxygen, hypobaric protection, and physical protection from aerodynamic forces. During reentry on June 29, 1971, *Soyuz 11* suffered a pressure equalization valve failure during module separation, and the crew was unable to isolate the leak before losing cabin pressure. Crewmembers were not in suits, and all three perished. Soviet Sokol suits were consequently added to the Soyuz system. During the Apollo–Soyuz mission on July 24, 1975, a valve configuration error resulted in a nitrogen tetroxide leak into the cabin during final descent. The crew elected to not wear pressure suits, and all were hospitalized with chemical-induced potentially fatal pneumonia. The final example involved the space shuttle *Challenger* accident on January 28, 1986. NASA determined that with a bailout system, some portion of the crew might have survived had they been wearing pressure suits. Accordingly NASA once again made wearing suits during launch and landing of the space shuttle a requirement. Over time wearing suits on the shuttle during reentry became optional due to the routine nature of this flight phase and general discomfort of wearing suits. The existing escape equipment on *Columbia* was capable of protecting the crew from rapid decompression via pressure suits, helmets, and either the orbiter oxygen or an individual emergency oxygen system for a limited time. However, the depressurization on *Columbia* was so rapid and the breakup forces so severe the crew was immediately incapacitated and would not regain consciousness as the crew module disintegrated and crashed to the ground in fiery fragments.

Kennedy Space Center—Landing Field
February 1, 2003, approximately 9:05 a.m. (EST)

The normally isolated airfield at KSC was buzzing with the family and friends of the *Columbia* astronauts and a few thousand spectators eagerly awaiting the shuttle return and landing. Family members holding flowers, friends, and VIPs were positioned in the front and center of the tightly controlled airfield. White posts roped the area off, and chairs littered the space. This would be the sixty-second time a shuttle mission landed at Kennedy. Landing sites at Edwards Air Force Base in California and White Sands in New Mexico served as alternates if the weather minimums at Kennedy were unacceptable. The weather, however, was clear and crisp. A cool sea breeze mixed with a subtle hint of jet fumes from NASA's T-38 chase planes and blew across the field. The atmosphere surrounding the landing was much different than for the launch. The crowds for launch day were much larger (approaching half a million people or more), and there was always the possibility the launch would be scrubbed for weather or mechanical reasons. Historically this happened more than 50 percent of the time, and it always left the crew, flight team, and crowd disappointed. On the other hand, landing is a certainty after the deorbit burn is executed, and the only thing left to chance is the weather and landing location. The spectators at the airfield are limited to mostly NASA employees and contractors, but there are some space enthusiasts that line the roads and beaches surrounding KSC and wait to experience the signature double sonic boom of the approaching shuttle and get a glimpse of the magnificent machine during its final flight phase. Nighttime landings are even more spectacular. The heat generated during reentry produces a phosphorescent glow and eerie trace across the sky as the shuttle maneuvers in the landing pattern.

Strategically positioned along the runway were approximately 160 space shuttle launch operations team members and their collective gear of fire engines, transport vehicles, and emergency equipment. Their job after the shuttle landed was to place the spacecraft in a safe configuration, remove the crewmembers and time-critical flight equipment, and tow the vehicle to the orbiter processing facility (OPF). Once inside the hangar, the vehicle would be completely inspected and prepared for its next flight. Several ground team members would wear self-contained atmospheric protective ensemble suits to protect them from toxic chemicals as they approached the spacecraft when it stopped rolling. Other ground team members would take sensor measurements to ensure the atmosphere in the vicinity of the spacecraft was not explosive. In the event of propellant leaks, a wind machine truck carrying a large fan would be moved into the area to create a turbulent airflow that would break up gas concentrations and reduce the potential for an explosion. It would take nearly an hour to perform the post-landing procedures and remove the crew. When the spacecraft landed at Edwards or White Sands, the same procedures and ground support equipment were used, but the orbiter would be mounted on the back of a modified 747 aircraft and ferried back to KSC. Then it would be removed and taxied to the OPF.

The shuttle was scheduled to land in approximately ten minutes when the alarming sound of multiple cell phone ringers and BlackBerry buzzers broke the excitement surrounding the landing and shattered the relief that another successful mission was coming to an end. As news of the shuttle's demise poured in, the airfield crowd quickly became quiet and stunned. Many broke down into tears. There would be no loud sonic booms or joyous celebration—just the sounds of sobbing and whistling wind. The human brain stores emotionally positive or negative memories in a particularly robust way. Many can easily recall where they were and what they were doing when President Kennedy was shot, *Challenger* exploded seventy-two seconds into flight, or the tragic events of 9/11 unfolded. The stunned crowd was shrouded in disbelief but understood another monumental tragedy had struck America's space program. The landing clock had no mercy as seconds and minutes clicked past the projected landing time. A cool early morning sea breeze was all that could be heard and felt as the families awaited the arrival of their loved ones. The lonely feeling of helplessness slowly succumbed to disbelief and shock. Before the shocked people dispersed and left to deal with their own emotions, the members of the astronauts' families were quietly ushered away. The flowers the families had once held now lay discarded on the airfield grass. *Columbia* would not be coming home.

CHAPTER 2

Picking Up the Pieces

**Reagan Airport, Washington, DC
February 1, 2003**

I recall watching the Apollo 11 moon landing on a black-and-white television in 1969, watching the *Challenger* explosion while attending graduate school at the University of Maryland in 1986, and hearing about the 9/11 terrorist attacks in 2001 while attending a conference in Dallas. It was easy to recall these significant events and the news about the *Columbia* disaster was no different. I was completing a Homeland Security assignment in Washington, DC, and was waiting at Ronald Reagan Washington National Airport for my flight home when news of the *Columbia* accident scrolled across a grainy airport terminal TV screen during a special CNN news report.

In the summer of 2002, the newly formed Transportation Security Administration tasked the Boeing Company with installing explosive detection equipment in all major US airports by the congressional deadline of December 31, 2002. I had just finished working my last assembly mission on the International Space Station and offered to help the Boeing team in Washington, DC, organize the effort. This involved major systems integration challenges similar to work I had performed on the station. The two weeks I planned to work in DC turned into a six-month effort. It was a seven-days-a-week job, and sixteen-hour days were not uncommon. I helped the Boeing team meet the congressional deadline with trained personnel and explosive detection equipment in place. Although the DC job was rewarding, my passion was human spaceflight, and I was eager to return home to Houston and start a new job as the systems engineering and integration manager on the space shuttle. While waiting in Reagan Airport for my flight home, I stood mesmerized in front of an airport television and watched the video footage of *Columbia* disintegrating during reentry. The final boarding call was announced. My wife and other friends called with the news of what I had just witnessed, and my BlackBerry was buzzing continually with more news about the accident as I boarded the plane. The flight was a blur. I sat deep in thought for three and a half hours, reflected on the calamity, and wondered if there could possibly be any survivors.

Houston, Texas
Boeing Building—Space Shuttle Program

After arriving in Houston, I drove straight to the Boeing Space Shuttle building to get more details about the accident and offer assistance with the rescue and recovery effort. The Boeing building was nestled on the Clearlake shoreline across from JSC. News crews now inundated the area. A large crowd had formed outside the space center's main entrance. I made my way to the main conference room, and I met about fifty shuttle engineers and managers wearing somber faces. Several people were sobbing and shaking their heads in disbelief. Even though it was a Saturday, most of the shuttle team was there either in the main conference room or in small huddles throughout the building. Even then the normally noisy hallways and offices were quiet. Any information was sketchy at best, and no one knew whether there would be any survivors or even where to start looking. Neither the cause of the accident nor the extent of the wreckage was known. The only known facts were that communications had been lost with the crew, the vehicle had disintegrated just south of Dallas, Texas, and instrumentation inside the left landing gear well under the left wing had indicated temperatures were elevated well above normal. Telemetry from *Columbia* showed the flight computers tried in vain to maintain control of the vehicle before the aerodynamic and thermal forces overwhelmed *Columbia* and caused the spacecraft to abruptly break apart. Nobody wanted to leave despite the late hour and the empty feelings of helplessness and diminishing hope crewmembers would be found alive. In the darkness of the evening, the news crews, helicopters, and large crowds were still swarming around the JSC entrance as I quietly drove home. It was an inconceivable way to start work on the shuttle program.

The large crowds that gathered at the entrance to the Johnson Space Center placed flowers and memorial gifts to honor the fallen *Columbia* crew. (Courtesy of NASA.)

Columbia disintegrated over remote parts of Texas starting about twenty-five miles south of Dallas. The debris path stretched for more than two hundred miles and was nearly ten miles wide. Local and federal emergency response teams were called out as standard procedure in an emergency event. Initially the biggest challenge was determining where to look for potential survivors in such a large search area. NASA barely had time to respond with their teams from JSC and Stennis Space Center before scavengers were selling purported shuttle wreckage on eBay. To prevent the plundering of hardware, federal agents and Texas authorities issued warnings to residents not to pick up any pieces, and they granted temporary amnesty to anyone who had taken space shuttle debris but brought it back. This reprieve came after two people were arrested in Nacogdoches and charged with stealing government property. Reportedly they walked off with shuttle fragments.

When *Columbia* disintegrated it scattered most of the debris from west of Dallas to Louisiana. Makeshift base camps were established in the small towns of Corsicana, Palestine, Nacogdoches, Hemphill, and Lufkin. Hemphill in east Texas served as the command post, and Lufkin (about fifty miles to its west) provided air operations support. Despite the tremendous efforts mounted by the National Guard, Texas Department of Public Safety, and emergency personnel from local towns and communities, the expanding bounds of the debris field quickly overwhelmed the search and recovery tasks. Faced with a search area several orders of magnitude larger than any previous accident site, NASA and Federal Emergency Management Agency (FEMA) officials activated US Forest Service wildland firefighters to serve as the primary search teams. Within two weeks the number of ground searchers exceeded three thousand. Within a month more than four thousand searchers were flown in. The searchers were drawn from across the United States and spread out to the base camps. They worked twelve hours per day on rotations of fourteen, twenty-one, or thirty days. GPS-equipped NASA and Environmental Protection Agency (EPA) personnel trained to identify and handle debris accompanied them. Navy dive teams searched Lake Nacogdoches and Toledo Bend Reservoir—two bodies of water located in the middle of dense debris fields. Sonar mapping of over thirty-one square miles of the lake bottom identified more than 3,100 targets in Toledo Bend and 326 targets in Lake Nacogdoches. Divers explored each target, but visibility was only a few inches in the murky water. Amid underwater forests and other submerged hazards, they recovered only one object in Toledo Bend and none in Lake Nacogdoches.

Houston, Texas
Mission Control Center—Johnson Space Center

In the mission control room at JSC, Leroy Cain, the bright young flight director and poster child for mission operations, ordered all doors locked. NASA officials impounded data, software, hardware, and facilities at NASA and contractor sites in accordance with the preexisting mishap response plan to preserve all material relating to STS-107 as evidence for the accident investigation. At KSC mission facilities and related hardware (including Launch Complex 39A) were put under guard or stored in secure warehouses. Similar actions were taken at other key shuttle facilities such as the Marshall Space Flight Center (MSFC) and the Michoud Assembly Facility (MAF) near New Orleans, Louisiana. Within minutes of the accident, the NASA Mishap Investigation Team was activated to coordinate debris recovery efforts with local, state, and federal agencies. The team initially operated out of Barksdale Air Force Base near Shreveport, Louisiana, and then in Lufkin, Texas, and Carswell Field in Fort Worth. The emergency response that began Saturday morning grew into a massive effort to recover and decontaminate debris strewn over an area that in Texas alone exceeded two thousand square miles. Decontamination of the debris was required to protect the recovery teams and general public from any toxic propellants, caustic chemicals, and explosive hazards from pyrotechnic devices and high pressure tanks.

Nacogdoches, Texas
Sunday, February 2, 2003

I volunteered to assist with the recovery operations and packed my bags for a trip to Nacogdoches, Texas where a makeshift command center had been established. Many questions were running through my mind on the drive up to the recovery command center. Most of my thoughts centered on the crew and their families—some I knew personally. I recalled Michael Anderson who had sat on my astronaut selection board in 2001 and was one of the board members who had questioned me during my interview. I had worked with Dave Brown on the International Space Station program. Dave was a US Navy attack pilot and doctor who had had a premonition of dying in a fiery crash. It was doubtful anyone could have survived the hypersonic breakup, and I wondered if anything would be left of the orbiter. A number of vehicles were parked along the road and in the field adjacent to the parking lot of the Nacogdoches command center, which was located at the approximate epicenter of the vast debris field. It was a dry, sunny day in the low eighties. This was somewhat warmer than normal but not unusual for a Texas winter. I had been involved with search and rescue missions flying with the Civil Air Patrol (CAP), an auxiliary part of the US Air Force, but this was different. CAP missions usually searched for crashed planes, lost climbers, or clandestine marijuana crops. Although such flights were conducted, volunteers did most recovery operations on foot.

The command center was little more than a large tent with dozens of people milling around. Mixed in the crowd were rescue crews, firefighters, ambulance squads, police, and media trucks sporting large mobile antennae on their roofs. Despite the large number of people, the mood was very solemn and serious. Most people stared aimlessly and silently at the ground or their recovery maps as they passed by. Everything about this major disaster in the middle of nowhere seemed surreal. Before I arrived President Bush declared East Texas a federal disaster area, and this enabled the dispatch of emergency response teams such as FEMA and the EPA. As the day wore on, county constables, volunteers on horseback, and local citizens headed into the pine forests and bushy thickets in search of debris and crew remains. National Guard units mobilized to assist local law enforcement officers guard the debris sites. Researchers from Stephen F. Austin University sent seven teams into the field with GPS units to mark the exact location of debris. The researchers and later searchers then used the GPS data to update the debris distribution on detailed geographic information system (GIS) maps.

I recognized a few familiar Boeing and NASA faces, and we grouped together to form five- to ten-person search teams. Each team contained at least one person familiar with the area and a shuttle program person to help identify debris and potential hazards. The teams were instructed to concentrate their initial search efforts on human remains in the debris corridor between Corsicana, Texas, and Fort Polk, Louisiana. They were also given a list of some twenty "hot items" that potentially contained crucial information regarding the accident. This included the orbiter's computers, film cameras, and modular auxiliary data system recorder—the shuttle's black box. NASA officials also warned the search teams of hazards posed by certain kinds of debris. *Columbia* carried highly toxic propellants such as monomethyl hydrazine and nitrogen tetroxide in pressurized tanks used to maneuver the orbiter in space. These propellants as well as concentrated ammonia used in the orbiter's cooling systems could severely burn lungs and exposed skin when encountered in vapor form. Other materials used in the orbiter such as beryllium were also toxic. Other hazard concerns were the various pyrotechnic devices that ejected or released items such as the K_u band antenna, landing gear doors, and hatches in an emergency. These pyrotechnic devices and their triggers were designed to withstand high heat and might have survived reentry, but they posed an explosive threat. Personnel trained in ordnance disposal had to remove them. Teams were given maps, assigned search grids, and provided contact information at the command center. As the teams embarked on their searches, someone shouted, "Watch for poisonous snakes, and drink plenty of water!"

Debris from the space shuttle *Columbia* in a hangar at Barksdale Air Force Base near Shreveport, Louisiana, on February 8, 2003. The debris was collected and cataloged prior to shipment to KSC. (Courtesy of NASA.)

Hemphill, Texas

The town of Hemphill (population 1,106) was transformed into a bustling, anxious depot of space shuttle debris. Hemphill residents welcomed the newcomers like long-lost cousins. Local residents such as Desa Gressett made thousands of sandwiches for searchers each day for several weeks. When asked about her experience, she wiped away her tears and said, "I'll always feel connected to the astronauts and their families."

More than two hundred families opened their homes to the scientists, state troopers, soldiers, and volunteers. This group quite outnumbered the local population. The whole town was involved in the debris hunt in one way or another. The local Veterans of Foreign Wars

(VFW) post was turned into a feeding hall lined with chips, pallets of bottled water, and tables loaded with chili dogs and cakes. This was where volunteers met before they stomped off into the soggy woods. A. L. Harper lived on the waters of the Toledo Bend Reservoir, and he was outside the moment the sky turned black with smoke and debris splashed down like belly flopping fish. Hopes were high that a significant clue to the *Columbia* disaster was lying on the reservoir bottom or in the nearby brush. Consequently more than five hundred military personnel were sent to this woodsy outpost a few miles from the Louisiana border. It turned out to have the largest concentration of material of any debris zone. Within a few days, large pieces of *Columbia* such as the nose cone and parts of the landing gear were being found in Hemphill's forests.

As more and more debris and human remains were found to have landed near Hemphill, the town's VFW post was established as the recovery command center. A top priority for NASA was expeditious recovery of those remains for fear wild animals would disturb, consume, or move the bodies. Hunters have long tracked fallen game by searching for the buzzard congregation, and that same technique was successfully used to locate the fallen crew. Early morning helicopter sorties were flown to coincide with buzzard flocks circling an area of interest. GPS coordinates would be immediately sent to ground crews to investigate the area and potentially locate the human remains. The NASA employee primarily responsible for the body recovery operations was Brent Jett, a fellow astronaut from the crew office. Brent had had stellar military and NASA careers. He was first of 976 in the 1981 US Naval Academy class and a distinguished graduate at the US Naval Test Pilot School. He was selected as an astronaut in March 1992 and served as a crewmember on three missions as a pilot and mission commander. However, nothing had prepared him for this. "When I arrived," he recalled, "I think everyone was looking to me for leadership because I was wearing my blue astronaut jumpsuit. I was unexpectedly thrust into that role. There were no procedures or manual for what to do next, so I stepped in quickly to establish a command center. I scheduled regular planning meetings to establish a routine and set some simple goals designed to provide direction among all the chaos and shock." Brent could, however, draw on his experience as a shuttle commander. "I witnessed firsthand how well-trained astronauts who had lifelong goals of flying in space essentially froze in fear before conducting space walks. I recognized the fear in their eyes and helped them through the preparation procedure. This allowed them to experience short-term successes to build their confidence, overcome their fears, and execute the space walks. Although the situations were totally different, there were some parallels that helped mobilize recovery efforts."

The remains of all seven members of *Columbia*'s crew were flown from Texas to Dover Air Force Base in Delaware within ten days of the accident for positive identification.

Area residents said they saw objects as big as cars plunge into the reservoir. Searchers in US Army helicopters spotted several large pieces of wreckage shining on the bottom. Unfortunately rains came and flooded the sixty-five-mile-long reservoir that straddles the Texas–Louisiana border. This turned already silty waters into murky darkness. The only indication of submerged objects were from sonar readings, but the lack of visibility made it impossible to recover any debris. Officials searched especially hard for a device that handled communications between the space shuttle and mission control. Instructions given to the search teams showed a box marked "Secret Government Property." In the first week, thousands of pieces were retrieved. This included the nose cone, sections of the fuselage, seats, harnesses, clothes, a shoe, windows, a rocket engine, fuel tanks, and countless strips of scorched metal. Remains from all the deceased astronauts had been located. Unfortunately crucial objects such as data recorders and wing pieces were still missing. Several weeks later *Columbia*'s flight data recorder was found near Hemphill on March 20. Nowhere was the activity more focused than around Toledo Bend—the largest human-made body of water in the South. The angler's paradise covers 185,000 acres and reaches to a depth of one hundred feet. The space shuttle ripped apart right above the reservoir. When the weather was good, police boats plied the waters and looked for debris with sonar and underwater cameras. More than 625 pieces of debris were found in the woods around Hemphill, but not a single one was recovered from the reservoir.

Searchers braved frigid rain, thick forest, and swamps to find debris and remains. Although the task was a somber affair for workers from across the country, the challenge often brought out the best in those who served in the field. "We learned firsthand that people love the space program and want to support it any way they can," said Ed Mango, a launch manager at KSC who served as NASA's recovery director for about three and a half months.

An East Texas high school was turned into a morgue as authorities collected the remains of the astronauts in an area between Hemphill and Jasper. A local funeral home assisted officials from the FBI and Defense Department in the grisly work of identifying the remains. Mango vividly remembered a local family's reaction on February 2 (the day after *Columbia* broke up) as he surveyed the debris field in a US Army Blackhawk helicopter. The group came across a farm field and spied what they later learned was the shuttle's left main tire and attached structure. "The family that owned the farm greeted the helicopter landing with open arms. All the men expressed sorrow for the loss, and the woman put her hand on my shoulder and said we were all in their prayers," Mango said. "The woman asked if we would fly the Shuttle again, and we knew the only answer she would accept was yes."

After the first two weeks, the search was organized by coordinating various FEMA-funded federal agencies. For the next three and a half months, teams of full-time and seasonal Texas and US Forest Service workers accompanied by space program and EPA workers scoured woods and fields for shuttle materials. The recovery groups wore chaps and other gear to protect themselves from briars and cottonmouth snakes in the woods. NASA and contractor team members who volunteered to work twelve-hour days and seven-day weeks in the field often did so with enthusiasm. Others not contacted sought the duty. They called it "an honor." Space program team members who worked intimately with *Columbia* on a day-to-day basis and those who knew the fallen STS-107 astronauts found their services in the field both painful and rewarding. Chris Meinert, a KSC team leader on the closeout crew—the team responsible for inserting the crew and closing the hatch at launch, remembered kidding with Israeli astronaut Ilan Ramon in the white room just before the launch of STS-107. Ramon presented a "boarding pass" the closeout crew had made for him. The closeout crew signed it, and Ramon took it aboard *Columbia*. Meinert thought about those moments while working with a recovery team in Nacogdoches. "I wondered if any of us would come across the boarding pass. It was such a personal item," Meinert said.

Some space program workers said their services in the field were life-altering experiences. Their faith in humanity was reaffirmed because they met and worked with so many who made sacrifices in support of the space program. KSC's Ronnie Lawson, lead at the Nacogdoches site, said he experienced many touching moments during his tour of duty. "It was so uplifting to me to see people of diverse cultures from all over the country work together for the common good," he said. "In camp I heard Native American and Hispanic music being played and felt in a very poignant way that no matter what our backgrounds we were all in this together."

Houston, Texas
Johnson Space Center
February 4, 2003

I assisted the search teams in Nacogdoches for most of two days before driving back to Houston Monday afternoon to ascertain my new shuttle duties. The next day was February 4, and President and Mrs. George W. Bush joined NASA administrator Sean O'Keefe in paying tribute to the brave heroes of the space shuttle *Columbia* during a special memorial service at JSC in Houston. The security was tight as thousands of people gathered for a private ceremony for family members, friends, invited guests, NASA employees, and contractors. It was a beautiful, crisp, clear day, and although I was across the lake and behind the JSC headquarters building, I could still see the podium. President Bush said the nation was blessed to have such men and women serving the space program. He affirmed that although NASA was being tested at this time, "America's space program will go on."

The magnitude of the ceremony moved me and many others. It concluded with a "missing man" aerial salute to the fallen crewmembers. President Bush's speech assured everyone associated with the Space Shuttle Program we would fly again, but first we needed to find the cause of the accident and fix the problem. Numerous teams began working around the clock to reconstruct the timeline of the final minutes of *Columbia*'s flight from the data being gathered from the wreckage and analyzed. For my first shuttle assignment, I was selected as a team lead on one of the many Boeing debris teams responsible for analyzing the massive amounts of data that would ultimately support the accident investigation.

On February 4, 2003, President George W. Bush addressed the crowd on the mall of the JSC during the memorial for the *Columbia* astronauts. Seated from the left are Captain Gene Theriot, Chaplain Corps (USN), NASA administrator Sean O'Keefe, and astronaut Kent V. Rominger, chief of the Astronaut Office. A portrait of the STS-107 *Columbia* crew is visible on the left. (Courtesy of NASA.)

Washington, DC
NASA Headquarters

On February 11, 2003, while the search and recover efforts continued in earnest, Admiral Hal Gehman held the first of many formal press conferences and introduced the Columbia Accident Investigation Board (CAIB). A joint hearing in the US Congress into the space shuttle *Columbia* accident followed this press conference the next day. NASA administrator Sean O'Keefe, Deputy Administrator Frederick Gregory, and Associate Administrator of Space Flight William Readdy provided testimony about the accident and the CAIB investigation CAIB. Senator John McCain, the ranking member of the Senate Committee on Commerce, Science, and Transportation, and Representative Sherwood Boehlert who chaired the House Science Subcommittee on Space and Aeronautics cochaired the hearing. One of the key questions asked concerned the International Space Station (ISS) contingency planning for the rest of the year and until the shuttle returned to operational flight status. All options to sustain a human presence aboard the station in the temporary absence of shuttle flights were being explored, but the only real option was using the Russian Soyuz to ferry crewmembers three at a time and the Russian Progress vehicle for ISS resupply. The next shuttle flight aboard *Atlantis* in March 2003 was slated to bring the Expedition 7 crew to the station and return the current resident crew to Earth.[2] This was no longer an option with the shuttle grounded. The next human journey to the station would involve an unprecedented landing by US astronauts in a Russian spacecraft. Earlier that same day, a Russian Progress resupply ship successfully docked to the ISS and delivered food, fuel, and supplies to Commander Ken Bowersox, Flight Engineer Nikolai Budarin, and NASA Science Officer Don Pettit of Expedition 6. Progress gave the station resident crew a solid supply of consumables. It was enough to sustain operations through at least late June until the launch planning and manifest could be modified.

---

[2] Expedition is a designation for a group of astronauts that remain aboard the ISS until the next Expedition crew replaces them.

Hemphill, Texas
VFW Post

After recovery of the crew remains, the search and rescue team's focus switched to strictly debris recovery. "I had been essentially in robot mode during the search for the bodies and did not realize how physically exhausted and emotionally spent I was," Brent Jett said.

Most volunteers were in the same state after nearly two weeks of intensive searching. One community leader suggested conducting a memorial service in the VFW Post for everyone involved and asked if Brent would lead the service. Brent agreed and felt it would be an honor to eulogize the fallen crewmembers and thank those who helped in the recovery effort. It would also help provide some closure to such a disastrous event. "The VFW hall was packed with many dressed in their dirty and wet search clothes, which seemed unusually fitting," recalled Brent. "I really had no time to grieve during the recovery operations, and the emotional toll finally caught up to me during the service. I broke down crying and struggled to muddle through the remainder of my speech."

Recovery operations lasted about three months before the effort began to wind down. Only about 40 percent of the vehicle was recovered and transported to KSC for investigation analysis. The wreckage was placed in a large, empty hangar, and people marked the floor into five-foot square grids to help organize the identification of thousands of fragmented pieces. Walking through the hangar was like walking through Arlington Cemetery. People strained to maintain their emotions as they surveyed the destruction's sheer expanse. Among the most gripping articles in the wreckage were the charred and tattered helmets and gloves worn by the astronauts. Those involved could not help but wonder about their last thoughts as their final moments transpired.

PICKING UP THE PIECES

Shuttle wreckage strategically placed in a hangar at KSC. The grid patterns and shuttle outline were used as dimensional reference points to aid in the reconstruction and accident investigation. (Courtesy of NASA.)

More than 25,000 people from 270 organizations took part in debris recovery operations. All told searchers expended over 1.5 million hours and covered more than 2.3 million acres—an area approaching the size of Connecticut. Nearly one-third of this acreage was searched by foot, and searchers found more than 84,000 individual pieces of orbiter debris. Collectively these weighed more than 84,900 pounds and represented 38 percent of the orbiter's dry weight. Though significant evidence from radar returns and video recordings indicated debris shedding across California, Nevada, and New Mexico, the westernmost piece of confirmed debris was the tile found in a field in Littleton, Texas. Heavier objects with higher ballistic coefficients (a measure of how far an object will travel in the air) landed toward the end of the debris trail in western Louisiana. The easternmost debris pieces included the space shuttle main engine turbo pumps, and these were found in Fort Polk, Louisiana.

Houston, Texas
Johnson Space Center

Major changes to the shuttle program's leadership and organization started well before the CAIB completed its investigation and reported its findings and recommendations. On April 23, 2003, Ronald D. Dittemore, a twenty-six-year NASA veteran, announced his intention to step aside as the Space Shuttle Program manager. Dittemore had served as the shuttle program manager for more than four years and was heavily criticized for his role in the accident and early dismissal of foam debris as the accident's cause. He originally intended to remain in his position until the CAIB finished its investigation and a complete Return to Flight path had been established. However, Dittemore stated, "The timing of my departure is based on what I believe will allow for the smoothest management transition possible as the pace of work to return the shuttle to flight begins to ramp up."

After Dittemore's announcement NASA selected William (Bill) Parsons as the new program manager on May 9, 2003. Parsons was a tall, commanding, and tactfully blunt former US Marine officer who had served as the center director at Stennis Space Center in south Mississippi since August 2002. "NASA is about the people who fly, fix, maintain, and design our vehicles, and I know we've found a terrific leader to help guide the team through this difficult time," stated Sean O'Keefe.

Once Parsons replaced Dittemore, he shook up the program in July by creating new positions and reassigning several managers five months after *Columbia* disintegrated over Texas. Gone were Ralph Roe, Linda Ham, and Austin Lambert, key members of the STS-107 mission. Before assuming the position of shuttle program manager in 1999, Dittemore held several positions. This included flight director on eleven shuttle missions, deputy assistant director of the Space Station

Program, and manager of Space Shuttle Integration. Roe, Ham, and Lambert were the key leaders involved in the discussion regarding whether the chunk of foam that flew off *Columbia*'s massive external tank and hit its left wing eighty-two seconds after launch could pose a danger to the orbiter and crew. Ultimately shuttle managers accepted an analysis that concluded the foam strike was not a safety of flight issue—a decision that later proved horribly wrong. Ralph, a tall man with a noticeable Southern accent, was removed as the shuttle program's chief engineer and reassigned to Langley Research Center in Virginia to head a new safety and engineering office. This organization was independent from the program and was to provide an additional level of safety oversight. It would later play a key role in debris testing and risk analysis. Ham, the ambitious and bright program integration manager and head of the Mission Management Team (MMT) that oversaw *Columbia*'s flight, was dismissed as the team's chair. Her talent and intellect had enabled her to became NASA's first female flight director. In 2001 she became the shuttle program's integration manager. She was one of six senior managers responsible for program operations. In this position Ham chaired the MMT meetings and reported directly to Dittemore. At the time of the *Columbia* mission, she was also serving as acting manager of shuttle launch integration. The CAIB later called this a dual role that permitted a conflict of interest. She was deemed the single-point failure regarding the wing damage problem in flight and was responsible for stopping the requests for telescopic pictures of *Columbia*'s wing in orbit by rationalizing that nothing could be done even if the photos showed damage. In the aftermath of the disaster, that decision was hotly debated, and ultimately NASA agreed the agency needed to obtain such photos on all future flights.

Austin Lambert, manager of the systems integration office, was replaced in July 2003. John Muratore, a fiery former USAF officer, NASA flight director, and X-38 program manager, replaced him. John had served twenty-eight years in the air force and with NASA. He started his career as an air force officer assigned to space launch duties at Vandenberg Air Force Base and Cape Canaveral Air Force Station. On loan to NASA, John served on the launch team for the first five flights of the space shuttle at KSC. He joined NASA permanently in 1983 as a flight controller, and he served in positions of increasing responsibility in JSC's mission control. He eventually worked his way up to a shuttle flight director. John then led the development effort to build, test, and put into operation the new MCC that supported the space shuttle and space station programs since 1995. Before taking over as the Systems Engineering and Integration (SE&I) manager, he led the X-38 Crew Return Vehicle project and brought a few of his top X-38 employees over to the shuttle program. Austin was effectively demoted for all that had gone wrong during STS-107 and even before that with the foam loss on STS-112 in October 2002. Within months of his reassignment, Austin resigned from NASA and hired on with Boeing as chief technical adviser for the shuttle program's systems engineering and integration team. Wayne Hale had just moved into the job overseeing shuttle launches at KSC in February, and he returned to Houston to become Parsons's deputy. Wayne, a longtime flight director at the JSC, was one of the people inside the shuttle program who had pushed for military satellites to take pictures of *Columbia* to check for damage but was denied.

Washington, DC
NASA Headquarters

On June 13, 2003, the NASA administrator appointed a task force to assess the agency's Return to Flight efforts and help implement the anticipated recommendations of the CAIB. Veteran astronauts Thomas Stafford and Richard Covey led the task force group, which was established to monitor NASA's compliance with the CAIB recommendations. The CAIB eventually released twenty-nine findings and recommendations for NASA. Fifteen needed to be addressed before returning the shuttle fleet to flight status. The job of this independent task force, which became known as Stafford–Covey, was to report the Space Shuttle Program's compliance with these recommendations to the administrator. In addition the NASA Engineering Safety Center

(NESC) was formed in July 2003 as an independently funded program with a dedicated team of technical experts who would provide objective engineering and safety assessments of critical, high-risk projects. Ralph Roe was named to lead the NESC and became very involved in the Return to Flight activities and resolving the shuttle's debris problems. Like many who grew up in the sixties and seventies, NASA's great accomplishments in the Apollo program inspired Ralph and made him want to work for NASA. He was hired directly out of college to test the space shuttle orbiter reaction control and orbital maneuvering subsystems at the KSC. Following the *Columbia* accident, Ralph and fellow NASA leaders Bryan O'Connor, Roy Bridges, and Theron Bradley developed the concept for the NESC in response to the CAIB's recommendation regarding the shuttle's safety program. What Ralph remembered most was Admiral Gehman's criticism and his quote about the safety program before the accident: "There is no there there." He was referring to how the shuttle safety organization lacked the funding, resources, and technical skills to offer the Space Shuttle Program a second perspective on tough technical issues such as foam loss. "After the accident I was devastated," recalled Ralph. "What helped me deal with my grief the most was leading the Space Shuttle Program accident investigation before I left to start the NESC in August 2003. This enabled me to devote all my energy into finding out exactly what happened. That difficult work—long hours every day for months following the accident—provided a constructive outlet for my grief. Without that opportunity to help find out what happened, I don't know how I would have dealt with my grief."

CHAPTER 3

Columbia Accident Investigation Board

The CAIB was similar the Apollo 204 Review Board, which investigated the Apollo 1 launchpad fire in 1967, and the 1986 Presidential Commission on the Space Shuttle *Challenger* Accident, known as the Rogers Commission. It was charged with the responsibility to determine the accident's cause and develop recommendations to fix the problem(s). From its inception the board considered itself an independent and public institution. It was accountable to the American people, the White House, Congress, the astronaut corps and their families, and NASA. With the support of these constituents, the board resolved to broaden the scope of the accident investigation into a far-reaching examination of NASA's operation of the shuttle fleet. This included the program's practices on shuttle safety. A staff of more than 120 and some four hundred NASA engineers supported the thirteen-member board. The board recognized early on that the accident was probably not an anomalous, random event. Rather it was likely rooted to some degree in NASA's history and the human spaceflight program's culture. Accordingly the board broadened its mandate to include an investigation of a wide range of organizational issues. This included political and budgetary considerations, past compromises, and evolving priorities over the life of the Space Shuttle Program. The board members' convictions regarding the importance of these factors strengthened as the investigation progressed.

Columbia's STS-107 Mission

Columbia was the first space-rated orbiter and made the Space Shuttle Program's first four orbiter test flights. Because it was the first of its kind, *Columbia* differed slightly from the orbiters that came after— *Challenger, Discovery, Atlantis,* and *Endeavour*. Built to an earlier engineering standard, *Columbia* was slightly heavier. Although it could reach the high-inclination orbit of the ISS, its payload was insufficient to make *Columbia* cost-effective for space station assembly missions. Therefore, *Columbia* was not equipped with a space station docking system. This freed up space in the payload bay for longer cargoes such as the science modules Spacelab and Spacehab. Consequently *Columbia* generally flew science missions and serviced the Hubble Space Telescope (HST) instead of space station missions.

The STS-107 flight was an intense science mission that required the seven-member crew to form two teams. This enabled around-the-clock shifts. Because the extensive science cargo and its extra power sources required additional checkout time, the launch sequence and countdown were about twenty-four hours longer than normal. Nevertheless the countdown proceeded as planned, and *Columbia* was launched from Launch Complex Pad 39A on January 16, 2003, at 10:39 a.m. (EST). At 81.7 seconds after launch, the shuttle was at about 65,820 feet and traveling at Mach 2.46 (1,650 miles per hour). At that time a large piece of handcrafted insulating foam came off an area where the orbiter attached to the external tank. At 81.9 seconds, it struck the leading edge of *Columbia*'s port wing. The crew did not detect this event, and the ground support teams did not see it until the next day's detailed reviews of all launch camera photography and videos. Although the crew was informed of the foam impact, this news had no apparent effect on the daily conduct of the sixteen-day mission. The crew met all its objectives. It was later determined that the accident really started two missions previous on STS-112 when the same bipod foam came off the external tank and dented the solid rocket booster. However, no one asked what would happen if it hit the orbiter.

RETURN TO FLIGHT

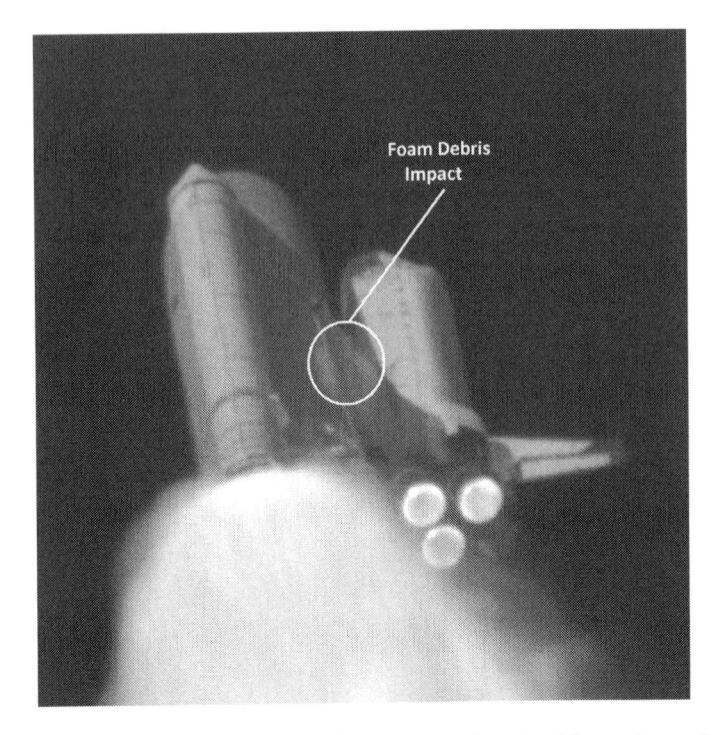

Foam Debris Impact

At 81.9 seconds after launch of STS-107, a chunk of foam from the external tank dislodged and struck the leading edge of *Columbia*'s left wing like an exploding snowball. The accident investigation team estimated the chunk of foam to be traveling between four hundred and six hundred miles per hour relative to the orbiter. (Courtesy of NASA.)

Determining the Cause of the Accident

During its investigation the board evaluated every known factor that could have caused or contributed to the *Columbia* accident. This included the effects of space weather on the orbiter during reentry and the specters of sabotage and terrorism. In addition to the analysis and scenario investigations, the board oversaw a NASA "fault tree" investigation. A group led by John Muratore constructed this. A fault tree is a common systems engineering tool that accounts for every chain of events that could possibly cause a system failure. NASA chartered six teams to develop fault trees. That meant one for each of the shuttle's major components or elements: the orbiter, space shuttle main engine, reusable solid rocket motor, solid rocket booster, external tank, and payload. A seventh "systems integration" fault tree team analyzed failure scenarios involving two or more shuttle components. These interdisciplinary teams included NASA and contractor personnel and outside experts. In all more than three thousand individual elements in the *Columbia* accident fault tree were examined. In addition the board analyzed the more plausible fault scenarios such as the impact of space weather, collisions with micrometeoroids or "space junk," willful damage, flight crew performance, and failure of some critical shuttle hardware. All these were eventually dismissed. When investigators ruled out a potential cascade of events (represented by a branch on the fault tree), it was deemed "closed." When evidence proved inconclusive, the item remained "open." Some elements could be dismissed at a high level in the tree, but most required delving into lower levels. Some elements would never be closed because there was insufficient data and analysis to conclude they did not contribute to the accident. For instance heavy rain fell on KSC prior to the launch of STS-107. This left investigators wondering if the abnormally heavy rainfall compromised the external tank bipod foam. Experiments showed the foam did not tend to absorb water, but the rain could not be ruled out entirely as a contributor to the accident.

There were two plausible explanations for the aerodynamic breakup of the *Columbia* orbiter. One, the orbiter sustained structural damage that undermined attitude control during reentry. Two, the orbiter maneuvered to an attitude in which it was not designed to fly. The former explanation dealt with structural damage initiated before launch, during ascent, on orbit, or during reentry. The latter considered aerodynamic breakup caused by improper attitude or trajectory control by the orbiter's flight control system. Telemetry and other data strongly suggested improper maneuvering was not a factor. Therefore, most fault tree analysis concentrated on structural damage.

One of the more prominent branches of the fault tree that was investigated as a cause of structural damage was the solid rocket booster "bolt catcher." Four large separation bolts connected each booster to the external tank. Three were at the bottom with a larger one at the top. Each weighed approximately sixty-five pounds. About two minutes after launch, the firing of pyrotechnic charges is meant to break each forward separation bolt into two pieces and allow the spent solid rocket boosters to separate from the external tank. Two bolt catchers on the external tank each trap the upper half of a fired separation bolt. The lower half stays attached to the solid rocket booster. In this way both halves are kept from flying free of the assembly and potentially hitting the orbiter. Bolt catchers have a domed aluminum cover containing an aluminum honeycomb matrix that absorbs the fired bolt's energy. The two upper bolt halves and their respective catchers subsequently remain connected to the external tank. This burns up on reentry. The lower halves stay with the solid rocket boosters that are recovered from the ocean. If one of the bolt catchers failed during STS-107, the resulting debris could have damaged *Columbia*'s wing leading edge. Although bolt catchers could be neither definitively excluded nor included as a potential cause of left wing damage to *Columbia*, the impact of such a large object would likely have registered on the shuttle sensors. The indefinite data at the time of solid rocket booster separation in tandem with the accumulation of overwhelming evidence related to the foam debris strike eventually led the board to conclude that bolt catchers were unlikely to have been involved in the accident.

Foam Debris Strike

At 81.9 seconds after launch of STS-107, a sizable piece of foam indisputably struck the leading edge of *Columbia*'s left wing. Visual evidence established the foam's source as the left bipod ramp area of the external tank. The widely accepted implausibility of foam causing significant damage to the wing leading edge system led the board to request independent tests to characterize the impact. While it was impossible to determine the precise impact parameters because of uncertainties about the foam's density, dimensions, shape, and initial velocity, intensive work by the board, NASA, and contractors provided credible ranges for these elements. NASA and the CAIB agreed tests would be required to validate an impact or breach scenario. Initially the board intended to act only in an oversight capacity in the development and implementation of a test plan. However, ongoing and continually unresolved debate on the size and velocity of the foam projectile convinced the board to take a more active role. This uncertainty was largely due to the Marshall Space Flight Center's insistence that, despite overwhelming evidence to the contrary, the foam could not have been larger than a football. In its assessment of potential foam damage, NASA also continued to rely heavily on the "CRATER" model. This was used during the mission to determine the foam-shedding event was nonthreatening. CRATER was an outdated debris model constructed from Apollo-era data. Another factor that contributed to the board's decision to play an active role in the test program was the Orbiter Vehicle Engineering Working Group's requirement that the test program be used to validate the CRATER model. NASA failed to focus on several key aspects of the testing such as assessing the physics-based pretest predictions, determining the most vulnerable reinforced carbon-carbon (RCC) panels, and incorporating appropriate test instrumentation. In discussions with the Orbiter Vehicle Engineering Working Group and the NASA Accident Investigation Team, the board ultimately provided test plan requirements that outlined the template for all testing. The board directed that a detailed written test plan with board signature approval be provided before each test.

Foam Debris Impact Testing

Scott Hubbard, one of the thirteen CAIB board members, took a personal interest in the foam debris impact testing. Scott was the director of the NASA Ames Research Center in California and the first Mars program director at NASA headquarters. He successfully restructured the program after a series of high-profile mission failures. He and some colleagues performed back-of-the-envelope calculations after viewing the bipod foam strike video, and he was convinced there was enough impact energy to damage the wing leading edge RCC panel and breach the thermal protection system. Unfortunately the size of the bipod foam debris exceeded the capability of the CRATER tool to predict the damage. Other branches of the fault tree (such as the bolt catcher) seemed more plausible, and many within NASA were skeptical the lightweight foam material could breach the RCC. Scott was determined to prove the skeptics wrong by firing an equivalent-size piece of foam into an RCC panel under the same impact conditions as encountered on *Columbia*'s ascent. Impact testing took place at the Southwest Research Institute in San Antonio, Texas. Scott's temperament matched the summer heat and his drive to show that foam falling off the external tank could create a hole big enough to eventually cause the vehicle's destruction. The biggest challenge was building a device that could accelerate a large piece of foam debris to a high velocity without causing the material to disintegrate before reaching the target. A nitrogen gas gun that had evaluated bird strikes on aircraft fuselages was modified to integrate a thirty-five-foot-long rectangular barrel. The target site was equipped with sensors and high-speed cameras that photographed at two thousand to seven thousand frames per second. The target was a leading edge structural subsystem. It was designed to accommodate the board's evolving speculation that RCC panels six through nine were the most likely points of impact.

Heading up the testing for NASA was Dr. Justin Kerr, a proud University of Texas graduate. He was always in a good mood and not someone to be challenged in a beer-drinking contest. He had Clark Kent looks. He was tall with dark hair, blue eyes, and a mischievous smile. Three days after the accident, Justin was assigned as the director of all the impact testing being performed at South West Research in San Antonio, Texas. A week later he and his team had constructed a test apparatus that enabled them to shoot large blocks of structural foam purchased from Home Depot that were similar to the size seen striking *Columbia*. Justin recalled being quite impressed at "how quickly we could pull together a team to solve a debris problem, run a test, and get top-notch people in the agency to support the effort. Holy cow. NASA and the shuttle program could move fast when all the normal bureaucratic limiters were gone. This enabled us to press full speed ahead."

Growing up in the Clearlake area minutes from JSC, Justin always wanted to be an astronaut, and he visited the center with his grandmother every chance he had. From 1991 through 1993, Justin worked as a NASA intern. This was how he was able to secure a job with the agency. "I actually turned down NASA initially to take a job with Lockheed working at NASA's arc jet facility. However, three months later the funding for that position ended, and I was left searching for another job. Fortunately friends of mine who followed my plight put me in touch with the NASA co-op coordinator, and through him I was given another shot at the NASA job. I took it and never looked back. Although the accident was a very tragic event, I was fortunate that everything I had done up to this point working for NASA prepared me for doing this type of work. Even though the test director position was much more critical and on a much larger scale, I felt I could contribute immediately because of all the hypervelocity impact testing and micrometeoroid debris impact assessments work I had already done for the shuttle program."

There were numerous setbacks and delays, and the test team was beginning to feel the strain of the summer heat, long hours, and schedule pressure to conduct the tests. Despite the obstacles Scott continued to push the team. RCC panel six was tested first. By the time of the test, though, other data had indicated RCC panel eight-left was the most likely site of the breach. Scott and the test team looked bewildered after the first shot. The impact only yielded a 5.5-inch crack on the outboard end of the panel that extended through the rib and a small crack through the web of the T-seal between panels six and seven. Without further crack growth, the specific structural damage this test produced would probably not have allowed enough superheated air to penetrate the wing during reentry and cause serious damage.

The eureka moment came on June 7 when nearly a hundred observers (including news crews and two astronauts) watched the second impact test of panel eight, which had flown twenty-six missions on *Atlantis*. Based on forensic evidence, sensor data, and aerothermal studies, panel eight was considered the most likely point of the foam debris impact on *Columbia*. The impact created a hole roughly sixteen inches by seventeen inches. This was within the range consistent with all the investigation's findings. The damage stunned the crowd. "I had very mixed emotions when we were able to replicate the RCC damage caused by the foam impact on the left wing," said Justin. "Initially I was ecstatic the test was a success and even celebrated by sharing a few high fives with coworkers. Literally within a few seconds, the elation turned to despair. I came to the scientific realization that the foam impact was responsible for the accident and loss of life. The mixtures of cheers and gasps made it an emotional experience and one shared by the stunned crowd of engineers and reporters who had just witnessed the test."

Scott Hubbard also experienced the same emotion and felt vindicated as he spoke to reporters. "We demonstrated for the first time that foam at the speed of the accident can actually break reinforced carbon wing pieces."

His feelings also became subdued with the knowledge that the foam debris strike was the accident's cause.

The test article from the foam debris impact testing at Southwest Research Institute's labs in San Antonio, Texas. A piece of foam comparable in size to the bipod foam and shot at similar impact conditions left a substantial breach (larger than a basketball) in the RCC panel. (Courtesy of NASA.)

John Muratore as Shuttle Integration Manager

John Muratore was leading the fault tree team, and he became the newly appointed systems integration manager at NASA in July. He was given the task of developing a series of tests to support the CAIB's investigation. John's New York accent and temperament complemented his expressive body language and gestures. His high energy, intensity, and enthusiasm were contagious for some and offensive to others in the normally meek and quiet NASA workforce. It was not uncommon for John to yell down the hall for some unsuspecting engineer to come see him in his office. As a systems engineer, he defined and paid meticulous attention to detail. John had an uncanny knack for simplifying complex concepts. Although demanding and sometimes short-tempered if work fell short of his expectations, John was quick to praise a job well done. From 1996 to 2003, he was the program manager for the X-38. Targeted as a space station rescue vehicle, the X-38 was an unpiloted demonstrator that performed a series of successful experimental flights at Edwards Air Force Base. He gathered a team of young, experienced, and highly motivated engineers to try to apply the "faster, better, cheaper" method advocated by NASA's previous administrator, Daniel Goldin, to human spaceflight. This was meant to allow NASA to obtain an affordable Crew Return Vehicle. Despite his efforts the X-38 program was canceled before John took over as SEI manager due to the ISS program's financial woes. However, he was able to bring most of his talented team over to the shuttle program. The CAIB requested that NASA and John perform several tests: one, an integrated vehicle aerodynamic test to assess the debris transport, air loads, and aeroheating characteristics during ascent; two, a wind tunnel or flight dynamics test to characterize the lift and drag characteristics of foam debris; three, thermal vacuum testing of foam panels to understand the foam failure mechanisms, and four, foam debris impact testing on the wing leading edge RCC panels to determine the extent of damage from a foam debris hit. John was eager to get started and assigned Bob Ess to lead what became known as the IA-700 (integrated aero) test. This was performed at Arnold Engineering Development

Center—a US Air Force research facility in Tullahoma, Tennessee. The F-15 foam flight dynamics test was performed at NASA Dryden. Denny Kross, the NASA external tank project manager at MSFC in Huntsville, Alabama, led the foam thermal vacuum testing. After July the lead for the impact testing was assigned to the orbiter project under newly appointed Steve Poulos and former flight director Chuck Shaw. Shaw took the lead for test operations and eventually took over as mission manager for the HST mission. John and Chuck were former coworkers and would argue tumultuously over the tests, but in the end the process was better for it. Their friendship stood the test of working through the arguments. Bob Ess was a young, focused, rising star, and John assigned him as the manager of the Engineering Integration Office within SEI. Bob was responsible for all shuttle ascent technical disciplines, and that included ascent debris assessment. Bob had spent three years working with John on the X-38 program, and before that he spent two years at Navy Test Pilot School at the Naval Air Station in Patuxent River, Maryland, where he developed exceptional data interpretation and analysis skills. I was Bob's counterpart at Boeing and helped develop, form, and execute the tests. In addition I was heavily involved in leading the debris team at Boeing. This team focused on the CAIB's recommendations and debris reconstructions of the bipod foam release. These tasks were the beginning of the debris odyssey on which I was about to embark.

CAIB Investigation Summary

Investigators examined more than thirty thousand documents, conducted more than two hundred formal interviews, heard testimony from dozens of expert witnesses, and reviewed more than three thousand inputs from the general public. In addition more than twenty-five thousand searchers combed vast stretches of the western United States to retrieve the spacecraft's debris. In the process *Columbia*'s tragedy was compounded when two debris searchers with the US Forest Service perished in a helicopter accident. After six months of investigation, the CAIB released its final report. It pinpointed the blame on a foam debris strike to the left wing at eighty-two seconds into flight. Using a combination of analysis and test results, the board concluded the foam strike observed during the flight of STS-107 was the direct, physical cause of the accident. During ascent foam liberated from the left bipod—part of the thick titanium structure that attached the external fuel tank to the forward attach point on the orbiter. It was a suitcase-sized cover used as an aerodynamic ramp designed to minimize the heating and aerodynamic forces at the tank attachment point. The foam struck the left wing leading edge with nearly three thousand pounds of force. This blasted a large hole in RCC panel eight, which was part of the structural protection against the searing 2,500°F reentry temperatures. The breach in the RCC thermal protection system allowed superheated air to penetrate the leading edge insulation and progressively melt the left wing's aluminum structure. This resulted in a weakening of the structure until increasing aerodynamic forces caused loss of control, wing failure, and orbiter breakup. At that point *Columbia*'s fate had been sealed. The crew and spacecraft were doomed.

On August 23, 2003, the CAIB released its final report and summarized its findings as a single accident with multiple causes—both physical and organizational. The board identified a number of pertinent factors, and these were grouped into three distinct categories: one, physical failures that led directly to *Columbia*'s destruction; two, underlying organizational weaknesses revealed in NASA's culture and history that effectively paved the way to catastrophic failure; and three, "other significant observations" made during the course of the investigation that might have been unrelated to the accident at hand. The report also concluded that left uncorrected any of these factors could have contributed to future shuttle losses. Rockets are complex and unforgiving vehicles. They must be as light as possible yet attain outstanding performance to reach orbit. Although engineers are getting better at building them, in the space program's early days, vehicles often exploded on or near the launchpad. That seldom happens anymore. The situation was not that different from early air flight. Airplanes tended to crash about as often as they flew. Aircraft seldom crash these days, but rockets still fail 2 to 5 percent of the time. This is true of just about any launch vehicle—Atlas, Delta, Soyuz, STS, and more. Regardless of what nation builds it or what basic configuration is used, they all fail at about the same rate. In the space shuttle's case, 7 million pounds of propulsive force found the vehicle's weakest link, and this was where failures occurred. Building and launching rockets is still a very dangerous business and will continue to be so for the foreseeable future while additional experience is gained. Launching a space vehicle will never be as routine an undertaking as commercial air travel—certainly not in this lifetime. Although scientists and engineers continually work on improved methods, if humans want to continue going into outer space, we must continue to accept the risks. It is an extraordinary challenge to manage high-risk technologies while minimizing failures. Complex space technologies are intricate and have many interrelated, fast-moving parts. Standing alone the components might be well understood and have anticipatable failure modes. When these components are integrated into a larger system, though, unanticipated interactions can occur, and unfortunately that can lead to catastrophic outcomes. The risk of failure in such complex systems is increased when complex organizations that can also break down in unanticipated ways produce and operate them.

CAIB Conclusions

The CAIB concluded the STS-107 left bipod foam ramp loss was the cause of the accident and failed due to a combination of factors. Investigators believed testing conducted during the investigation (including the dissection of flight hardware and testing of simulated defects) showed conclusively that preexisting defects in the foam were a major factor in and a necessary condition for foam loss. However, analysis indicated that preexisting defects alone were not responsible for the foam loss.

The basic external tank deployed on the space shuttle was designed more than thirty years before and used a design process that was substantially different than today's. In the 1970s engineers often developed particular facets of a design (e.g., structural or thermal) independently from one another and in relative isolation from other engineers. Today engineers usually work collaboratively on all design aspects as an integrated team. The bipod fitting was designed first from a structural standpoint to withstand the orbiter attachment loads. Application processes for foam to prevent ice formation and super lightweight ablator (SLA) to protect from high heating were developed separately. Unfortunately the structurally optimum fitting design and the geometric complexity of its location posed many problems in the foam and SLA application. This eventually led to foam ramp defects.

The bipod foam application also had its root in a previous accident during Skylab, which MSFC designed. When flutter and turbulence associated with a protrusion caused loss of one of the Skylab solar arrays during its launch, MSFC became concerned. In order to eliminate the aerodynamic flutter problem with protuberances, bipod ramps were developed.[3] Engineers and designers used the best methods available at the time. This involved testing the bipod and foam under as many severe combinations as could be simulated and then interpolating the results. It was—and still is—impossible to conduct a ground-based, full-scale simulation of the combination of air loads, airflows, temperatures, pressures, vibrations, and acoustics the external tank experiences during launch and ascent. Consequently the testing did not truly reflect the combination of factors the bipod would experience during flight.

Further complicating this problem, foam does not have the same properties in all directions, and there is also variability in the foam itself. Because it consists of small hollow cells, it does not have the same composition at every point. This combination of properties and composition makes foam extremely difficult to model analytically or characterize physically. The way foam was produced and applied

---

[3] A protuberance is an external part or piece of the spacecraft structure that protrudes or "sticks out" above the surface into the airstream.

(particularly in the bipod region) also contributed to its variability. Foam consists of two chemical components that must be mixed in an exact ratio and sprayed according to strict specifications. Foam is applied to the bipod fitting by hand, and this process is the primary source of foam variability. Board-directed dissection of foam ramps revealed defects such as voids, pockets, and debris. These were likely due to a lack of control of various parameter combinations in spray-by-hand applications. The complexity of the underlying hardware configuration exacerbated this. These defects often occurred along "knit lines"—the boundaries between each layer formed by the repeated application of thin layers. Multiple defects can combine to weaken the foam along a line or plane or near a geometric boundary.

Cryogenic effects such as cryopumping and cryoingestion were theorized as among the processes contributing to foam loss from larger areas of coverage. Cryopumping occurs if cracks in the foam lead through the foam to voids at or near the surface of the liquid oxygen and liquid hydrogen tanks. Cracks of this type allow air chilled by the extremely low temperatures of the cryogenic tanks to liquefy in the voids. As propellant levels fall and aerodynamic heating of the exterior increases during launch, the temperature of the trapped air increases. This leads to boiling and evaporation of the liquid with concurrent buildup of pressure within the foam. If the pressure buildup exceeds the fracture toughness of the foam, it will fail and break away from the external tank in an explosive fashion.

Cryoingestion follows essentially the same scenario, but it involves gaseous nitrogen seeping out of the intertank and liquefying inside a foam void. The intertank separates the upper oxygen tank and lower hydrogen tank, and nitrogen fills it to prevent condensation and prevent liquid hydrogen and liquid oxygen from combining. If pooled liquid nitrogen contacts the liquid hydrogen tank, it can solidify because the temperature of liquid hydrogen (−423°F) is cooler than the freezing temperature of liquid nitrogen (−348°F). As with cryopumping cryoingested liquid or solid nitrogen could also flash evaporate during launch and ascent and cause the foam to fail.

There was no evidence to suggest defects or cryo-effects alone caused the loss of the left bipod foam ramp from the STS-107 external tank. Calculations showed that during ascent the hydrogen tank's outer wall near the bipod ramp did not reach the temperature at which nitrogen boils until 150 seconds into the flight. This was too late to explain the bipod ramp foam losses. Tests conducted at the MSFC revealed flight conditions could have permitted ingestion of nitrogen or air into subsurface foam, but they would not have permitted flash evaporation and a sufficient subsurface pressure increase to crack the foam. When conditions were modified to force a flash evaporation, the only thing observed was a crack that provided pressure relief rather than explosive cracking. The evidence suggested the flight environment itself must also have played a role. Aerodynamic loads, thermal and vacuum effects, vibrations, stress in the external tank structure, and myriad other conditions might have contributed to the growth of subsurface defects and a weakening of the foam ramp until it could no longer withstand the flight conditions and liberated. NASA lacked an integrated structural model that enabled conducting a more thorough analysis that would provide insight into the failures. John recognized this shortcoming and vigorously pursued building one during the RTF activities.

In addition to identifying the physical failure, the CAIB concluded that management decisions made during *Columbia*'s final flight reflected missed opportunities, blocked or ineffective communications channels, flawed analysis, and ineffective leadership. Perhaps most striking was that the Space Shuttle Program's management—the MMT, mission evaluation room (MER), flight director, and mission control—displayed little interest in understanding the foam problem and its implications. Managers failed to avail themselves of the wide range of expertise and opinion necessary to achieve the best answer to the debris strike question of whether this was a safety of flight concern. Some Space Shuttle Program managers failed to fulfill their implicit contracts to do whatever was possible to ensure the crew's safety. Their management techniques unknowingly imposed barriers that kept both engineering concerns and dissenting views at bay, and they ultimately helped create blind spots that prevented them from seeing the danger the foam strike posed. These organizational and leadership issues would have to be corrected before the space shuttle could fly again.

CAIB Recommendations

It was paramount NASA apply the board's recommendations (based on the insights gained while investigating the loss of *Columbia* and its crew) to continue the nation's journey into space. The recommendations spanned three time frames: one, short-term—NASA's Return to Flight (RTF) after the *Columbia* accident; two, intermediate—what was needed to continue flying the space shuttle fleet until a replacement means for human access to space and other shuttle capabilities became available; and three, long-term—future directions for US space programs. The objective in each case was to maintain a human presence in space but with enhanced safety of flight. In all the CAIB developed twenty-nine recommendations. Fifteen had to be implemented before the shuttle could be returned to operational status. The twenty-nine specific recommendations to NASA to improve the safety of future shuttle flights included eliminating foam debris from the external tank, initiating better preflight inspection processes, increasing the quality of images available during ascent and flight, recertifying all shuttle components, developing a repair and contingency capability, and establishing an independent technical engineering authority that would be responsible for technical requirements and all waivers. This independent unit would be responsible for building a disciplined, systematic approach to identifying, analyzing, and controlling hazards throughout the life cycle of the shuttle system.

John Muratore Mobilizes His Team

Even before the CAIB released its final report, John Muratore was mobilizing his team toward RTF activities. The first thing he changed was the name from shuttle integration to shuttle systems engineering and integration (SEI). This sent a clear message that things would be done differently. After the scathing assessment in the CAIB report, NASA personnel and contractors began to focus all their attention on addressing the CAIB recommendations and returning the shuttle to flight status. NASA designated the next scheduled flight (STS-114) as the Return to Flight mission and responded to the board by making significant organizational changes in terms of people, processes, and procedures. NASA also reclassified the Space Shuttle Program as "experimental" and restricted flights to only ISS assembly missions. This gave the crew a safe haven in the event a rescue mission was needed. Shortly after the CAIB report was released, John presented his RTF plan to the program managers. He recognized that a debris field (environment) was generated as the vehicle flew, and he concluded the shuttle must be engineered to successfully fly through that environment. RTF was not a new shuttle design but an exercise in maximizing the safety of an existing shuttle design. John was committed to closing the gaps regarding debris threats. This included: one, incomplete identification of all ascent and liftoff debris sources; two, incomplete or nonexistent flight rationale for each debris source on each vulnerable structure; three, limited verification debris generation did not exceed maximum specification; four, very limited verification of minimum impact tolerance by test or analysis; five, limited validation of debris transport models; and six, incomplete debris hazard analysis.[4] Closing these gaps was his primary objective for certifying vehicle safety vis-à-vis the debris environment. Certification was the standard systems engineering approach taken for all shuttle systems to ensure they could successfully perform mission operations. In other words John wanted to eliminate all debris sources that could lead to catastrophic damage while simultaneously hardening the vulnerable structures to withstand debris strikes at any time during ascent. John's certification approach to the debris environment had four key

---

[4] Flight rationale is the justification needed to proceed with another launch. Flight rationale is typically derived from test data, analysis, flight history, operational controls, flight rules, mitigations, engineering judgment, or a combination of factors.

elements. Ideally, he wanted to eliminate debris sources by redesign or mitigation. If that was not possible, he wanted to understand the debris transport mechanism and impact tolerance for any debris that could reach the shuttle's vulnerable structures. Finally, John wanted an on orbit inspection and repair capability to determine if reentry was possible. If not, he recommended implementing contingency shuttle crew support (CSCS) procedures using the ISS as a safe haven until a rescue shuttle was launched.

The first certification action involved redesign of the ET bipod and Solid Rocket Booster (SRB) bolt catcher and identification of all potential debris sources on the vehicle—not just the external tank. The bolt catcher contained the fragments from the explosive bolts used to separate the SRB from the ET. John ordered a comprehensive audit of all potential debris sources and wanted them classified as "unexpected" or "expected" debris. Unexpected debris was not inherent in the design if the design was implemented properly and was operated within its certification limit. Examples include things like thermal tiles or orbiter windows. The vehicle itself generated expected debris. This debris was inherent in the system design to liberate and included items such as foam and ablators. The vehicle would have to be able to fly through the expected debris environment when operating in its certified flight envelope. Numerous debris generation tests (including ice characterization and liberation testing) were performed to understand the failure mechanism(s) associated with all expected debris.

John also ordered a complete overhaul and enhancement of the imagery assets to detect all falling debris and possible damage to the shuttle during liftoff and ascent. As in the 1986 *Challenger* accident, when *Columbia* lost its foam, ground camera failures resulted in the loss of vital data to assess the accident. The plan was to place 107 cameras on and around the launchpad to film the shuttle during liftoff. These cameras included various infrared, high-speed digital video, high definition TV, 35 mm, and 16 mm ground cameras. In addition ten sites within forty miles of the launchpad would be equipped with cameras to film the shuttle during ascent. Two WB-57 aircraft (modified U-2s) would be outfitted to film the shuttle from high altitude as it ascended. This was in the event cloud cover obscured ground cameras. Three radar tracking facilities, one with C-band radar and two with Doppler radar, would also monitor the shuttle to detect debris releases or orbiter impacts. Members of the external tank project took responsibility for the new digital video cameras to be installed on the external tank to monitor the orbiter's underside and relay the data to ground control through an antenna installed in the tank. The Solid Rocket Booster project office took responsibility for installing cameras on the solid rocket booster noses to monitor foam debris from the external tank. The shuttle crew would be provided with new handheld digital cameras to photograph the external tank after separation. These images would be downloaded to laptops on the orbiter and then transmitted to ground control. Areas of foam failure on the tank could then be determined. Under the leadership of Steve Poulos and the orbiter project, engineers and technicians installed sixty-six tiny accelerometers and twenty-two temperature sensors in the leading edge of both wings on the orbiter. These devices were designed to detect the impact of any debris hitting the orbiter's wings and RCC panels. In combination the imaging, radar, and wing sensors would enable confirmation of the expected debris environment and possible damage from debris strikes. This allowed engineers to analyze these images and make recommendations to the crew during the mission to fly as is, make a repair, or declare a rescue mission.

The second certification action involved understanding debris transport for all generated expected debris. Prior to STS-107, the external tank and bipod ramp were not tested in the complex flight environment, and fully instrumented external tanks were never launched to gather data for verifying analytic tools. The accuracy of the analytic tools used to simulate the external tank and bipod ramp were verified only by using flight and test data from other space shuttle regions. Although significant analytic advancements had been made since the external tank was first conceived—particularly in computational fluid dynamics—more work was needed. This shortcoming had resulted from the development of three generations of the ET—the original, lightweight, and super lightweight tank. The bulk of the testing had been done on the original tank, and analysis had largely qualified the two succeeding designs. Although the underlying structure had been extensively tested, no wind tunnel testing had been carried out since the early 1980s. Computational fluid dynamics (CFD) include a computer-generated model that uses fluid flow physics and software to create predictions of flow behavior and stress or deformation of solid structures. The CFD solutions enabled engineers to develop a comprehensive debris transport model. This was needed to determine where the debris was going to travel if released and characterize the impact conditions if the debris hit the orbiter.

John was interested in developing a debris transport model, but he also knew the bipod ramp removal changed the aerodynamic and thermal environments around the bipod attachments. The bolt catcher redesign required a new thermal environment be developed for the design process. Both changes occurred in complex flow fields where multiple aero and thermal interactions were happening. The bipod region was home to a complex flow field. During ascent the solid rocket booster and orbiter shocks impinged forward of the bipod and on the intertank. There were also numerous protuberance interactions between the bipod structure and liquid oxygen feed line as well as attached and separated flows. Verification of the new designs and verification of the debris transport model required an integrated vehicle wind tunnel test. This would be performed at the Arnold Engineering Development Center (AEDC) in Tullahoma, Tennessee. NASA Ames used Columbia—the fourth-largest supercomputer in the world—to perform the majority of CFD computations. The wind tunnel test was designated IA-700. The 700 series designated all post-*Columbia* accident aerodynamic wind tunnel tests in support of RTF. Specific IA-700 test objectives included: one, verify air loads for the bipod foam closeout redesign; two, validate the CFD flow field in support of debris transport analysis; three, support the aerothermal analysis of the bipod foam closeout redesign; and four, validate the aeroacoustic environment. Direct heating and pressure measurements on the bipod assembly and the bolt catcher were performed at the Calspan-University of Buffalo Research Center (CUBRC) in Buffalo, New York.

The third certification action required understanding the impact tolerance for any expected debris that could be generated and reach the orbiter. This was accomplished by full-impact testing of foam, ice, ablators, and other debris on RCC panels, thermal tiles, and windows. The goal was to determine the damage characteristics under different impact conditions and when the structure failed. Ideally for certification an impact-tolerant structure able to withstand an impact without change of capability or configuration was desirable. If a structure demonstrated the ability to complete a mission even though impact had altered the design configuration or capability, it was considered "damage tolerant." The impact data were used to develop damage maps and determine the set of conditions considered a catastrophic failure that would necessitate a repair or rescue mission. John worked closely with Steve Poulos and the orbiter project after they initiated a nondestructive evaluation (NDE) of RCC panels and thermal tiles to determine the effect of such factors as age and thermal cycling on structural integrity and impact resistance. There was also a set of tests designed to evaluate different hardening methods and ways to increase the impact tolerance of vulnerable structures.

The fourth certification action involved developing contingency plans in the event a debris impact compromised the orbiter's thermal protection system. Deputy Program Manager John Shannon took the lead on the effort with the support of Muratore's team. The astronauts would use digital cameras to inspect the orbiter while in orbit. The Canadian Space Agency constructed a fifty-foot-long extension—the remote manipulator system/orbiter booster sensor system (RMS/OBSS)—that would attach to the shuttle's existing robotic arm. This extension allowed the RMS to reach the orbiter's underside. Cameras mounted on the extension could photograph the wing leading edge and orbiter underside for damage. NASA also formulated ideas about how to repair damaged tiles or RCC panels in orbit. This included applying pre-ceramic polymers to small cracks and using small mechanical plugs made of carbon-silicone carbides to repair damage up to six inches in diameter. If it was determined the heat-shielding tiles and RCC panels were damaged beyond repair capabilities, NASA could launch a rescue

RETURN TO FLIGHT

mission designated as STS-300—launch on need (LON). A CSCS plan was developed for such a rescue flight, and it could be launched within forty days of being called up. During that time the damaged (or disabled) shuttle's crew would have to take refuge on the space station. The space station would be required to support both crews for around eighty days. Oxygen supply would be the limiting factor. Initially the plan was for the damaged shuttle to be abandoned and burned up on reentry. However, the shuttle had the capability to autonomously perform entry and landing, and if a five-billion-dollar asset could be saved, why not try? The prime landing site for an autonomous orbiter reentry was Vandenberg Air Force Base in California. Vandenberg was selected due to its proximity to the Pacific coast. This allowed the shuttle to be ditched in the ocean should a problem develop that would make landing dangerous. Although the CSCS plan placed an extra logistical burden on the shuttle team (two vehicles had to be ready for launch instead of one), it was unanimously accepted.

CHAPTER 4

The Amazing Shuttle Program

The Space Shuttle Program was enormous and involved support from multiple NASA centers and contractors throughout the country. Due to the program's massive size, it is imperative to understand the fundamental organizational structure regarding technical authority and operations. NASA, like many federal agencies, is primarily a matrix organization. This means the lines of authority are not necessarily straightforward. At the basic level, three major types of entities are involved in the human spaceflight program: NASA field centers, NASA programs carried out at those centers, and the industrial and academic contractors. Each entity corresponds to a specific hierarchy or level. NASA headquarters is "Level 1." The program office at the JSC is "Level 2." The project offices that manage each element of the shuttle are "Level 3," and project support contractors are "Level 4." NASA's ten field centers provided the buildings, facilities, and support services for the agency's various programs. The programs, along with field centers and headquarters, hired civil servants and contractors from the private sector to support various aspects of the enterprise. In 2004 NASA had a total workforce of more than eighteen thousand employees, forty thousand contractors, and an annual budget around sixteen billion dollars.

The head of NASA at the time of the *Columbia* accident was Sean O'Keefe. President George W. Bush nominated him, and the US Senate confirmed him on December 21, 2001, as the agency's tenth administrator. O'Keefe joined the Bush administration on inauguration day and served first as the deputy director of the Office of Management and Budget. In this capacity he helped oversee the preparation, management, and administration of the federal budget and government-wide management initiatives across the executive branch. Prior to joining the Bush administration, O'Keefe held a number of leadership positions in academia and served as the secretary of the navy from 1992 through 1996 and as the comptroller and chief financial officer of the Department of Defense from 1989 to 1992. Known more for his budgetary and fiscal expertise, many within the agency viewed Sean as a bureaucratic outsider who lacked the technical knowledge needed to run the agency effectively. However, his critics did not deter the stern but mild-mannered O'Keefe, and he eliminated a $5 billion cost overrun in the ISS construction before helping NASA cope with the trauma of the space shuttle *Columbia* accident and its aftermath. One of Sean's most controversial decisions came in January 2004 when he canceled an upcoming shuttle mission to service the aging Hubble Space Telescope. (This was later overturned.) In light of the *Columbia* accident, he claimed the mission would be too risky due to the inability to reach the space station as a safe haven if debris damaged the shuttle. Before the technically astute Michael Griffin replaced him in April 2005, Sean had the shuttle program preparing for a Return to Flight, and he reorganized NASA to start work on President Bush's 2004 Vision for Space Exploration—a plan to send humans to the moon and Mars in what became known as the Constellation program.

Washington, DC
NASA Headquarters

NASA headquarters was responsible for leadership and management across five strategic enterprises: aerospace technology, biological and physical research, Earth science, space science, and human exploration and development of space. NASA's organizational structure changed over the decades to make the agency more efficient and adaptable to changes in its direction and priorities in the post-Apollo era. After the *Columbia* accident, this structure reflected the strategic plan formulated in 2005 to carry out the goals set forth in President Bush's plan. NASA in 2004 had four main offices, or mission directorates, through which it worked to accomplish its goals and provide the strategic management for the Space Shuttle and ISS programs: one, the Aeronautics Research Mission Directorate conducted research and development for safe, reliable flight vehicles and aviation systems; two, the Exploration Systems Mission Directorate developed technologies to support human and robotic exploration of space; three, the Science Mission Directorate was chartered to explore the Earth, sun, solar system, and universe through missions the principal investigators (scientists) within NASA and academia proposed; and four, the Space Operations Mission Directorate oversaw all launches, operations, and communications for all spacecraft (such as the space station and shuttle) both in Earth orbit and beyond. The mission directorates were all located in NASA headquarters and formed the core of the agency's efforts and coordination activities within its various centers across the country. Bill Readdy, a former pilot astronaut with three spaceflights, served as the associate administrator of the Space Operations Mission Directorate at the time of the accident. Readdy chaired the Space Flight Leadership Council, which was charged with oversight of NASA's successful Return to Flight STS-114 mission. He retired from NASA in October 2005 after William Gerstenmaier (Gerst) replaced him. Through the remainder of the shuttle program, Gerst was responsible for assessing the numerous debris challenges, evaluating the debris risk, and making the final launch decisions as chair of the shuttle program's flight readiness reviews.

RETURN TO FLIGHT

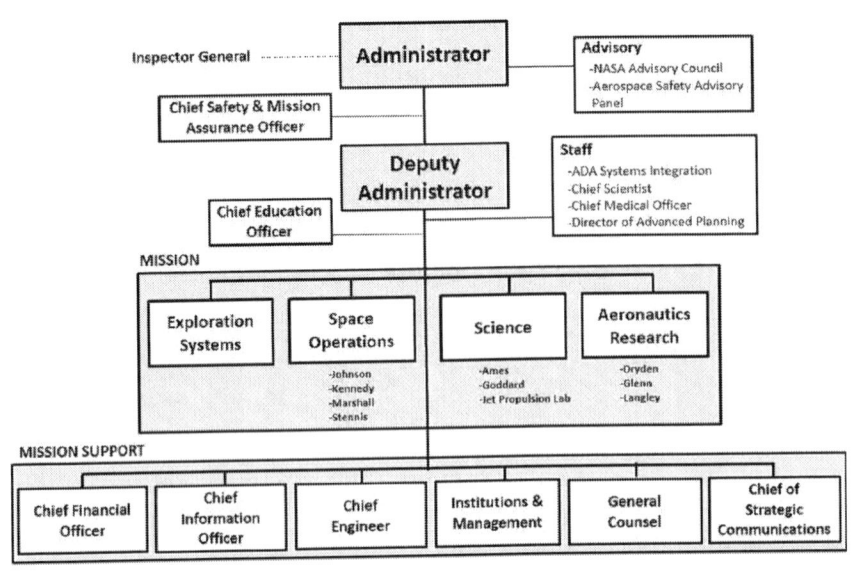

NASA headquarters organizational chart in 2004. (Courtesy of NASA.)

Although NESC was based out of Langley, it reported directly to NASA's chief engineer and served as an independent organization dedicated to promoting safety through engineering excellence while remaining unaffected and unbiased by the programs it evaluated. The NESC's basic goal was to help solve NASA's most difficult issues by bringing together some of the best engineers in NASA and partnering with industry, academia, and other government agencies. Ralph Roe's biggest concern at the outset was whether he could actually recruit and hire some of NASA's best engineers to participate in this new organization. As it turned out, NESC received an overwhelming response owing to strong support by NASA leadership all the way up to the administrator and the safety-conscious environment created after the accident. "We had over nine hundred applicants for the first fifteen technical leadership jobs," recalled Ralph. "I was proud that several years later I could happily report the concept had worked very well. By 2011 NESC had conducted nearly four hundred engineering and safety assessments and shown that when you bring solid technical expertise to help solve the problem, it breaks down a lot of barriers, and you often get more robust solutions."

Johnson Space Center
Houston, Texas

The Lyndon B. Johnson Space Center in Houston, Texas, established in 1961 as the Manned Spacecraft Center, has led the development of every US human spaceflight program. The JSC has been home to the offices of the Space Shuttle Program, the ISS program, and the shuttle's orbiter project. JSC's facilities have included the training, simulation, and mission control centers for the space shuttle and space station. It also had flight operations at neighboring Ellington Field, where they stationed the training aircraft for the astronauts and support aircraft for the Space Shuttle Program. JSC also managed the White Sands Test Facility in New Mexico where hazardous testing was conducted. The JSC is a 1,620-acre site located twenty-five miles southeast of Houston. After the site selection was announced in September 1961, personnel began moving their offices from Langley Field in Virginia to Houston. For more than four decades, NASA's Johnson Space Center has led the way in the continuing adventure of human space exploration, discovery, and achievement. In 2004 the JSC civil service workforce consisted of about three thousand employees. The majority were professional engineers and scientists. Of these approximately 110 to 130 were astronauts. About fifty companies provided contractor personnel to the center, and more than twelve thousand contractors worked on-site or in nearby office buildings and other facilities. The JSC was responsible for the training of space explorers from the United States and other ISS partner nations. It was, therefore, the principal training site for both space shuttle crews and ISS Expedition crews. The center's famed mission control center is often referred to as the nerve center for the US human spaceflight program. Since 1965 and *Gemini IV*, men and women working in MCC have been instrumental to the success of every crewed spaceflight. Today the MCC supports all space station missions and simulations.

Kennedy Space Center
Cape Canaveral, Florida

The John F. Kennedy Space Center (KSC) is America's premier spaceport. It was created in the Project Mercury days to launch the Apollo missions to the moon. The KSC provided launch and landing facilities for the space shuttle and has been the launch facility for all NASA manned missions. Rocket stages and payloads (including the space shuttle and Delta rocket) were assembled in the Vehicle Assembly Building and transported to one of several launchpads. The Launch Control Center at KSC conducted all operations at the launchpad after the space shuttle was rolled into position on the mobile launch platform. The center is located on Merritt Island, Florida, and is adjacent to Cape Canaveral Air Force Station, which also provided support for the Space Shuttle Program and was the site of the earlier Mercury and Gemini launches. KSC personnel supported launch and landing operations and maintenance and overhaul services for the orbiters. They also assembled and checked the integrated vehicle prior to launch, and they operated the Space Station Processing Facility, where components of the orbiting laboratory were packaged for launch aboard the space shuttle.

Marshall Space Flight Center
Huntsville, Alabama

The Marshall Space Flight Center (MSFC), in Huntsville, Alabama, is home to NASA's rocket propulsion efforts. The Space Shuttle Project Office located at Marshall was organizationally part of the Space Shuttle Program Office at JSC and managed the manufacturing and support contracts of Boeing Rocketdyne for the space shuttle main engine (SSME), Lockheed Martin for the external tank (ET), and ATK Thiokol Propulsion for the reusable solid rocket motor (RSRM)—the major piece of the solid rocket booster. MSFC was also involved in microgravity research and space product development programs that flew as payloads on the space shuttle.

NASA's Other Shuttle Testing and Research Centers

The Stennis Space Center in Bay St. Louis, Mississippi, is the largest rocket propulsion test complex in the United States. Stennis provided all the testing facilities for the space shuttle main engines and external tanks. The solid rocket boosters were tested at the ATK Thiokol Propulsion facilities in Utah.

The Ames Research Center at Moffett Field, California, has evolved from its aeronautical research roots into the Center of Excellence in Information Technology. Ames's primary importance to the Space Shuttle Program, however, lay in wind tunnel and arc jet testing, development of thermal protection system concepts, and utilization of NASA's supercomputer for analytical computations.

The Langley Research Center at Hampton, Virginia, is the agency's primary center for structures and materials. It supported the shuttle program in these areas and in basic aerodynamic and thermodynamic research.

Space Shuttle Program Office

In 2004 the two major human spaceflight efforts within NASA were the Space Shuttle Program and International Space Station. Both programs had headquarters at the Johnson Space Center but reported to the deputy associate administrator of space operations at NASA headquarters in Washington, DC. The Space Shuttle Program Office at JSC was responsible for all aspects of developing, supporting, and flying the space shuttle. To accomplish these tasks, the program maintained a large workforce at the various NASA centers that hosted the facilities the program used. The program office was also responsible for managing the spaceflight operations contract with United Space Alliance. This provided most of the contractor support at JSC and KSC and a small amount at MSFC.

The Space Shuttle Program employed a wide variety of commercial companies to provide services and products. Among these were some of the largest aerospace and defense contractors in the country—United Space Alliance (USA), Boeing Company, Lockheed Martin, and more. USA, a joint venture between Boeing and Lockheed Martin, was established in 1996 to perform the spaceflight operations contract and conduct day-to-day space shuttle operation. Headquartered in Houston, Texas, USA employed more than ten thousand people at Johnson, Kennedy, and Marshall. Rockwell International, located in Downey and Palmdale, California, designed and manufactured the space shuttle orbiter. In 1996 Boeing purchased Rockwell's aerospace assets and later moved the Downey operation to Huntington Beach, California, as part of facilities consolidation. As a subcontractor to USA, Boeing provided support to orbiter modifications and operations. Work was performed in California and at Johnson and Kennedy Space Centers. Before the Boeing Company acquired them, the Rocketdyne Propulsion and Power Division of Rockwell International was responsible for the development and manufacture of the SSMEs. The Space Shuttle Project Office at MSFC managed the main engines contract, and most of that work was performed in California or at Stennis and Kennedy. ATK Thiokol Propulsion in Brigham City, Utah, manufactured the RSRM segments—the propellant sections of the SRBs. Lockheed Martin Space Systems developed and manufactured the external tank at the NASA Michoud Assembly Facility near New Orleans. The external tank was the only disposable part of the space shuttle system, so new tanks were continually under construction. Lockheed Martin's Missiles and Fire Control division made the RCC panels used on the orbiter's nose and wing leading edges.

Near the end of the Apollo program, NASA officials were looking at the US space program's future and considering whether to continue using disposable rockets. A reliable, less expensive, and perhaps reusable rocket was needed. The idea of a reusable space shuttle that could launch like a rocket but land like an airplane was appealing and would be a great technical achievement. NASA began design, cost, and engineering studies on such a space shuttle. Many aerospace companies also explored the concept. In 1972 President Nixon announced NASA would develop a reusable space shuttle—the Space Transportation System (STS). NASA decided the shuttle would consist of an orbiter attached to solid rocket boosters and an external fuel tank, and they awarded the prime contract to Rockwell International. The shuttle's two SRBs and three main engines provided the nearly 7 million pounds of thrust needed to lift the 4.5-million-pound shuttle into low Earth orbit. The main engines provided nearly 30 percent of the total thrust needed to lift the shuttle off the pad and into orbit. The main engines burned liquid hydrogen and oxygen stored in the external fuel tank at an amazing rate. Fuel was consumed at an equivalent of ten thousand pounds every second. This would empty an average family swimming pool every ten seconds! The external tank was 158 feet long, had a diameter of 27.6 feet, and held a half million gallons of cryogenic propellant. The original shuttle design had a reusable first stage, but the inclusion of the SRBs and ET was a design compromise NASA accepted to minimize development cost. Both shuttle accidents (*Challenger* and *Columbia*) resulted from implications of that original compromise.

The program office at JSC was the tip of the iceberg in the mammoth shuttle program. All the program requirements, directives, and procedures were contained in the multiple volumes of NSTS 07700 documents. These define the fundamental organizational responsibilities and requirements for space shuttle operations and key leadership positions. The shuttle program manager controls these documents, and any changes, waivers, or deviations to these requirements have to be approved through a manager-chaired Program Requirements Control Board. The management and organizational responsibilities and relationships are described for the various phases of the space shuttle operational activities involved in the preparation for and execution of each mission. This includes a description of the requirements and responsibilities for management of the launch commit decision process. All elements of the shuttle program must adhere to these baseline requirements.

Along with the program manager, other key leadership positions on the shuttle program included the deputy program manager and launch integration manager. The Program Manager delegated key responsibilities for flight preparation, mission execution, and landing activities

RETURN TO FLIGHT

to the deputy program manager at JSC. The deputy also chaired the MMT during a mission. This was done after the 1986 *Challenger* accident to serve as a check on decision-making to prevent the possibility of groupthink. The MMT had representation from every project office and program support organization, and it was the program decision-making body responsible for programmatic trades and decisions associated with launch countdown and in-flight activities. The MMT was the interface between real-time flight operation activities flight directors led and senior NASA management at headquarters and various centers. The Mission Operations Directorate at JSC managed the actual mission operations. This body was responsible for crew training, the MCC, and simulation and training facilities. It also provided the teams in mission control. Senior experienced personnel called flight directors led the MCC teams. *Apollo 13* made this role legendary. The film depicted Flight Director Gene Kranz and his team's epic fight to bring the Apollo 13 crew home after an explosion in space.

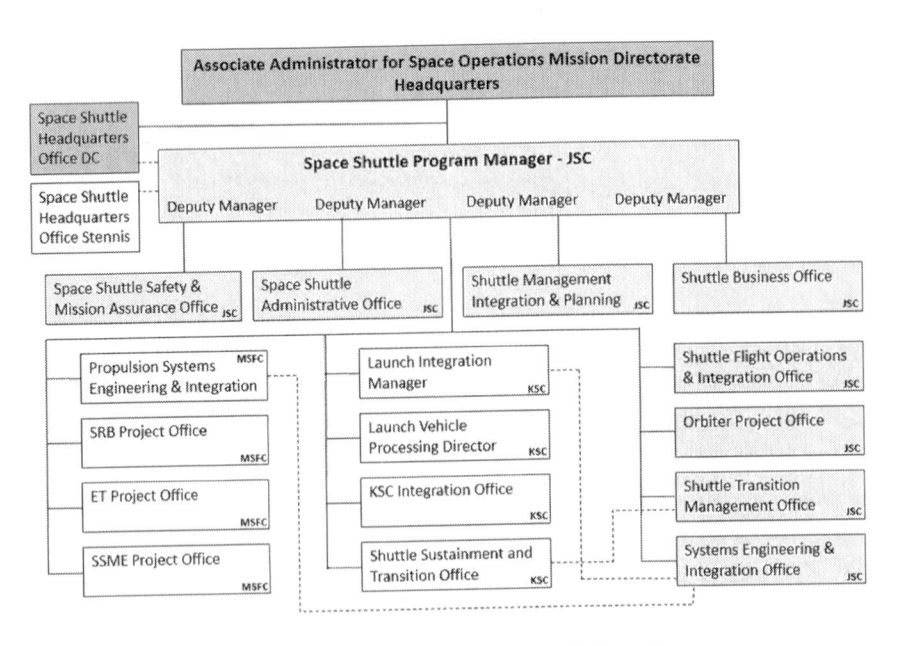

Space shuttle program organizational chart in 2004.
The dotted lines represent dual reporting paths.

The Program Manager assigned all the launch operations to the launch integration manager at KSC. Once the vehicle cleared the launch tower, flight directors from the JSC Mission Operations Directorate were responsible for conducting real-time shuttle flight operations in accordance with established mission plans and procedures. This responsibility included integrating all activities that supported the flight control team in the Mission Operations Center. The processes defined in NSTS 07700 ensured the successful performance of the flight and ground systems processes were interwoven into all program activities with shared responsibility throughout all program levels.

Preparing a shuttle for flight was an enormous undertaking and required a regimented process that included a series of systematic reviews starting at "Level 4" and working up to the major "Level 1" reviews. The program followed a documented flight preparation process that outlined the steps leading to the completion of a flight readiness review (FRR). This provided the certification of flight readiness (CoFR) for a specific shuttle mission. The major elements of the flight preparation process (in order) were the project milestone review, external tank/solid rocket booster mate milestone review, orbiter rollout review, and flight readiness review. The program FRR was done two weeks prior to launch to assess relevant documentation and data in sufficient detail to ensure flight hardware discrepancies, anomalies, and launch constraints had been fully evaluated and resolved. At this review's conclusion, NASA and shuttle contractors signed the CoFR. This indicated the program was ready to proceed with the launch and start the official countdown. Two days prior to launch (L minus two), the MMT was activated to assess any changes since the FRR and provide a go-no-go decision regarding the countdown. The L minus two review included the closure status of any actions assigned at the FRR and resolution of any constraints to launch identified at that review. During the countdown the MMT provided recommendations (as necessary) to support programmatic decisions outside the responsibility or authority of the launch director at KSC or flight director at JSC. At T minus nine minutes the launch MMT was polled for a go-no-go launch decision. The final go-no-go launch decision was the responsibility of the launch integration manager at KSC.

After the shuttle cleared the launchpad and during the mission, various project offices supported the flight directors at JSC and MMT. These offices included the Shuttle Program Office, mission evaluation room (MER), Huntsville Operations Support Center, and certain elements of the Launch Control Center and Customer Support Room. The MER provided real-time engineering services (as required) to support prelaunch, launch, orbit, and landing activities. Services included but were not limited to subsystem evaluation, anomaly resolution, analysis in support of mission action requests, and launch commit evaluation. The Huntsville Operations Support Center at MSFC provided technical support for the final launch preparation activities and recommendations for the disposition of in-flight anomalies and/or off-nominal operation of the space shuttle main engines, external tank, and solid rocket boosters. Once a shuttle landed, the entire process was repeated for the next flight.

Building 4N
Johnson Space Center

In 2004 former flight directors from the Mission Operations Directorate or astronauts from the crew office held most leadership positions at JSC and in the various program offices. The flight directors, astronauts, and managers in the program offices played the largest roles in driving the direction of the human spaceflight program, and each group had its unique way of asserting influence. Although the flight directors spent most of their time in Building 30 (MCC), they had offices located on the top floor in Building 4N. Even though the building was constructed in the mid-1960s, pictures and patches from previous crewed spaceflight missions decorated the walls. It had a nostalgic triumphal charm from decades of successful space endeavors. Over the years the Mission Operations Directorate has been the breeding ground for leaders such as Gene Kranz, Chris Kraft, Glynn Lunney, and more.

Building 30, Mission Control Center Johnson Space Center

Since 1965 mission control has been at the helm of America's human spaceflights, and since assembly of the International Space Station began in 1998, the center has become a focal point for human spaceflight worldwide. The teams that work in mission control have been vital to every US human spaceflight since the 1965 Gemini IV mission. That included the Apollo missions that took humans to the moon and more than 130 other space shuttle flights from 1981 to 2011. With a permanent human presence in space aboard the ISS, flight control teams of about a dozen experienced engineers and technicians are on duty seven days a week, twenty-four hours a day, 365 days a year. Flight directors and controllers keep a constant watch on the crew's activities and monitor spacecraft systems, crew health, and overall safety as they check every system to ensure operations proceed as planned. These highly trained flight controllers have the skills needed to closely monitor and maintain increasingly complex missions and respond to unexpected events.

About twenty flight controllers staffed the space shuttle's Flight Control Room when in operation. These "front rooms" are supported by dozens of experts working in the Mission Evaluation Room and back rooms located around the perimeter of the main control rooms. Mission control's focal points are the two Flight Control Rooms. These are where flight controllers get information from console computer displays or from projected displays that fill the wall at the front of the room. Atop each console is a sign with abbreviations for its respective function. Each console has a call sign—the name the flight controller uses when talking to other controllers over the various telecommunication circuits. Televised pictures of flight controllers working at their consoles with headsets in place are often shown on network news and the NASA channel. There are two functionally identical Flight Control Rooms. One is on the second floor, and one is on the third. Either room can be used for mission control, or both can be used to control separate, simultaneous flights. More often one team of flight controllers conducts flight operations while a second team participates in highly realistic training for a future mission. Flight controllers who work in the Flight Control Rooms represent only a small percentage of the total staffing in the MCC. Each of the twenty to thirty flight controllers who sit at consoles in the Flight Control Room has the help of many other engineers and flight controllers who monitor and analyze data in nearby staff support rooms.

The historic Apollo Flight Control Room in Building 30 has been preserved and designated a national historical landmark. The Apollo Room from which controllers guided humankind's first landing on the moon was last used operationally in 1995 for control of space shuttle missions. John Muratore led the effort to revamp Building 30 in the mid-1990s with the newest generation of control room hardware and computer workstations. This change from the old Apollo-style mainframe computer-based rooms to the current architecture reduced the cost of operating and maintaining mission control and dramatically increased the center's mission support capabilities. During the shuttle missions, MMT meetings were held in a large, secure room on the first floor. Several theater-sized viewing screens and high-tech video equipment to communicate face-to-face with all NASA centers outfitted the room.

Building 4S
Johnson Space Center

Adjacent to Building 4N and connected by only a first-floor corridor is the newer 4S building. This is home to astronaut offices on the top floor. The building was constructed in the early 1990s for the engineers and contractors working on the space station. The term "astronaut" (derived from the Greek words meaning "space sailor") refers to all who have been launched as crewmembers aboard NASA spacecraft bound for low Earth orbit and beyond. The vast majority of astronauts have been current or former military officers, and since the selection of the Mercury Seven astronauts in 1959, they have been the most public, featured attraction at NASA. There is a special camaraderie and tight allegiance among the crewmembers who routinely put their lives on the line for the love of spaceflight. Of all the NASA organizations, the crew office was one of the few that could exercise veto power over any objectionable decisions. After all, it was their lives at risk, and if push came to shove, they wanted to be in control of their destinies. Getting consensus among astronauts was sometimes difficult, and there was even a running joke among program managers—ask any two astronauts their opinions, and you'll get three. Over the years NASA has taken great strides to protect their astronauts' images and reputations. Many times inappropriate or unacceptable behavior was dismissed, overlooked, or simply covered up. In addition NASA rarely removed astronauts from missions. Even with all the astronauts' successes, some within the NASA community criticize the crew office as being elitist. This is given their locked office doors and special privileges. Fortunately most have a good working relationship with the crew office, and I personally have had a rewarding experience working for former astronauts both at Boeing and NASA.

Building One, JSC Headquarters
Johnson Space Center

The JSC headquarters and program offices are located in Building One. At nine stories it is the tallest structure on the center's grounds. It was built in the early 1960s, and Robert Gilruth, JSC's first center director, declared it the headquarters building. Gilruth was a legend in the flight test community and had written the bible on aircraft handling capabilities while at the National Advisory Committee for Aeronautics (NACA). Gilruth often developed techniques while flying in cramped backseats behind NACA test pilots. With its tight linkage of engineering, medical, mission control, and flight crew, JSC takes much of its personality from its first leader. The center director of the JSC resides on the top floor, and the Shuttle Program and

System Engineering Offices are located on the fifth and seventh floors respectively.

Following the release of the CAIB findings, NASA implemented significant organizational changes in the shuttle program. This included replacing key members, restructuring the SEI team, and broadening representation from safety, engineering, and project teams in the decision-making process. As part of NASA's response to the loss of *Columbia* and its crew, Space Shuttle Program Manager Bill Parsons restructured and strengthened the role of the existing SEI Office. The restructured office was responsible for reviewing and integrating all elements of the program. The SEI duties were also expanded to include all the Return to Flight debris work with emphasis on the importance of rigorous engineering testing and analysis as key ingredients in the systems engineering function. John Muratore (the new SEI manager) reported directly to Bill, and this established SEI within the space shuttle organization as the program's integrator of all space shuttle elements. The SEI Office had unprecedented power and responsibility that extended to all aspects of systems engineering, integration, performance, and safety of the space shuttle vehicle. This included all ground and flight activities where multiple project elements were involved.

John was responsible for overseeing all multi-element, integrated analysis relating to certification of flight readiness (CoFR), and he was a designated member of the Program Requirements Control Board and Mission Management Team. Review and approval of element recommendations and actions to ensure appropriate integration of activities in the shuttle program were conducted at the Systems Integration Control Board (SICB). John chaired the SICB, and all project elements and NASA centers directly involved in the Space Shuttle Program (most notably MSFC and KSC) answered to him. A member of the SICB who had a major role supporting debris activities was Helen McConnaughey. She was from the Propulsion Systems Engineering and Integration (PSEI) Office. She was a tall, slender woman with striking blue-green eyes and blond hair. She looked more like a model than a NASA project manager. Her PhD in applied mathematics and combustion with a minor in fluid mechanics from Cornell made her well suited for a career at NASA and with propulsion systems. When Helen joined NASA, she spent eighteen years in engineering doing computational fluid dynamics, propulsion testing, and vehicle systems design before joining the shuttle program to build the PSEI organization. Although very happy in her position as Vehicle and Systems Department manager, Marshall's Center Director Dave King told her they needed someone who was a strong leader and who knew about propulsion systems. She saluted and took over as the PSEI manager in September 2003. "I think one of the biggest challenges initially was dealing with resentment and antagonism from all the elements," she recalled, "because we were viewed as bad guys for getting into their business. It made it difficult to do our integration job, but it also made it discouraging on the personal, human level."

Very few people were technical matches for John, but Helen's background and experience complemented his. Although she didn't have John's experience with flight systems, she was a natural, intuitive engineer on whose judgment John relied heavily. Since most rework on shuttle RTF involved MSFC, and because MSFC and JSC had a traditional rivalry, Helen served a key role in trying to keep the centers working smoothly together. Helen and her team supported all the debris testing conducted at MSFC and formed a small debris team that Louise Strutzenberg headed. This team tackled liftoff debris associated with debris sources from the launchpad. The SICB was given the authority initially to baseline, dispose of, and implement changes to program requirements and documents. It was further authorized to prepare and maintain systems integration plans in support of those authorized changes. Only changes affecting the program cost, safety, or schedules were forwarded to the Program Requirements Control Board (PRCB) for final disposition. Although the multielement SEI teams at the various centers performed unique technical activities local to their individual centers, the entire SEI organization reported technically to and received direction from John.

Space Shuttle Systems Engineering and Integration Office Johnson Space Center

The SEI office at JSC was organized into three offices. The Engineering Integration Office provided the fundamental systems engineering capability and managed the core technical disciplines necessary to produce and assess the natural and induced vehicle environments. The Element Integration Office provided insight into the daily operation of Level-3 elements to ensure the integrated effects of design and operational changes in the system were properly evaluated and resolved, and the Flight Software Office was responsible for approving flight software loading and managing configuration control[5] of the flight software. It did not take John long to make a few changes to the SEI organization after he took over. Rick Schmidgall remained as John's deputy, and Darryl Stamper was still head of the Flight Software Office. However, John replaced Brian Anderson and Rod Wallace in Element Integration with Kim Doering and Bob Ess respectively. Bob was given the primary debris responsibilities and led the, multi-discipline teams in developing technical solutions for debris challenges. A rising star within NASA, Bob was young and focused. He had worked for John on the X-38 program, and before that he had spent two years at PAX River Navy Test Pilot School as a flight test engineer. This provided him exceptional skills of analysis and interpretation of technical data. Even though he had gone through the astronaut interview as a mission specialist candidate and was not selected, this did not slow his career. Kim Doering was a Notre Dame graduate and a rising star in her own right. She had all the attributes of the character Kate Jackson from *Charlie's Angels*. She was attractive but with the brains to match. Darryl Stamper was one of the few post-*Columbia* employees to remain following the SEI reorganization. Rod Wallace was another. He had spent the majority of his NASA career on the space shuttle and was

---

[5] Configuration control is a process that ensures all changes to a complex system are performed with the knowledge and consent of management.

promoted to chief engineer of the SEI Office when Bob Ess joined the team.

The daily activities worked by the SEI Office included requirement management, environmental changes and assessments, integration of multielement changes, integrated risk and hazard analysis, combined element verification, flight anomaly investigation, and flight software. In addition to supporting all major program reviews, Bob Ess and the engineering integration team monitored all the technical panels and working groups. Kim Doering and her element integration team monitored all Level-3 Chief Engineer Control Boards. During a mission the SEI Office provided back room support during launch day and worked to resolve any real-time problems.

Both element integration and engineering integration teams reviewed and assessed all design center non-conformances being tracked in the shuttle Problem Reporting and Corrective Action (PRACA) system to determine the implications to the integrated vehicle and other project elements. If a PRACA-reportable nonconformance occurred during the mission, it was classified as an in-flight anomaly (IFA). The project office or design identified IFAs and reported them to the SICB. From there they were evaluated to determine whether the problem should be elevated to an integrated in-flight anomaly (IIFA). An integrated IFA had the potential to affect multiple elements as well as integrated system performance, or to invalidate an integrated hazard. All identified integrated IFAs were maintained in a tracking matrix in support of the MMT during a mission and reported to the PRCB afterward. It was SEI's responsibility to lead the resolution of all IIFAs and validate ("close") the integrated hazards before the next mission could proceed. The integrated hazards were reports used to identify hazards to personnel, the vehicle, and other systems associated with the integrated shuttle. There was a hazard report for each technical discipline and major vehicle subsystem. Each hazard report contained a fault tree to assess all potential hazard conditions, document controls and verification methods for all hazard conditions, and specify the acceptance rationale for each hazard.

The hazard reports were a focus of the CAIB report because one of the hazard reports at the time of *Columbia* identified foam as low risk. John led a massive multicenter effort that reestablished baselines for all the hazard reports prior to RTF. In order to get this work done, John and his assigned lead, Steve Bauder, held hazard report review meetings for thirteen consecutive Saturdays during early 2004. SEI led the technical coordination with all the applicable program elements and presented the report to the Integrated Safety and Engineering Review Panel (ISERP) before going to the PRCB. Integrated debris risk assessments, however, were only part of the overall ISERP system. The ISERP served as an independent adviser to the program office to integrate all system safety risk assessment products in accordance with program hazard analysis techniques and reporting requirements. After the *Columbia* accident, the ISERP was given more representation and authority on the shuttle program through the hazard report approval process. A shuttle flight was essentially grounded until the ISERP approved or "closed" all hazard reports. Roy Glanville led the ISERP. He was a meticulous and dogmatic man with a reserved sense of humor. NASA Engineers tend to be creatures of conservatism, and Roy was the king. Preparing for the ISERP was like drinking from a fire hose when assessing the entire integrated risk environment—not just the debris risk. The CAIB criticized NASA for insufficient safety oversight, and Roy was going to ensure his safety and mission assurance organization was fully engaged.

Shortly after becoming the SEI manager, John implemented a weekly status meeting called "TopX" to discuss the latest status and planning associated with the top ten critical issues or priority tasks impacting the Space Shuttle Program and SEI. Debris was number one on the list and would remain there for the next five years. The TopX meeting was held every Monday to avoid interfering with the Tuesday SICB and Thursday PRCB meetings. They would sometimes last for several hours. The objective was to ensure senior SEI management was cognizant of the forward plans associated with designated issues and risks, but John often used the meetings as technical forums to resolve issues, identify work-arounds, or reestablish priorities. Many complained the TopX meeting took up too much time and was unnecessary because of redundancies with what the SICB covered. John disagreed and threatened to conduct it on the weekend to allow more time for SICB preparation. Eventually the complaints stopped.

CHAPTER 5

External Tank Overview

Volumetrically the external tank was the largest element of the space shuttle. Because it was the common element to which the solid rocket boosters and the orbiter were connected, it served as the main structural component during assembly, launch, and ascent. It also carried the cryogenic propellant for the space shuttle main engines.

External Tank Construction

The basic and lightweight models of the external tank were constructed primarily of aluminum alloys. A special aluminum–lithium alloy was used for the super lightweight version of the tank. Steel and titanium fittings were used for attach points, and some composite materials were used in fairings and access panels. The external tank measured 153.8 feet in length and 27.6 feet in diameter and comprised three major sections: the liquid oxygen tank, the liquid hydrogen tank, and the intertank area between them. Both the liquid oxygen and liquid hydrogen tanks were welded assemblies of machined and formed panels, barrel sections, ring frames, and dome sections. They held 143,351 gallons of liquid oxygen at −297°F in the forward (upper) tank and 385,265 gallons of liquid hydrogen at −423°F in the aft (lower) tank. The cryogenic propellant flowed from each tank through a seventeen-inch-diameter feed line out of the external tank and into the orbiter through an umbilical line to the shuttle's main engines. Through these lines liquid oxygen flowed at a maximum rate of 17,600 gallons per minute, and hydrogen flowed at a maximum rate of 47,400 gallons per minute.

A minimum one-inch-thick coating of insulating foam covered the external tank to keep the super-cooled propellants from boiling and to prevent ice from forming on the outside of the tank while it sat on the launchpad. This insulation was so effective the external tank's surface felt only slightly cool to the touch, even though the liquid oxygen was stored at subzero temperatures. Insulating foam also protected the tank's aluminum structure from aerodynamic heating, which could reach temperatures in excess of 600°F in the first one hundred seconds of flight. Although generally considered the least complex of the shuttle's main components, the external tank was a remarkable engineering achievement. In addition to holding more than 1.5 million pounds of cryogenic propellants, the tank had to support the orbiter's weight while on the launchpad and absorb the 7.3 million pounds of thrust generated by the two solid rocket boosters and three space shuttle main engines during launch and ascent. The external tanks were manufactured at the MAF and transported by barge to the KSC in Florida for final vehicle assembly. Unlike the reusable solid rocket boosters that were retrieved, the external tank was discarded during each mission and burned up in Earth's atmosphere over the Indian Ocean after being jettisoned from the orbiter.

A feed line ran externally along the right side of the liquid hydrogen tank up and into the intertank. Its main purpose was to transport the liquid oxygen from the upper tank to the umbilical feed in the aft tank. The feed line was attached to the tank with five brackets that resembled an L-shaped boomerang. These brackets allowed movement ("articulation") of the feed line to compensate for propellant flow during fueling on the launchpad and during detanking.[6] Engineers

---

[6] The massive external tank was filled with super-cold liquid oxygen and hydrogen fuel just as it was on launch day to test upgrades made to the tank in advance of an upcoming flight. Detanking drained the tank.

RETURN TO FLIGHT

designed the brackets to accommodate external tank thermal expansion and contraction during launch. The feed line also contained three bellows (joints) that allowed the feed line to move or flex when the tank was assembled, when it was fueled, and during liftoff and ascent. The feed line brackets and bellows were very difficult to insulate due to their geometric complexity, and they were one of the major sources of ice formation on the external tank. Running beside the feed line were two five-inch-diameter pressurization lines—one to supply hydrogen gas to the liquid hydrogen tank and the other to supply oxygen gas to the liquid oxygen tank.

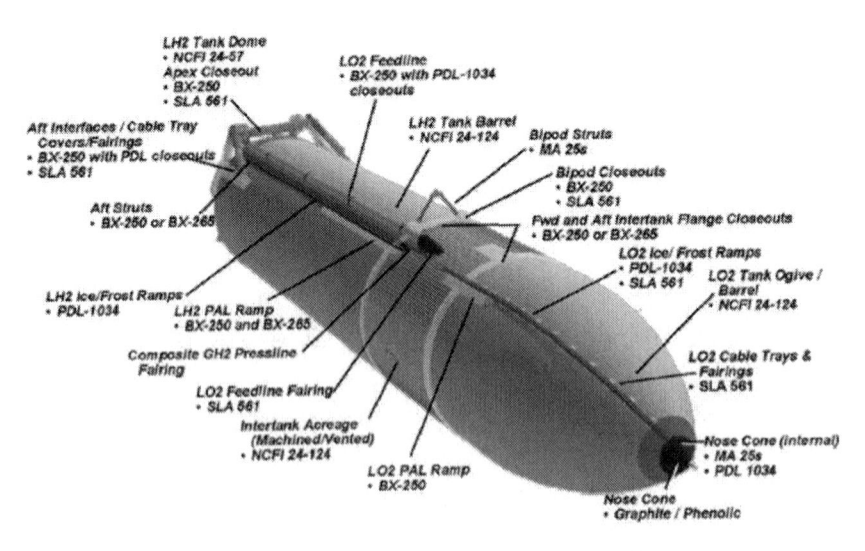

Location of key external tank components and their respective foam applications. (Courtesy of NASA.)

During launch the pressurization lines were used to maintain the ullage pressure—the pressure of the unfilled space in the container of liquid. Maintaining this pressure was critical because it forced the propellant into the massive turbopumps that fed the SSMEs. Loss of pressure would have led to loss of propulsion and could have caused the turbopumps to rev to the point of catastrophic failure. Adjacent to the pressurization line was a four-inch cable tray that routed all the sensor

EXTERNAL TANK OVERVIEW

data from the oxygen and hydrogen tanks through the umbilical and into the orbiter. Metal support brackets attached the pressurization lines and cable trays along the tank's length at multiple locations. Foam that was poured and formed into specific shapes protected these metal brackets from ice formation and frost during tanking operations. These foam shapes were called ice/frost ramps (IFRs). Thirty-four IFRs were on the tank: twelve on the liquid oxygen tank, six on the intertank, and sixteen on the liquid hydrogen tank. The size of the IFRs depended on their locations. The smaller ramps on the liquid oxygen tank were roughly 1.5 feet long by 1.5 feet wide by 5 inches high. Each weighed about twelve ounces. The larger ramps on the liquid hydrogen tank were roughly two feet long by two feet wide by one foot high. They weighed approximately 1.7 pounds apiece.

The intertank connected the oxygen and hydrogen tanks. The intertank was a 22.5-foot-long hollow cylinder made of eight stiffened aluminum alloy panels bolted together along longitudinal joints. Two of these panels were the integrally stiffened thrust panels. They were called this because they supported the solid rocket booster thrust loads and were located on the sides of the external tank where the solid rocket boosters were mounted. Each consisted of single slabs of aluminum alloy machined into panels with solid longitudinal ribs. The thrust panels were joined across the inner diameter by the intertank truss—the major structural element of the external tank. During propellant loading nitrogen was used to purge the intertank, prevent condensation, and prevent liquid oxygen and liquid hydrogen from combining.

Bolts and fittings on the thrust panels and near the aft end of the liquid hydrogen tank attached the external tank to the solid rocket boosters. Two umbilical fittings at the bottom and a "bipod" fitting located at the top attached the orbiter to the external tank. These fittings also contained fluid and electrical connections. The bipods were titanium forgings attached to the external tank by fittings at the right and left of the external tank centerline. Each forging contained a spindle that attached to one end of a bipod strut and rotated to compensate for external tank shrinkage during the loading of cryogenic propellants.

Michoud Assembly Facility, New Orleans

Although NASA's external tank program office was based at Marshall Space Flight Center, the external tank manufacturing facility and Lockheed Martin program managers were located at the Michoud Assembly Facility on the eastern side of New Orleans. The property boasted a long, colorful history dating back more than 240 years when it was part of an original thirty-five thousand-acre site first deeded to a Louisiana soldier–statesman by the king of France in 1763. In 1961 NASA acquired the Michoud facility through a congressionally approved property transfer from the Department of Defense. The site was used to support the nation's space program by serving as a final assembly facility for the design, development, and manufacture of large launch vehicles requiring water transportation to launch sites. Early development efforts culminated in July 1969 when Neil Armstrong and Buzz Aldrin visited the dusty surface of the moon. A Michoud-built Saturn 1C booster powered the crew of that Apollo mission on their flight. In 1973 Martin Marietta Aerospace was awarded a contract to design, develop, and manufacture the external propellant tanks for the space shuttle system. Lockheed Martin became the external tank contractor in 1995 when the merger of Lockheed Corporation with Martin Marietta formed the new company. The 833-acre Michoud facility contained one of the largest production buildings in the nation. At its height it had more than two thousand workers supporting the tank production.

External Tank Project Leadership

Shortly after the *Columbia* accident, the external tank project experienced more changes in management than any other shuttle organization on either the NASA or contractor side. At MSFC NASA named Sandra Coleman to manage the external tank project office. James Reuter was the deputy manager, and Neil Otte (the previous deputy) was the chief engineer. Coleman, Reuter, and Otte were responsible for fixing the foam debris problems in addition to the normal design, manufacturing, certification, and testing of the external tank. Sandra joined NASA in 1965. She worked in Marshall's Saturn Program Office and supported the effort that launched American astronauts to the moon. In 1969 she became a member of the Space Shuttle Task Team. From 1972 to 1997, Sandra served in three of the four shuttle propulsion project offices in increasingly responsible positions. This included as integration subsystem manager for the Solid Rocket Booster Project and deputy manager for the Reusable Solid Rocket Motor Project. Sandra and her team represented the commitment, experience, and leadership necessary to be successful.

Ron Wetmore of Lockheed Martin was the vice president of the external tank program that was responsible for ET assembly at the time of the *Columbia* accident. He joined Lockheed Martin in 1981 as a field engineer and progressed through a series of engineering positions. He was eventually named director of Kennedy Space Center operations in 1995. He served in that position until he was promoted to vice president of the external tank project in 2000. After the accident Ron served as the RTF manager for the ET project until Wanda Sigur replaced him. Wanda assumed the position days after Hurricane Katrina hit New Orleans in 2005. When the levees broke, eight feet of water flooded her home. "Maintaining focus was our biggest challenge. We had displaced employees, and several of those employees slept in the factory for days."

This was the first of many challenges Wanda would face. She began her Lockheed Martin career as a materials engineer in 1979, and she gained experience in a number of engineering positions. In 2000 she was promoted to director of the Engineering and Technology Laboratories. After the shuttle accident, Wanda led the investigation of the external tank's role in the mishap. Her job was to help investigators identify the problem, work to resolve the deficiencies, and regain the trust of the shuttle program managers. She was unassuming and quiet and relied heavily on Dan Callan, the director of mission success, Mike Quiggle, the chief engineer, Phil Kopfinger, the lead engineer for materials testing, and others to work the redesigns and debris issues.

Another key leader for the ET project was Jeff Pilet. (He eventually became the chief engineer.) The ET manager gave him the incredible responsibility of certifying the tank's thermal protection system for flight. This became a nearly impossible task due to the inherent flaws in the foam and foam application processes, all the analytic uncertainties, and limitations of the historical test data. Jeff's father started working on the ET project in 1977 as an hourly technician and was always very proud of the work he did and what it represented. "I still remember him coming home from work and trying to explain to my family why the tank had a liquid oxygen and liquid hydrogen tank. Of course, I didn't quite get it at the time because I was only in eighth grade, but the conversations sure stuck with me and got me interested in the space program."

Jeff still has a dollar bill his dad had John Young and Bob Crippen, the first shuttle astronauts, sign when Young and Crippen visited MAF prior to their STS-1 flight.

After the *Columbia* accident, the ET project either ran or was involved with a number of debris-related tests. This included aerodynamic and thermal testing of the bipod, aerodynamic and vibroacoustic testing of the integrated vehicle, thermal vacuum testing of the foam to investigate the failure modes, ice characterization and separation testing, foam dissection analysis, and more. Jeff and several of his engineering colleagues such as Jimmy Doll and Mike McBain were instrumental parts of the tests and were heavily involved with the data analysis. Jimmy focused on the data reduction and analytic model development. Mike led most of the material property and failure mode testing as well as the process development and validation for the new foam applications.

External Tank Thermal Protection System

There were two primary materials that served as the thermal protection system and coated the external tank: dense composite ablators for dissipating heat and low-density closed-cell foams for high insulation efficiency. Closed-cell foam materials were sprayed onto the tank, and they consisted of small pores filled with air and blowing agents that were separated by the thin membranes of the foam's polymeric component. The external tank thermal protection system was designed to maintain an interior temperature that would keep the oxygen and hydrogen in a liquid state and maintain the temperature of external parts high enough to prevent ice and frost from forming on the surface. The adhesion between spray-on foam insulation and the external tank's aluminum substrate was actually quite good—provided the substrate had been properly cleaned and primed. Large areas of the aluminum substrate were also usually heated during foam application to ensure the foam cured properly and developed maximum adhesive strength. The interface between the foam and aluminum substrate experienced stresses due to differences in how much the aluminum and foam would contract when subjected to cryogenic temperatures and the stresses on the external tank's aluminum structure while it served as the shuttle stack's backbone. While these stresses were certainly not trivial, they did not appear to be excessive. Very few of the observed foam loss events indicated the foam was lost down to the primed aluminum substrate.

Most of the external tank was insulated with three types of spray-on foam. NCFI 24-124, was used on most areas of the liquid oxygen and liquid hydrogen tanks. NCFI 24-57, was used on the lower liquid hydrogen tank dome. BX-250, was used on domes, ramps, and areas where the foam was applied by hand. The letters represented the chemical compound and the numbers the composition and mixture ratios. Throughout the external tank's history, factors unrelated to the insulation process have caused foam chemistry changes. These included Environmental Protection Agency (EPA) regulations and material availability. The most recent changes resulted from modifications to governmental regulations of chlorofluorocarbons. The foam types changed on external tanks built after external tank ninety-three—the one used on STS-107. Metallic sections of the external tank insulated with foam were first coated with an epoxy primer. In some areas, such as on the bipod's hand-sculpted regions, foam was applied directly over ablator materials. Where one foam layer was applied over cured or dried foam, a bonding enhancer called Conathane was first applied to aid the adhesion between the two foam coats.

After foam was applied to the intertank region, the larger areas of foam coverage were machined to about a one-inch thickness. Since controlling weight was a major concern for the external tank, this machining served to reduce foam thickness while still maintaining sufficient insulation. The insulated region where the bipod struts attached to the external tank was structurally, geometrically, and materially complex. Because of concerns that foam applied over the fittings would not provide enough protection from the high heat of exposed surfaces during ascent, ablators coated the bipod fittings. BX-250 foam was sprayed by hand over the fittings (and ablator materials), allowed to dry, and manually shaved into a ramp shape. The foam was visually inspected at the MAF and KSC, but no other nondestructive evaluation was performed.

Since the shuttle's inaugural flight, the shape of the bipod ramp has changed twice. The bipod foam ramps on external tanks one through thirteen originally had a forty-five-degree ramp angle. On STS-7 foam was lost from the external tank bipod ramp. Subsequent wind tunnel testing showed shallower angles were aerodynamically preferable to reduce the air loads. The ramp angle was changed from forty-five degrees to between twenty-two and thirty degrees on external tank fourteen and later tanks. A slight modification to the ramp was implemented on external tank seventy-six and was the last ramp geometry change. Beginning with the STS-6 mission, a lightweight ET was introduced. Although future tanks varied slightly, each weighed approximately sixty-six thousand pounds inert. The last heavyweight tank (flown on STS-7) weighed approximately seventy-seven thousand pounds inert. For each pound of weight reduced from the ET, the cargo-carrying capability of the space shuttle spacecraft increased almost one pound. Eliminating portions of stringers (structural stiffeners running the length of the hydrogen tank), using fewer stiffener rings, and modifying major frames in the hydrogen tank contributed to this weight reduction. Significant portions of the tank were also milled differently to reduce thickness, and using a stronger, lighter, and less expensive titanium alloy reduced the weight of the ET's aft solid rocket booster attachments. Earlier several hundred pounds had been eliminated by deleting the anti-geyser line. The line paralleled the oxygen feed line and provided a circulation path for liquid oxygen to reduce accumulation of gaseous oxygen in the feed line while the oxygen tank was filled before launch. After propellant loading data from ground tests and the first few space shuttle missions was assessed, the anti-geyser line was removed for STS-5 and subsequent missions. The total length and diameter of the ET remain unchanged.

External Tank Integration

After arriving by barge at the Kennedy Space Center, the external tank was then towed to the Vehicle Assembly Building, mated with the solid rocket boosters, and stacked on the mobile launcher platform. The space shuttle orbiter (after refurbishing) was then mated, and all connections were made. The integrated vehicle was then checked as a system before the ordnance was finally installed. Once the entire stack was fully mated structurally and electrically, all the interfaces were tested, and the ordnance was installed, the shuttle was readied for rollout to the pad. Testing involved opening the crew module, applying power to the vehicle, performing all interface (orbiter–ET–SRB) checks, testing the solid rocket booster flight controls, and checking for leaks in the main propulsion system (MPS).

CHAPTER 6

Orbiter: The Debris Target

The Orbiter

The orbiter was commonly referred to as the "space shuttle," and it was about the size of a small commercial airliner. It normally carried a crew of seven that included a commander, pilot, and five mission or payload specialists. It could accommodate a payload the size of a school bus weighing between 38,000 and 56,300 pounds, depending on the orbiter parameters. The crew compartment had three decks: the uppermost flight deck, middeck living quarters, and lower deck. The orbiter's upper flight deck contained seats for the commander, pilot, and two specialists in the aft. Equipment for flying and maneuvering the vehicle and controlling its remote manipulator arm filled the upper deck. The middeck contained the living quarters and stowage lockers for food, equipment, supplies, and experiments. It also included a toilet, a hatch for entering and exiting the vehicle on the ground, and—in some instances—an air lock for doing space walks in orbit. During liftoff and landing, four crewmembers would sit on the flight deck, and the rest would be on the middeck. The lower deck or equipment bay contained all the life-support equipment, electrical systems, and additional payload stowage. Each orbiter had three main engines mounted at the aft fuselage. These engines used the most efficient propellants in the world—liquid oxygen and liquid hydrogen—at a rate of half a ton per second. At 100 percent power, each engine produced 375,000 pounds of thrust—four times that of the largest engine on commercial jets. To provide steering control during ascent, the large bell-shaped nozzle on each engine could swivel 10.5 degrees up and down and 8.5 degrees left and right.

Orbiter Project Leadership

Like the ET project, the orbiter project experienced substantial management turnover after the *Columbia* accident. Most leadership changes occurred within NASA. On July 2, 2003, Steve Poulos became manager of the orbiter project at JSC. He later recalled, "I became interested in NASA just like every other kid from my generation—watching the Apollo moon landings. When I first started working as a contractor in 1984 with Rockwell International on cargo integration, I became more interested in the shuttle program and later joined NASA in 1989. The orbiter project manager position was offered to me after working in the Engineering Directorate and as the deputy to the NASA Accident Investigation Team. It was a job I could not pass up since most of my career was dedicated to project management. This was a dream job, but my biggest concern taking over the orbiter was lacking experience in a number of technical discipline areas: propulsion, avionics, and aero to name a few."

Adept at integrating humor and high energy into his work, Steve was quick to understand a problem and explain it to others. His sense of humor would often be pushed to the limit when disagreements arose with John Muratore and SEI over risk assessments, debris risk classification, or orbiter impact capability. "I certainly agree I was frustrated in a number of the debris meetings," Steve said. "My general approach to leading a team relies on the eighty-twenty rule. If we have about eighty percent of the information, then make a decision. John typically wanted much more data than I did. Also, as we were approaching RTF (about six months before flight), my mind-set moved away from ascent debris to the models and analyses we would need to perform in order to show the orbiter was safe for reentry."

Steve was able to surround himself with good, qualified people such as his deputy manager, Ed Mango, who had recently served as the director for *Columbia*'s debris recovery effort in East Texas. Ed joined NASA at KSC in 1986 after serving in the air force at Vandenberg Air Force Base. He had held numerous positions in the agency. That included shuttle project engineer, deputy director for shuttle process engineering, and shuttle launch manager. By the time Ed completed his work on the shuttle, he had supported the launch team for more than fifty launches. More than a dozen came as assistant launch director. Steve was also responsible for offering Justin Kerr a job on the orbiter project working debris and impact testing. Justin recalled, "It was great working for Steve. He had a reputation for taking care of his people. Steve was big on commitments and doing what you said you were going to do. You knew he was going to honor his, and you did not want to be the one who fell short of yours. He backed all the testing I did with funding or allowing more time to get the job done or extending the testing a little bit further to capture additional data."

Steve also put Justin in charge of the orbiter's damage assessment team (DAT). This was responsible for all the on-orbit, real-time thermal protection system inspections and debris damage assessments during a mission. His team was made up of NASA, United Space Alliance, and Boeing employees such as Mike Stoner, Paul Parker, and Ban Bell since they were most familiar with the damage models. This dedicated team met the debris challenges of every mission through the end of the Space Shuttle Program.

Orbiter's Thermal Protection System

To protect its thin aluminum structure during reentry, various materials collectively referred to as the thermal protection system (TPS) covered the orbiter. Different parts of the orbiter were subjected to dramatically different temperatures during reentry. The nose and leading edges of the wings were exposed to superheated air temperatures of 2,800 to 3,000°F, depending on the reentry profile. Other areas on top of the fuselage were so shielded from superheated air that ice sometimes survived through landing. The TPS comprised three major components: various types of heat-resistant tile, blankets, and RCC panels on the leading edge of the wing and nose cap. The RCC panels most closely resembled high-tech fiberglass. The panels are layers of special graphite cloth molded to the desired shape at very high temperatures. Reusable black tiles protected most other areas of the orbiter exposed to medium and high heating. They were 90 percent air and 10 percent silica (similar to common sand). Produced in varying strengths and sizes depending on the orbiter areas they were to protect, they were designed to withstand either 1,200° or 2,300°F. In a dramatic demonstration of how little heat the tiles transfer, one can fire a blowtorch at one side of a tile and place a bare hand on the other. The white blankets were capable of withstanding 700° to 1,200°F. They covered regions of the orbiter that experienced only moderate heating.

The TPS consisted of various materials applied externally to the outer structural skin of the orbiter to maintain acceptable temperatures—primarily during the mission's entry phase. The orbiter's outer structural skin was constructed primarily of aluminum and graphite epoxy. During entry the TPS materials protected the orbiter's outer skin from temperatures above 1,221°F (the melting point of aluminum). These materials were reusable for one hundred missions with refurbishment and maintenance. They performed in temperature ranges from −250°F in the cold expanse of space to entry temperatures of nearly 3,000°F. The TPS also sustained the forces induced by deflections of the orbiter airframe as it responded to the various external environments. Because the thermal protection system was installed on the outside of the orbiter skin, it determined the aerodynamics over the vehicle in addition to acting as the heat sink.

The white low-temperature reusable surface insulation (LRSI) tiles protected areas where temperatures were below 1,200°F. These tiles provided better thermal protection on-orbit and were used in selected areas of the forward, mid, and aft fuselage, vertical tail, upper wing, and orbital maneuvering system/reaction control system (OMS/RCS) pods. After the initial delivery of *Columbia,* an advanced flexible reusable surface insulation (AFRSI) was developed to reduce the weight, decrease the installation time, and improve producibility and durability. This material consisted of composite quilted fabric and insulation batting between two layers of white fabric that were sewn together to form a quilted blanket. AFRSI was used on *Discovery* and *Atlantis* to replace the vast majority of the LRSI tiles. Following its seventh flight, *Columbia* also was modified to replace most of the LRSI tiles with AFRSI. White blankets made of coated Nomex felt protected areas below 700°F. This included the upper payload bay doors, portions of the fuselage sides, and upper wing surfaces. Additional materials were used in the windowpanes and operable penetrations such as the umbilical and landing gear doors.

At launch the three space shuttle main engines were fed liquid hydrogen fuel and liquid oxygen oxidizer from the external tank. Once it had been verified the engines were operating at the proper thrust level after ignition, a signal was sent to ignite the solid rocket boosters. At the proper thrust-to-weight ratio, initiators (small explosives) at eight hold-down bolts on the SRB fired to release the space shuttle for liftoff. All this took only a few seconds. Maximum dynamic pressure was reached early in the ascent—normally sixty seconds after liftoff. Approximately sixty seconds later the two SRBs had consumed their propellant and were jettisoned from the external tank. Flight operations up until SRB separation were referred to as the first stage of powered flight.

The boosters briefly continued to ascend, and small booster separation motors (BSMs) fired to carry the booster away from the space shuttle. Eventually the boosters turned and descended. At a predetermined altitude, parachutes were deployed to decelerate the SRBs for a safe splashdown in the ocean. Splashdown occurred approximately 141 nautical miles (162 statute miles) from the launch site. The boosters were then recovered for reuse. Meanwhile the orbiter and external tank continued to ascend, using the thrust of the three SSMEs. Approximately eight minutes after launch and just short of reaching orbital velocity, the three engines were shut down in a critical operation known as main engine cutoff (MECO). The external tank was then jettisoned on command from the orbiter. The second stage of powered flight referred to the time between SRB separation and MECO.

The forward and aft reaction control system engines provided attitude (pitch, yaw, and roll) and the translation of the orbiter away from the external tank at separation. The external tank continued on a ballistic trajectory, entered the atmosphere, and disintegrated over the Indian Ocean. Normally two thrusting maneuvers were performed using the two OMS engines located at the aft end of the orbiter. They were used in a two-step thrusting sequence—to complete insertion into Earth orbit and to circularize the spacecraft's orbit. The OMS engines were also used on-orbit for any major velocity changes. This phase of shuttle operations was called the post-insertion coast.

In the event of a direct-insertion mission, only one OMS thrusting sequence was used. The orbital altitude was dependent on that mission with the nominal altitude varying between 100 to 217 nautical miles.

Orbiter Reentry

For a successful return to Earth and space shuttle landing, dozens of things had to go just right. First, the orbiter had to be maneuvered into the proper position. This was a crucial factor for a safe landing. When a mission finished and the shuttle was halfway around the world from the landing site, mission control gave the command to come home. This prompted the crew to close the cargo bay doors. In most cases they were flying nose-first and upside down, so they then fired the RCS thrusters to turn the orbiter tailfirst. Once the orbiter was tailfirst, the crew fired the OMS engines to slow the orbiter down and initiate the fall back to Earth. It took about twenty-five minutes before the shuttle reached the upper atmosphere. During that time the crew fired the RCS thrusters to pitch the orbiter over so its bottom faced the atmosphere (about forty degrees) and the craft was again moving nose-first. Finally they initiated a burn of leftover fuel from the forward RCS as a safety precaution because they would encounter the highest heat of reentry in this area.

As it moved at about seventeen thousand miles per hour (twenty-eight thousand kilometers per hour), the orbiter hit air molecules and built up heat from friction. In a successful reentry, the orbiter encountered the main air of the atmosphere and was able to fly like an airplane. The orbiter was designed from a lifting body design with swept-back wings. With this design the orbiter generated lift with a small wing area. At this point flight computers took over flying the orbiter. They directed the orbiter to make a series of S-shaped, banking turns to slow its descent as it began its final approach to the runway. The commander picked up a radio beacon from the runway (tactical air navigation system) when the orbiter was about 140 miles (225 kilometers) away from the landing site and 150,000 feet (45,700 meters) high. At twenty-five miles (forty kilometers) out, the shuttle's landing computers gave up control to the commander. He or she flew the shuttle around an imaginary cylinder (18,000 feet or 5,500 meters in diameter) to line the orbiter up with the runway and drop the altitude. During the final approach, the commander steepened the angle of descent to negative twenty degrees (almost seven times steeper than the descent of a commercial airliner). After landing the main objective was to "safe" the vehicle for the ground personnel and crew. This was because there were a lot of hazards present such as toxic gases, active pyrotechnics, hot surfaces, and fire potential. Approximately 160 space shuttle launch operations team members supported spacecraft recovery operations at the nominal end-of-mission landing site. Ground team members wearing self-contained atmospheric protective ensemble suits that guarded them from toxic chemicals approached the spacecraft as soon as it stopped rolling. The ground team members took sensor measurements to ensure the atmosphere in the vicinity of the spacecraft was not explosive. In the event of propellant leaks, a wind machine truck carrying a large fan would be moved into the area to create a turbulent airflow that would break up gas concentrations and reduce the potential for an explosion.

An air-conditioning purge unit was attached to the right-hand orbiter umbilical so cool air could be directed through the orbiter's aft fuselage, payload bay, forward fuselage, wings, vertical stabilizer, and OMS/RCS pods to dissipate the heat of entry. A second cooling unit was connected to the left-hand orbiter umbilical Freon coolant loops to provide cooling for the flight crew and avionics during the postlanding and system checks. The spacecraft fuel cells remained powered up at this time. The flight crew then exited the spacecraft, and a ground crew powered down the spacecraft. Flight crew's actions following a landing included "safing" the OMS/RCS, positioning the space shuttle main engines to allow rain drainage, and securing the auxiliary power units. Ground crews deployed instruments and sniffed for hazardous gases, connected ground purge and coolant equipment, and assisted with opening the hatch. The astronauts egressed, time-critical flight crew equipment was removed, and the orbiter was then towed to the obiter processing facility. If the spacecraft landed at Edwards AFB, the same KSC procedures and ground support equipment were used after the orbiter stopped on the runway. The orbiter and ground support equipment convoy moved from the runway to the orbiter mate and demate facility at Edwards. After detailed inspection the spacecraft was prepared to be ferried atop a modified 747 shuttle carrier aircraft from Edwards to KSC. Although never needed, landing at an alternate site involved a crew of about eight team members assigned to the landing site to assist the astronaut crew in preparing the orbiter for loading aboard the shuttle carrier aircraft for transport back to Florida. For a landing outside the United States, personnel at the contingency landing sites were provided minimum training on safe orbiter handling with emphasis on crash rescue training and how to tow the orbiter to a safe area.

Orbiter Processing

Once the orbiter left the runway or returned from ferry flight, it was immediately taken to one of the three OPFs. Typically it spent about two months in the OPF getting inspected, tested, and prepared for the next flight. Actual time in the OPF depended on the launch schedule. Launch rate, major modifications, and resolution of any vehicle problems drove this schedule. Once in the orbiter processing facility, the payload (if any) was removed, and the orbiter payload bay was reconfigured for the next mission. Any required maintenance and inspections were also performed while the orbiter was in the OPF. Depending on the mission, the payload for the particular orbiter's next mission was installed either in the orbiter's payload bay in the OPF or in the payload bay at the launchpad.

Once inside the OPF, support structure completely surrounded the vehicle. The shuttle was jacked, leveled, and configured for access by opening hatches and removing all panels and covers. Vehicle inspections were conducted, and the vehicle was prepared for its next flight. The spacecraft was then towed to the Vehicle Assembly Building where all necessary connections were made to mate it to the external tank. The mobile launcher platform moved the entire space shuttle system on four crawlers to the launchpad. Connections were then made, and servicing and checkout activities began. If the payload was not installed in the orbiter processing facility, it was installed at the launchpad. Prelaunch activities followed. The most significant prelaunch activity was the terminal count demonstration test—a launch dress rehearsal done once per mission. This test was a verification of launch countdown processing, checkout of vehicle system, flight software, and launch procedure rehearsal. It was an integrated simulation using all the flight hardware and launch facilities collectively. However, no hydraulics, fuel cells, or payload systems were operated. The firing room, MCC in Houston, and flight crew were all involved. Terminal count demonstration tests were shorter than the regular launch count, but the terminal countdown was nearly identical to the actual launch sequence. At the end of the exercise, the astronauts also practiced an emergency egress. Only then was the vehicle considered ready for flight.

CHAPTER 7

Hooked on the Space Program

Toledo, Ohio

The space program had enthralled me ever since watching the Apollo 11 lunar landing on a black-and-white TV set. My father was a big fan of the space program as was the rest of an enthusiastic nation captivated by the Mercury, Gemini, and Apollo missions. Dad would occasionally take off work and let me skip school to watch a launch. The most impressionable event was watching Apollo 11 land on the moon and the historic walk on the lunar surface with the family gathered around the television. It sparked my childhood fantasy of one day becoming an astronaut. Even at this early age, I knew I was destined to someday work for NASA and be part of the space program. In those days any special event involving an astronaut inspired "rock star" hysteria. The grand opening and dedication of the Armstrong Air and Space Museum in Wapakoneta, Ohio, in 1972 was no exception. Wapakoneta was the birthplace and hometown of Neil Armstrong, and it is located a short drive south of Toledo—where I grew up. My poor mother dragged four rambunctious, sugar-fueled, anticipatory kids to the event. We all hoped to see a live astronaut and climb aboard a spaceship. We did manage to get close enough to an astronaut (presumably the late Neil Armstrong, who was by then retired from NASA and working as a professor at the University of Cincinnati) to ask him how to become

RETURN TO FLIGHT

one. "Do well in school, learn to fly, go to college, get an engineering degree, and consider serving in the military," he replied.

Those words of wisdom stuck in my young, impressionable mind and served as a blueprint for decisions I would make throughout my life.

I was born in Toledo, Ohio, on September 29, 1963, to proud parents William and Melba Peters. I was the second of four kids, and a mere five years separated us. Toledo was a populous Roman Catholic, blue-collar city nestled on the southwest corner of Lake Erie. Once the glass capital of the world, it was home to the Toledo Mud Hens, Wolfman Jack, and Tony Packo's Café. Toledo had a limited downtown skyline and several loading docks along the murky Maumee River. This allowed Great Lakes freighters to carry raw materials for glass production, salt, and coal and snake their way from Lake Erie to the downtown ports of call for unloading. Other cargo ships carried away the grains grown by Ohio farmers and stored in the large silos that lined the river and accented the skyline. My Roman Catholic parents named me after three saints—James, Francis, and Peter (close enough). Unfortunately for them I did not live up to the billing and was in constant trouble. This resulted in frequent accidents and trips to the hospital. When I was a baby, I rolled off the changing table and fractured my skull. As a five-year-old, I had a rather lengthy streak of several ER visits ranging from a toy's metal axle rod piercing through my ankle to a near overdose of orange-flavored aspirin. The hospital staff, suspicious of child abuse, sent out a Social Services representative to investigate. While my mother sat with the interviewer, my older brother and I managed to venture into my dad's fishing tackle, and I ended up snagged with a fishing lure. Bleeding and screaming, I ran into the kitchen in frantic search of first aid. My mother calmly looked at the social worker and asked, "Do you want to take him to the emergency room, or do you want me to?" The social worker never returned.

I loved sports, but during my first year in junior high, I did not fare well in football or basketball due to my small size. I ended up mostly sitting on the bench. I was still rambunctious and often instigated fights or played practical jokes on fellow students. I pushed my luck one too many times, however, when I started a cafeteria food fight. I found myself headed to the principal's office for detention and a probable suspension. Principal Richard Barton recognized a case of unchanneled energy and gave me two options—take a three-day suspension or attend wrestling practice for two weeks. Failure to complete the two weeks of practice would result in the suspension. Finish the two weeks, and all would be forgiven. As it turned out, Mr. Barton had several sons involved in wrestling, and this made it easy to spot the untapped potential that flourished in front of him. It was a no-brainer. I did not want to face my father's wrath along with additional grounding and chores at home. I told my parents that evening I was going out for wrestling and not to worry about picking me up from practice. I would walk home. Having successfully engineered the school suspension cover-up, I showed up for wrestling practice the next day and proceeded to fall in love with the sport. Wrestling was pure. It matched one person against another of equal weight. No judging. No bias. No ties. I liked best that wrestlers were in control of their own destinies. Although conditioning, size, technique, and strength were advantages, they were not prerequisites for victory. The one factor that tended to win out more than any other was desire or mental toughness. Mental toughness drove the champion wrestler to work on all aspects of the sport, strengthen resolve, and minimize or eliminate any weakness. When others gave up, the champion kept going. I excelled at the sport, the cloud of trouble over me cleared up, and I never again visited the principal's office. Mr. Barton had given me a chance. Perhaps he came to regret it, though, as I dominated a few of his sons in local competitions.

I competed in every event I could enter, and three years later I worked my way to the Junior Nationals and a slot on the World Schoolboy team where I won the Junior World Championship. My wrestling career continued through high school, but despite having won the World Schoolboy Championship, my success did not immediately carry over. I recognized early on one was only as good as one's last match. Fortunately losses only strengthened my resolve to get better.

Clay High School was known for successful football teams, and although it always had had good wrestling teams, they never seemed to beat the top schools or compete well in the state tournament. Only in college did I learn the real difference between "good" and "great" teams. Like all good Buckeyes, I had my sights on attending Ohio State University. As a top student and wrestler, I was counting on an athletic and/or academic scholarship. My dad had been out of work off and on at General Motors during my senior year due to the struggles of the American automotive industry with rising Japanese companies. Without a scholarship my chance to attend Ohio State and wrestle for the Buckeyes was in jeopardy. This added a little extra performance pressure. My record was fifty and one as I headed into the state tournament, and I was ready and confident. My dad had stretched the finances and made a special trip to Columbus—the site of the state championships. Unfortunately I lost my first match on a penalty point for "slamming" my opponent, and my championship dream was over.

Two weeks after the state tournament, I received a call from Chris Ford—the Ohio State wrestling coach. Although still interested in recruiting me, he already had a number of athletes at my projected weight class. One wrestler on the team was Dane Tussel, and he had been Mr. Ohio and state tournament MVP the year before. Coach Ford stated he would only be able to manage a half-ride scholarship for academics. I knew this was not enough, and I replied I would not be going to Ohio State. I thanked him and hung up the phone. Fortunately my high school principal, John Wolfe, asked if I would be interested in the Naval Academy. John was a captain in the US Navy Reserves, and he had followed the wrestling team and my academic standing in the class. He knew about the troubles at General Motors and heard about my conversation with Chris Ford and the Ohio State scholarship situation. I was unfamiliar with the academy or even where the campus was located. None of that mattered, though. It was my best option, and having grown up near Lake Erie, I loved the water. Principal Wolfe helped me with the application process and securing a congressional appointment. The acceptance letter soon followed, and at age sixteen

I was packing my bags for Annapolis, Maryland. I had never even visited the Naval Academy campus. Despite the rigors of military indoctrination and the strictly regimented life as a first-year midshipman (or "plebe"), I knew I had made the right decision.

United States Naval Academy
Annapolis, Maryland

The camaraderie and highly competitive environment of the Naval Academy appealed to me. It also had a sensational wrestling team loaded with talent and nationally ranked wrestlers. I had yet to fully mature physically and easily made the lowest 118-pound weight class fully clothed. As a freshman I was at the bottom of the depth chart and well behind the senior team captain, Guy Zanti, and freshman prodigy, Sal Lacorte. Sal had pummeled me in wrestle-offs and routinely challenged Guy for the top spot. I kept working hard and eventually broke into the starting lineup at the end of my sophomore year right before the annual Navy–Army match. Guy had graduated, Sal had been dismissed from the academy for grades, and I had slowly edged my way into the number one spot. At the time Navy's team had seniors in eight of the ten weight classes, and they all wanted to end their distinguished regular season careers with a shutout over our archrival. I was not as confident as the rest of the team as we finished our warm-ups and huddled together for the "Go Navy" cheer. I was up against Army's senior team captain and lost the match by a point. I was more disappointed for my teammates losing the shutout than I was for myself losing a close match. The disappointment grew stronger as the bouts wore on. I was the only Navy wrestler to lose that day. From that day forward, I vowed never again to be the one to let my teammates down.

The most significant advantage we had as wrestlers and midshipmen was our attitude of expectation. Navy expected to win. The Naval Academy demanded high standards of conduct and performance and expected graduating officers to lead the navy and nation. This attitude of high expectations drove me and my teammates to commit and then pay the price to be successful. This attitude molded my desire to succeed and strengthened my mental toughness in wrestling and other areas of my life. This was especially true at NASA, where failure was not an option. I began to excel my junior year and was winning the majority of my matches. At one point during the season, we wrestled Syracuse, and my opponent was Sal Lacorte. I won the match, and that momentum carried me forward the rest of the year and into the NCAA championships held at the Meadowlands in New Jersey. I entered the tournament as an unheralded, unseeded wrestler and drew the Ohio State competitor Dane Tussel in the first round. Dane was a senior and had continued his brilliant wrestling career. He was a top seed in the tournament and had won the Big Ten Championship and All-American honors. I faced him the first day of competition. Much to my surprise, my parents made the long trip from Ohio to watch the competition. I was a prohibitive underdog, but even against the odds, I expected to win. My strategy was to focus on what I did best—the same things that had helped me earn a spot on the team, beat Sal, and make it to the NCAA tournament. I won the match by pinning Dane in the first period and went on to earn All-American honors. The debacle of my high school senior year during the state wrestling tournament and Ohio State scholarship loss now seemed insignificant. In hindsight it was a blessing in disguise. However, the aura of invincibility would not last long. After a successful run during the NCAA championships and earning All-American honors, I tore the ACL in my left knee during the finals of a regional tournament. Although I won the match, the damage was done, and surgery was required. I was devastated and wondered if I would be able to compete my senior year. The Naval Academy did not redshirt athletes, and the recovery time from reconstructive surgery was months. It would be deep into my senior season before I could step on the mat again. I opted for a partial reconstruction that required a special brace for practice and competition. The knee injury disqualified me from becoming a navy pilot after graduation, so I chose nuclear submarines as my service selection. Acceptance into the navy's nuclear program involved screening and an interview with Admiral McKee at

Naval Reactors in Washington, DC. Although the choice pleased me, I intended it to be temporary, as I had planned a full knee reconstruction after the wrestling season ended. I planned to rehabilitate the knee while attending graduate school at the University of Maryland, pass the flight physical, and put in a designator change to switch from a navy "nuke" to a fighter pilot. The injury had definitely dealt a blow to my psyche, and I tended to be more cautious and concerned about damaging my knee further. When I resumed wrestling, I had to deal with the self-doubt that came with not being at peak performance. I was worried about being too protective of my knee instead of aggressive and attacking—my normal style. As it turned out, living through this experience made me better prepared to deal with the same cautiousness and emotion I encountered years later working the first Return to Flight mission after the *Columbia* accident. Athletic tape and painkillers enabled me to grapple through the season and make it back to the NCAA championships. I lost my last match in overtime and did not place in the tournament. Afterward I found a quiet place in the locker room and cried with the recognition my competitive wrestling career was now over. It was the same type of sadness I felt when the shuttle flew its last mission.

Studies at the Naval Academy came easily, and I was especially good at mathematics. I did not burden myself with memorizing formulas. I would just derive them. While others were cramming for quizzes or exams, I would work out, watch TV, or make it an early night and head off to bed. "Well rested, well tested," was my motto—much to the disdain of others. My frequent pranks helped me earn the nicknames "Practical Joke Peters" and "Peterhead." These followed me through graduation and the rest of my military career. I did not care and reveled in the displeasure my clever pranks created. I was young, and it felt good to be invincible, but I always had a backup plan.

The first time the Army–Navy football game was played on the West Coast was in 1983 in the Rose Bowl. I knew the Playboy Mansion was in Los Angeles, and I had the bright idea to write Hugh Hefner a letter informing him about the historic event and inviting Playmates to the game. With the help of three fellow midshipmen, we wrote the letter and received a rather quick response approving our request. We were invited to the Playboy Mansion to meet Hugh and receive a personal tour of the Playboy Modeling Agency along with some Hollywood sightseeing. Other midshipmen thought we had concocted another "whopper" prank and refused to give our story any credence. They were not going to fall for another gag—especially one so far-fetched. On the flight to LAX, I developed a backup plan if the Playboy invitation letter turned out to be a fake and the weekend hobnobbing with Playmates was just a fanciful fantasy. It was entirely possible someone would try to prank us, but if the Playboy visit failed to materialize, we would declare the whole deal just another practical joke and mock those who had believed otherwise. I breathed a large sigh of relief after the mansion gates opened to invite us in. It was the start of a joyous weekend that ended with Navy beating Army in the annual grudge match.

Navy Nuclear Power Program

After graduating from the Naval Academy, I finished graduate school at the University of Maryland a year later. I underwent successful reconstructive knee surgery and passed my flight physical to be a naval aviator. Unfortunately the "designator change" to complete the last step of my master plan was disapproved. The response letter stated, "There would be no voluntary disenrollment from the US Navy Nuclear Power Program." Admiral Rickover had signed it. So it was off to "Nuke School" in Orlando, Florida, rather than to pilot training in Pensacola. The nuclear training was so intense I would rather have gone through another four years of the academy than another year of it. I struggled early and started poorly by failing the first four tests. One more failure would lead to dismissal and possible forfeiture of my nuclear bonus. I attributed the slow start to the rejected designator change letter and an episode involving a regrade on an exam I initially passed. All exams were four hours long and based on a 400-point scale. A student needed to earn 250 points to pass and could request a regrade if he or she believed a mistake had been made in the grading. I originally scored a 260 on one of the tests, and I believed I should have earned more points on a few of the questions. I submitted for a regrade. I was not aware at the time, but the entire exam was regraded—not just the answers in question. The regrade resulted in a failing score of 242. As I was trying to regroup and get serious about making it through Nuke School, I noticed a few students taking wagers on who would earn the highest test scores. A competition was exactly what I needed, and based on my poor performance, the other students were eager to take my wager. The winner could earn anywhere between fifty and one hundred dollars per test, depending on the wager and number of students betting. Midterm and final exams would go even higher. I earned a rare perfect score on the first exam I took after getting into the competition. This flabbergasted students and instructors alike, and some even accused me of cheating. As I collected the winnings, I joked I had hustled everyone by failing the first four tests. I laughed as I carried off the winnings. The competition inspired me to focus, and because the scores now counted for more than just grades, I wanted to win. After the six-month classroom training in Orlando, I was assigned prototype training in Idaho Falls, Idaho. During this training phase, students received hands-on experience running an actual reactor plant. I planned to spend the Christmas holidays with my parents in Ohio before making the long drive west to Idaho. During a New Year's Eve party that year, I met Jennifer Peth. (We eventually married.) She was a beautiful green-eyed blonde who captivated the party revelers with her outgoing personality and Valley girl impressions. Her fun-loving charm went well with the impersonations and sloe gin fizzes I was mixing up at the party. We shared a lot in common, but it was nearly a year before we met again. She had a boyfriend and was busy attending the University of Toledo, and I was off to finish my nuclear training in Idaho. My nuclear training journey came to an end after successfully completing seven months of prototype training and three months of submarine training in Groton, Connecticut.

USS *Hyman G. Rickover,* Nuclear Fast Attack Submarine
Norfolk, Virginia

My first duty station as a twenty-three-year-old lieutenant was on the USS *Hyman G. Rickover*—a nuclear-powered fast attack submarine stationed in Norfolk, Virginia. The sub's namesake, Admiral Rickover, was the "father" of the nuclear navy. *Rickover* had silenced all the critics in 1958 by successfully deploying the USS *Nautilus*—the first nuclear submarine. The *Rickover* was the marquee submarine in the fleet, and with it came high expectations and frequent visits by dignitaries and periodic visits by the admiral himself. Although naval aviation was my preferred service selection, the extreme technological environment of submarine operations fascinated me.

I was eager to embark on my first tour of duty and pleased I was on the *Rickover*. It was one of the newest Los Angeles class submarines in the fleet and already had established itself as the best, or "E," boat in the Norfolk submarine squadron. Captain Jay Cohen was a distinguished elite commander despite his extensive diesel boat background. I also had the privilege of teaming up with a great crew that included Charles Sims, my Naval Academy roommate. I felt confident Charles would mentor me through the transition of life aboard the sub and assist with my submarine qualifications. Both officers and enlisted people were referred to as NUBs (nonuseful bodies) until they passed their first qualifications. This enabled them to stand watch. Until that time NUBs were deemed practically worthless because they were breathing the valuable oxygen qualified watch standers needed. At a minimum it took a year of intensive training, exams, and board reviews to reach submarine qualifications. At that point the captain proudly awarded the gold "Dolphins." This was a significant milestone that officially welcomed sailors into the family of submariners, and the unofficial transition from a NUB to a Nuke was complete.

Although I was scheduled to report on a Monday, I paid the boat an early visit over the weekend to scout out the Norfolk Naval Base and meet some of the crew. I was wearing civilian clothes when I made my first request for permission to come aboard. Weekends were typically slow days, and just the topside watch was there to greet visitors. To the topside watch, I was an unexpected visitor and was asked to show my identification while someone called the ship's duty officer to grant boarding permission. As I was waiting for his reply, the chief of the boat (the senior enlisted man aboard) appeared on deck and leisurely wandered over. I introduced myself by name only, and he responded by asking which person aboard the sub was my dad. He apologized after I told him I was the new lieutenant reporting for duty. This was how my shipboard life began.

During my years at the Naval Academy, I had enthusiastically followed the Space Shuttle Program from the inaugural flight in 1981. In 1984 President Reagan announced during his State of the Union Address that NASA would embark on the construction of an Earth-orbiting space station named *Freedom*. The shuttle was originally designed for space station construction, but in the absence of an assembly mission, the shuttle had been used to launch, retrieve, and repair satellites, conduct limited microgravity research, and prove out technologies that would be essential for space station construction. NASA assessed several space station designs for the final configuration selection. Everything was progressing well until the *Challenger* disaster on January 30, 1986. I was attending graduate school at the University of Maryland and watching the launch during a break between classes when the explosion occurred. I was beside myself and could not muster the fortitude to attend my remaining classes that day. Instead I watched countless video replays of the accident on the news until late evening and wondered what caused the explosion. The ISS program quickly lost momentum as NASA focused on the investigation and returning the shuttle back to flight status. The post-*Challenger* RTF effort took more than two years and diverted much-needed development funds and labor away from the space station. Despite the setback NASA was still committed to moving forward with the shuttle and Space Station *Freedom* programs. I followed NASA's RTF efforts and kept a keen eye on the space station development. I knew someday I would be involved with this program. Many submarine systems (most notably the environmental control and life-support systems) were similar or identical to those eventually used on the space station. Knowledge of those systems and how they operated could work to my advantage. I planned on resigning from my commission and applying for the astronaut program as a mission specialist. I intended to finish my military obligation in a little over two years and then work in the space industry.

Not long after I reported aboard the *Rickover,* I received a call from my brother Aaron. He was a sophomore at the Naval Academy, and he inquired if I would be interested in a date with Jennifer. As it turned out, Aaron's girlfriend, Gwen, was good friends with Jennifer. Gwen and Jennifer planned to drive from Ohio to meet my brother and me in Annapolis for a long weekend and to watch one of my brother's wrestling matches. Although I had not seen Jennifer for more than a year, I was excited she remembered me and wasted no time in accepting Gwen's weekend proposition. Although the *Rickover* was late getting into port from a short deployment the day we were scheduled to meet, I grabbed a few belongings from the ship and sped off toward Maryland for my long-anticipated date. I made the two hundred-mile drive to Annapolis in less than three hours on a Friday evening. I arrived at the hotel at eight o'clock and ran to the room to greet my brother, Gwen, and Jennifer. Jennifer opened the door, and she was even more beautiful than I remembered. She was wearing jeans and a pink sweater. She looked a little surprised because I was still in my uniform, but after saying hello to my brother and Gwen, I grabbed a piece of leftover pizza and excused myself for a shower. It was a remarkable weekend, and from then on I knew I was going to marry Jennifer.

Life aboard the submarine was challenging with the long hours, extended time at sea, divisional responsibilities, and constant work on qualifications. However, a lot of these difficulties were offset by the camaraderie, level of responsibility, and submariner sense of humor, which was often brash, sarcastic, honest, and typically offensive—just the way I liked it. No one was spared from the possibility of a well-planned practical joke or humor-filled comment—not even the captain.

While on the bridge, the new captain, Bruce Lemkin, had ashes from a burial at sea blow in his face. As he descended from the bridge into the control room, the first words spoken came from the chief of the boat. "Captain, that color makeup does not match your outfit."

Everyone, including the captain, burst into laughter. Humor was one of the few things that could help one forget about the inherent risks of tunneling through the ocean depths on a nuclear-powered machine carrying nuclear weapons and looking for enemy subs. My sense of humor eventually served me well during my NASA career. This was especially true when used to break the ice of a tense situation or difficult decision.

A submarine, like a spacecraft, had inherent risks associated with operations in a harsh, unforgiving environment. The ship's nuclear reactor was actually comparatively safe relative to other hazards. The biggest risks were flooding, fire, and toxic gas—nearly identical to the depressurization, fire, and toxic gas threats aboard a crewed spacecraft. Collision threats, oxygen generator explosion, hot run torpedo, battery explosion, major steam leak, high-pressure hydraulic rupture, and a reactor problem rounded out the top ten risks. While at sea I spent my free time working out in the engine room. I rode an exercise bike in "shaft alley" and used hydraulic lines to do chin-ups and other strength exercises. While on watch I broke the monotony by challenging the watch team to solve difficult problems or scenarios. For instance, I asked how many atoms in a fusion reaction had to be split to take a warm five-minute shower. It turned out to be several million given all the assumptions we used to solve the problem. While in port it was common to combine the mandatory ship's duty officer and engineering duty officer watches into one. On several occasions I was the lone officer aboard one of the most destructive weapons humans ever built. The operations tempo on the *Rickover* was intense, and it was fairly common to see the crew strung out from lack of sleep, supporting an extended exercise, constant training, or troubleshooting periodic problems. After a few days of stumbling around in a zombielike state from borderline exhaustion, a crewmember would inevitably develop a blank, emotionless stare that submariners referred to as "sub face."

At that point everyone would drift in and out of focus with thoughts about better places. My thoughts always focused on Jennifer and the next time we would meet.

Life as a Civilian
Lexington, Kentucky

Not long after I was discharged from the navy on April 28, 1989, Jennifer and I were married and started our lives as newlyweds. I had a job offer from Rockwell International in Downey, California, to work on the space shuttle, but I turned that down in favor of a higher-paying job with E-Systems designing aircraft mission enhancement modifications for the USAF Special Forces. The job was in Lexington, Kentucky, and only a four-hour drive away from our hometown, families, and friends in Ohio. The opportunity to work in the space program would have to wait. I was making over five times what I made in the military, and we were living well in Lexington. We bought new cars, moved into the best apartment complex in the city, joined the most exclusive athletic club, and attended all the major sporting and social events. Jennifer transferred to the University of Kentucky to finish her degree, and my job was going extremely well. Our inner circle of friends quickly grew, and many of them referred to us as Ken and Barbie.

I had kept up with the shuttle's RTF after the *Challenger* accident and started to focus more on the progress of Space Station *Freedom*. It was now gaining momentum and advertising operational capability by the mid-1990s. Although I loved how things were going with my new job and life in Lexington, I often wondered about the shuttle job I had declined and how I might get into the spaceflight business. The development of the space station fascinated me, and I submitted my résumé to Boeing and McDonnell Douglas and applied for a space station position. At that time Space Station *Freedom* comprised three major work packages. Boeing was the prime contractor for all the pressurized modules, McDonnell Douglas was responsible for all truss segments, and Rockwell International was the prime contractor for all solar arrays and power truss elements. Boeing responded and invited me down to Huntsville, Alabama, for an interview to work on the environment control and life-support systems (ECLSS) and thermal control systems (TCS). I was excited about the prospect of working on the *Freedom* and soon received an offer I was not about to pass up. Although it paid less than what I was making, I accepted and began work in the space industry on September 29, 1991—my twenty-eighth birthday.

CHAPTER 8

Prelude to Disaster

Debris and Flight History

Through the 113 missions prior to STS-107, the orbiter had sustained upward of 15,000 debris hits with an average of nearly 130 total impacts per flight. Each flight sustained about twenty impacts with damage size to the thermal protection system greater than one inch and approximately two to three impacts per flight with damage size greater than five inches. The worst flight in terms of debris impact was STS-27R. It sustained more than seven hundred impacts with three hundred creating damage size greater than one inch. Many hit counts and damage sites were poorly documented because the damage was treated as an orbiter processing issue, and the tiles or impact sites were either replaced or repaired with special putty. Although the exact numbers were never confirmed, more than three hundred tiles had to be replaced after *Columbia*'s inaugural 1981 flight along with thousands of tile putty repairs for minor damage. Despite the limitations to the damage documentation, more than two thousand damage sites had postflight pictures and mission parameters. Approximately three hundred of the two thousand damage sites were well documented and became known as the platinum data set. This data set had the largest impacts from each flight and contained the detailed dimensions of the damage site along with the possible cause based on forensic analysis. In addition the external tank's observed foam losses from each mission were merged with the tile damage site database. This became known as the Dead Sea Scrolls because the summaries were printed on ten-foot-long pieces of continuous computer paper so they could be examined for patterns. This was the most comprehensive source of debris performance and orbiter damage from each mission where imagery and postflight inspection data were recorded. Only about 60 percent of the flights prior to the *Columbia* accident had documented foam losses at tank separation. This was due to imaging limitations such as night launches or poor lighting conditions or other imagery restrictions such as insufficient viewing angles. Even with these limitations, the Dead Sea Scrolls contained a wealth of information related to debris performance, and as the CAIB found out, it did not take much effort to show a significant correlation between foam loss and orbiter damage. The challenge when assessing the debris performance was determining what specific debris source was responsible for the orbiter damages. It was like starting with a pile of hamburger and trying to determine the original shape of the cow.

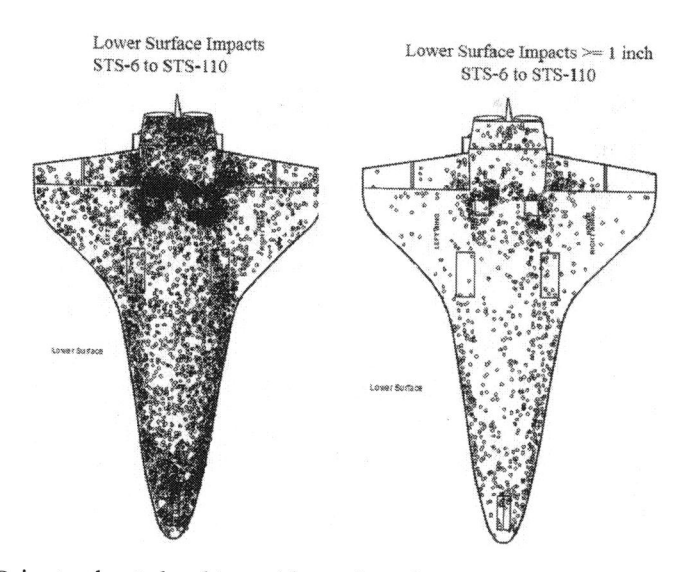

Prior to the *Columbia* accident, the orbiter had sustained more than fifteen thousand hits. On the left is the total number of lower surface impacts, and on the right are those impacts greater than one inch in diameter. Notice the concentration of hits to the nose and around the umbilical doors. (Courtesy of NASA.)

Given the thousands of foam debris releases per flight, it was a seemingly impossible task to connect specific foam loss and impact points. Eventually debris tools were developed to determine "probability zones." These showed the most likely foam debris release locations. This prediction ability coupled with the extensive imagery and radar data were used to pinpoint the exact location and time of release.

Close Calls

Before STS-107 the estimated risk for crew loss was around one in four hundred. The risk drivers were main propulsion system failures or an on-orbit micrometeoroid or debris strike. Considering the losses of *Challenger* and *Columbia*, the historical program risk was one in fifty-seven. If the *Challenger* accident (a propulsion failure) was hypothetically dismissed, the historical odds for crew loss due to debris alone was 1 in 114. However, the real risk from a debris strike was much greater due to the extensive number of debris hits and documented close calls. Close calls were a small subset of damage cases selected from the platinum data and chosen based on severity (both length and depth), thermal degradation after reentry, and extent of the remaining structural integrity of the tile or RCC panel. These close calls were deemed large enough to have caused catastrophic failure had the depth been a fraction of an inch deeper or the damage site been a few inches away. The close calls spanned much of the vehicle forward to aft, and all occurred on the right side of the vehicle. Although flight history showed total orbiter hit counts above three hundred were rare, there was no correlation between high hit counts and close calls. Although there were multiple close call candidates, the qualitative analysis identified four close calls on the tile and one close call to the RCC panel. The close calls suggested the debris risk was much greater than the historical performance showed.

During the postlanding inspection of STS-45 in March 1992, workers found two gouges on the upper portion of the right wing leading edge at RCC panel ten. Damage was not repairable, and they removed and replaced the panel. The most probable cause of impact damage was from debris on-orbit because traces of gold were found in the crack. The likely debris source, therefore, was a small part of a thermal protection blanket from some other spacecraft or launch vehicle. Debris damage during prelaunch or ascent was not ruled out. RCC panel ten suffered two major impact damages, and both were located approximately twenty inches above the leading edge. The larger damage site measured 1.9 inches by 1.6 inches and was 0.17 inches deep. The smaller damage site measured about 0.8 inches by 0.3 inches and 0.07 inches in depth. Due to the depth of the impacts, the carbon-impregnated cloth substrate was exposed with some evidence of oxidation of the carbon fibers. This was characteristic of charring during reentry. Before the next flight, an in-flight anomaly assigned to the orbiter project was closed as "unexplained [but]...most likely orbital debris." The Safety and Mission Assurance Office expressed concern as late as the prelaunch MMT meeting two days before the launch of the next flight—STS-49. However, the mission was cleared for launch on May 7, 1992.

Although three of the four close calls to the tile were from "unexplained" causes, most experts agreed the most likely cause was from large pieces of foam during ascent. This assessment was based on the damage size and shape, which was consistent with observations from foam-on-tile impact testing. The first close call was a large damage site measuring 11.0 inches by 3.8 inches by 0.8 inches. It occurred on STS-73. Although the cause of tile damage was unknown, it was most likely a foam impact that could have compromised the structure and hydraulic lines had the damage been deeper or the impact been to thinner tiles located a few feet away. The debris penetrated the hard, black outer coating and exposed the white silica insulation—the only barrier between the structure and searing reentry temperatures.

Another unexplained cause of tile damage occurred on STS-58. That fracture measured 5.0 inches by 2.0 inches by 0.75 inches and extended through the thickness of the tile. Discoloration and charring of the structure occurred where the damage was deepest. The structure and critical components in the surrounding area would have been damaged had a larger cavity penetrated to the structure.

RETURN TO FLIGHT

Tile damage on STS-26R was attributed to a piece of debonded cork insulation from the right-hand SRB. Although the damage characteristics were similar to foam debris, a large piece of missing cork insulation from the recovered SRB was consistent with the debris transport to this damage site. This debris event produced the largest documented case of tile damage and measured an incredible 18.0 inches by 6.0 inches by 1.5 inches. It also reinforced the fact that debris sources other than foam from the external tank could be just as damaging.

The most damaging close call occurred on STS-27R after an entire tile was found missing along with scorched structure. Although the adjacent primary structure was not severely damaged, analysis showed the structure was subjected to 550° to 575°F temperatures for about a fifteen-minute period. The only things that prevented a burn-through were the relatively mild reentry thermal environment for this location and the thin carbon-epoxy plate used as a substrate to bond the silica insulation material and attach the tile to the structure. STS-27R damage was shown to be definitively caused by a bad ablator application on the nose of the SRB.

Missing tile close call from a possible foam strike on STS-27R. (Courtesy of NASA.)

PRELUDE TO DISASTER

The biggest limitation on identifying all the close calls was the difficulty of determining the debris threats that *missed* the orbiter. All the documented close calls were based on debris impacts to the orbiter. The key question no one could answer was how many missed the orbiter or hit the ET or SRB instead. With all the imagery limitations before the *Columbia* accident, there was no way to tell which near misses could have led to catastrophic damage. There were some debris risk estimates to the shuttle as high as one in five. These tried to account for the near misses, but the analysis was based on unsubstantiated assumptions driving the final answer. One such near miss occurred on STS-112, which had flown two missions prior to *Columbia*. Thirty-three seconds after STS-112 launched on October 7, 2002, ground cameras observed a large (four inch by five inch by twelve inch) piece of bipod ramp foam liberate from the external tank and impact the solid rocket booster and external tank attachment ring. The foam failure was from the exact location that doomed *Columbia*. Post-mission inspection of the SRB confirmed a dent to the steel structure on the forward face of the attachment ring. This was consistent with the crew's post-ET separation photography that confirmed a substantial corner of the left bipod ramp was missing.

After the *Columbia* accident Roy Glanville became the program chair of the Independent Safety and Engineering Review Panel for the remainder of the Space Shuttle Program. He was part of the team that influenced the events leading to the loss of STS-107. The team actions prior to STS-107 demonstrated that problem resolutions had to be comprehensive in order to break the disaster chain. Overextended engineering debris models and assumed impact capability provided a false sense of safety. In other words the foam debris issues prior to STS-107 had been essentially dismissed. The problem assessment for the STS-112 bipod foam release that occurred two missions prior to STS-107 considered only what happened and not what could have happened. "NASA tragically learned that failure of the imagination by focusing on the obvious and not the possible precedes catastrophic failures," summarized Roy.

When dealing with problems, the use of such terminology as "in-family" or "within experience base" must be avoided to prevent the normalization of deviance, which dulls risk perception and results in unwarranted risk acceptance.[7]

STS-113 Flight Readiness Review October 31, 2002

During the next flight readiness review for STS-113 on October 31, 2002, the external tank project reported that, "Foam loss over the life of the Shuttle Program had never been a Safety of Flight issue and that more than 100 external tanks have flown with only three documented instances of significant foam loss on a bipod ramp." The CAIB's review later concluded that bipod foam loss occurred on six missions prior to STS-107. The number of bipod releases eventually climbed to nine as the imagery analysis techniques were improved and historical performance was reassessed. The STS-113 flight rationale given by the ET project stated, "1) Current bipod ramp closeout design had not been changed since STS-54; 2) There had been no design/process/equipment changes over the last 60 flights; 3) All ramp closeout work was performed by experienced practitioners all with over 20 years' experience each; 4) Ramp foam application involved craftsmanship in the use of validated application processes; 5) There were no changes in Inspection/Process control/Post application handling, etc.; and 6) The probability of a bipod ramp loss was no higher/no lower than previous flights." They concluded that overall the, "ET was safe to fly with no new concerns (and no added risk)."

With each successful landing, NASA engineers and managers increasingly regarded foam-shedding as inevitable and as either unlikely to jeopardize safety or presenting only a minimal risk. They believed this even though the risk had not been quantified. The distinction between foam loss and debris events also appeared to blur.

---

[7] "In-family" is a term used to describe an observed phenomenon or event consistent with previous flight history performance.

NASA and contractor personnel came to view foam strikes as a maintenance or orbiter turnaround issue rather than a safety of flight issue. In other words it was just a matter of repairing or replacing tile during the orbiter processing for the next flight. During its investigation the CAIB discovered a consistent and seemingly complacent pattern of downplaying debris events as safety concerns. What was originally considered a serious threat to the orbiter during the first few flights came to be treated as in-family in FRR documentation, MMT minutes, and IFA disposition reports. Debris failures and impacts were reportable problems within the known experience base. They were believed to be understood and not regarded as safety of flight issues. Engineers stated during the accident investigation that they knew in advance the external tank was going to produce a debris shower, and it was going to be difficult clearing even *Columbia*'s inaugural flight. The engineers were correct as foam shed from the external tank to various degrees on every flight from 1981 until the *Columbia* accident. Boston College Sociology Professor Diane Vaughan called this behavior of ignoring foam debris as a safety of flight concern the "normalization of deviance." NASA managers were getting so used to the debris warning signals they were no longer much of a warning.

The first time the "low fuel" gauge lights up, one frets about running out of gas. When one makes it to the gas station easily, that person isn't as concerned the next time. So it goes until one day, ignoring the fuel light, the car runs out of gas.

Foam Debris and CAIB Findings

Although workers conducted tests to develop and qualify foam for use on the external tank, there were large gaps in NASA's knowledge about this complex and variable material. Before the accident, testing conducted at Marshall Space Flight Center and later under the auspices of the CAIB indicated that mechanisms previously considered a prime cause of foam loss (cryopumping and cryoingestion) were not feasible in the conditions experienced during tanking, launch, and ascent. However, during the accident investigation, dissections of foam bipod ramps on tanks not yet launched revealed subsurface flaws and defects. Only now were these being discovered and identified as contributing to the loss of foam from the bipod ramps. Over the years, NASA engineers and management did not appreciate the scope (or lack of scope) of the hazard reports involving foam-shedding. Qualification from thirty years before had utilized methods that were rudimentary compared to today's standards. Testing had been done with essentially solid foam blocks that had not been sprayed onto the complex bipod fitting geometry. Hence the practice violated NASA's "fly what you test, and test what you fly" philosophy. The external tank and bipod ramp were not tested in the complex flight environment, and fully instrumented tanks were never launched to gather data for verifying the analytic tools used to certify the design.

In addition to the close calls, there had been a number of foam loss events that caused significant debris and tile damage and thus required responses. When the intertank area started to lose significant foam on flights in the early 1990s, a "tiger team" was convened that came up with a way to mitigate intertank foam loss by making small perforations in the foam.[8] The small perforations relieved the internal stresses causing the failure and minimized the foam debris size in the event of a failure. When the Environmental Protection Agency mandated a change of material in the late 1990s, large foam loss and damage resulted, and another tiger team was convened to improve the process. Until that fateful day in February 2003, though, shuttle program managers had largely become insensitive to the debris risk. There was a famous story circulated at NASA regarding how to boil a frog. Place a frog in boiling hot water, and it will hop out—so the legend goes. Put it in warm water, though, and slowly bring it to boil. By the time the frog realizes the water is too hot, its muscles no longer function, and the frog boils. Because all the foam-damaged orbiters survived, the agency got increasingly comfortable in the hot water. When a crisis finally arose on STS-112, NASA leadership didn't have

---

[8] A tiger team is a specialized team or group of experts assigned to investigate and/or solve technical or systemic problems.

the muscles to stop flying and resolve the issue. At one point during the RTF redesign after *Columbia*, John Muratore wondered aloud, "If the bipod ramp came off early on STS-112, and the ramp was supposed to protect against flutter and heating at higher speeds, why didn't we just remove the ramp after STS-112? Clearly it wasn't needed."

Rod and John surmised the ramp was kept because of residual concerns about heating and aerodynamic flutter on other trajectories. It actually took two years of effort to convince everyone it was safe to fly without the bipod ramp.

The CAIB identified fourteen flights with significant tile and RCC panel damage or major foam loss, including six foam losses from the left bipod. They also discovered that, "Space Shuttle Program personnel knew that the monitoring of tile damage was inadequate and that clear trends could be more readily identified if monitoring was improved, but no such improvements were made."

Shuttle program managers confused the notion of foam posing an "accepted risk" with foam not being a safety of flight issue. The process for closing IFAs was not well documented and appeared to vary from flight to flight. The CAIB noted that, although there was a process for conducting hazard analyses when the system was designed and a process for reevaluating them when a design was changed or the component replaced, no process addressed the need to update a hazard analysis when anomalies occurred. In the final analysis, the debris threat was known but ignored, and the unknown debris risk was accepted. Ultimately the CAIB pinpointed the blame of the *Columbia* accident on a foam debris strike to the left wing at eighty-two seconds into flight. The foam came from the left bipod, which was part of the thick titanium structure that attached the external fuel tank to the forward attach point on the orbiter. The foam was a suitcase-sized cover used as an aerodynamic ramp designed to minimize the heating and aerodynamic forces at the tank attachment point. The foam struck the wing leading edge with nearly three thousand pounds of force. This blasted a large hole in the RCC panel used to protect the structure against the searing 2,500°F reentry temperatures. At that point, *Columbia*'s crew and spacecraft were doomed.

Ballistic Coefficient and Foam Debris

The key question the CAIB investigators asked was how a lightweight material such as foam could impact the orbiter with a velocity high enough to puncture a hole in the RCC panel. Just prior to separating from the external tank, the left bipod foam was traveling with the shuttle stack at about 1,570 miles per hour. Visual evidence showed the foam debris impacted the wing approximately 0.161 seconds after separating from the external tank. That was 81.9 seconds into flight. In that split second, the velocity of the foam debris slowed from 1,570 miles per hour to about 1,025 miles per hour. Consequently the orbiter hit the foam with a relative velocity of about 545 miles per hour. That was the equivalent of a bowling ball striking the orbiter's wing at 175 miles per hour. The foam debris slowed, and the orbiter did not, so technically the orbiter ran into the foam. The bipod foam slowed rapidly because low-density and high-drag objects such as foam have low ballistic coefficients. This means their speeds rapidly decrease when they lose their means of propulsion. A ballistic coefficient is a function of mass (M), area (A), and drag coefficient (C_d), and it measures an object's ability to overcome air resistance in flight. A bullet has a very high ballistic coefficient and a very slow decrease in speed after the main charge is fired. The bipod ramp was a piece of foam debris with a low mass, large size, and high drag. This produced a low ballistic number and hence a high deceleration. Throwing a baseball out of a moving car reveals a gradual deceleration compared to an empty shoe box, which has a much lower ballistic coefficient.

CRATER Model Limitations

Without the video footage of the bipod release, it would have been difficult to determine the exact nature of the wing damage using the existing CRATER model. In 1966 during the Apollo program, engineers developed an equation to assess impact damage ("cratering") by micrometeoroids. The equation was modified between 1979 and 1985 to enable the analysis of impacts to the orbiter's tile. The modified equation became known as CRATER, and it predicted possible damage from sources such as foam, ice, and launch site debris. It was most often used in the day-of-launch analysis of ice debris falling off the external tank. CRATER had been correlated to actual impact data using results from several tests. However, the test projectiles were relatively small (maximum volume of three cubic inches) and targeted only single tiles rather than groups of tiles as were actually installed on the orbiter. No tests were performed with larger debris objects because it was not believed such debris could ever impact the orbiter. This resulted in a very limited set of conditions under which CRATER's results were empirically validated. Although CRATER was designed and certified for a very limited set of impact events, the results from CRATER simulations were generated quickly. During STS-107 CRATER was used to model an event well outside the parameters against which it had been empirically validated. For instance, while CRATER had been designed and validated for projectiles up to three cubic inches in volume, the bipod foam was estimated to be 1,200 cubic inches—four hundred times larger. Due to the limitations of CRATER, the debris analysis tools that existed prior to STS-107 had to be completely rewritten.

CHAPTER 9

Debris Analysis 101

Release-Transport-Impact Model

After the *Columbia* accident, an alternative analytic tool was clearly needed to replace CRATER. The newly developed debris assessment method known as the release-transport-impact model would be it. This model contained three major components: the release model, transport model, and impact damage model. It analyzed hundreds of various debris source–impact target pairs of expected debris sources. If the debris damage was less than the target's minimum impact tolerance, the debris strike was considered cleared "deterministically." The orbiter was considered "certified" if all foam debris strikes cleared their respective targets deterministically. This was the original approach used to clear the STS-114 orbiter for all ET foam sources. However, due to the low impact tolerance of the orbiter thermal protection system and high-energy debris releases, "certification" was rarely the case.

The fundamental parameters or inputs into the release model for a particular debris type and shape included the following: mass of the liberated debris, time of release, release rate, and release location. The impact targets included the RCC panels, tiles, windows, and special tiles. "Special" tile was defined as a tile within a two-inch boundary around the landing gear and ET umbilical doors. These tiles had thin tips covering three redundant seals around the doors that protected the inner cavity from the searing reentry temperatures. Damage to a special tile could have resulted in high-temperature flow directly into a critical region if the seals became compromised. Much of the effort leading up to STS-114 involved developing the analytic tools needed to assess the debris release, transport, and impact characteristics for all expected debris sources.

Debris Release Modeling

The "release model" portion of the release-transport-impact debris assessment characterized the debris environment. Debris release models for foam, ice, and other debris characterized the release in terms of debris type, failure mode, release location, time of release, release angle, velocity, and debris mass and shape. These release parameters defined the input conditions for the transport model, which described the aerodynamic characteristics of the debris and established the initial transport conditions. Release models for specific failure modes were typically derived analytically or from empirical test data. Unfortunately debris releases from several sources were either too infrequent and/or small to be characterized accurately with a statistical model. Thus empirical models were restricted to regions where meaningful release data existed. In cases where adequate release data were unavailable, flight history or a model based on engineering first principles was used to construct a release model for a specific debris source such as ice and other nonfoam debris. The external tank project at MSFC and Lockheed Martin at MAF were responsible for the majority of foam and ice release model development work. While MSFC provided the large testing facilities and technical oversight, test execution and analysis of test results were Lockheed's responsibilities. Wanda Sigur relied heavily on Dan Callan, Mike Quiggle, Phil Kopfinger, along with the lead engineer for materials testing, and dozens of technicians, engineers, and analysts to characterize the foam and ice failure modes and release models. Eventually the other shuttle elements such as the orbiter and SRB would be integrated into characterizing release models for various types of debris other than foam and ice shedding from the hardware.

Debris Transport Analysis Modeling

The debris transport analysis model replicated the movement of liberated debris through the space shuttle flow field during all flight phases. Doing so required representing the two principal forces acting on the debris: drag and lift. Determining the drag and lift of any shape involved solving the Navier–Stokes equations. These equations are nonlinear partial differential equations with the nonlinearities corresponding to turbulence—the toughest phenomenon to characterize. Because of the nonlinearity, the Navier–Stokes equations are very difficult or impossible to solve unless simplifying assumptions are made such as laminar flow (flow without turbulence). Analytic tools developed using simplified assumptions are called "classical methods," and engineers frequently use them to solve relatively simple problems. A solution of the Navier–Stokes equations is called a velocity field or flow field. This is a description of the velocity of the fluid at a given point in space and time. Once the velocity field is known, the lift and drag quantities for any piece of debris can be calculated. All the existing analytic tools and classical methods used to determine the shuttle's lift, drag, and aerodynamic loads were very limited when considering the multitude of debris types and countless debris shapes and sizes. Due to the complexity of the debris problem, an alternative debris transport analysis model had to be developed. The new debris transport analysis (DTA) model yielded a mathematical trajectory that provided the position as a function of time for each piece of debris in the predicted debris environment. The model spanned from release until debris impacted or cleared the space shuttle vehicle. NASA developed the debris transport model by conducting a multitude of aerodynamic wind tunnel tests and using the supercomputer named Columbia at the Ames Research Center (ARC). In 2004 Columbia was the second-largest supercomputer in the world. The wind tunnel testing was also used to validate the complex computational fluid dynamic (CFD) model. This enabled thousands of debris scenarios to be evaluated on Columbia in minutes or hours. Computational fluid dynamics is a branch of fluid mechanics that uses numerical methods and algorithms to solve and analyze problems involving fluid flows. While wind tunnel testing provided limited data at select flow conditions, CFD simulations provided detailed information about many conditions used to simulate the effect of debris transport on the space shuttle launch vehicle. High-speed supercomputers were needed to perform the extensive calculations required to simulate the flow field interactions with surfaces defined as boundary conditions. The program ("code") used to solve the three-dimensional Navier–Stokes equations and better understand the aerodynamic transport forces acting on the vehicle was named OVERFLOW. This was developed as part of a collaborative effort between NASA's Johnson Space Center and Ames Research Center. The driving force behind this work was the need to evaluate the flow around the space shuttle launch vehicle. NASA's Pieter Buning, Dennis Jespersen, and others originally developed this in the early 1990s. The code was an outgrowth of earlier codes and a result of the ARC's long history of flow-solver development. Before OVERFLOW could be used for debris transport purposes, it had to be validated using an integrated vehicle wind tunnel. This drove NASA to conduct one of the largest integrated vehicle tests ever conducted on the shuttle program since its inception in the late 1970s.

No one knew shuttle aerodynamics better than Ray Gomez. He began work at NASA's Johnson Space Center in 1985 after completing a bachelor's degree in mechanical engineering at Rice University. Well-groomed and known for wearing a bow tie to major shuttle meetings, Ray spent the majority of his career on the shuttle program. He worked on a wide range of CFD applications ranging from simulations of wind loads on the shuttle prior to launch to hypersonic reentry flow fields. His main responsibility working for JSC's Engineering Directorate was to serve as the technical panel chair for shuttle aerosciences—otherwise known as the "aero panel." In addition to reviewing all integrated aerodynamics products for the Space Shuttle Program, he led the development of wind tunnel tests, CFD simulations, and in-flight shuttle debris transport assessments. Ray had a reputation for being very meticulous and thorough. Everything had to be right and double-checked. When it came to aerodynamics, CFD, and debris transport, Ray was the program's technical conscience, and the shuttle program managers trusted his work. Ray's counterpart at Ames was Stuart "Stu" Rogers—a studious and intellectual engineer in his own right. Stu was also a CFD expert and worked with Ray to refine the code and perform the runs on the Columbia supercomputer. Before the first RTF mission, the shuttle program conducted more than a billion separate debris transport runs and monopolized the majority of Columbia's processing time.

Debris Impact Modeling

Overall the debris impact model transformed the transport model output (the debris environment) into an impact table that enabled damage assessment from a debris strike. Due to the lengthy development time, impact damage capability models were only developed for foam, ice, and ablator impacts to the orbiter windows, RCC panels, tiles, and special tiles. The orbiter project and Justin Kerr's team of NASA, United Space Alliance, and Boeing engineers and analysts developed the impact models for John Muratore and SEI. Justin had worked with John Muratore before on the X-38 and was familiar with his leadership style. "Although John was demanding, I enjoyed working with him and the SEI team," recalled Justin. "Looking back on the experience, a few of the things that impressed me most about all the debris work, mission support, and Return to Flight preparations was how well people got along and how quickly things could be accomplished. Working with John and SEI was no exception. It was as if all the normal bureaucratic constraints and organizational resistance was somehow removed, and everyone operated in a purely collaborative environment and worked toward a single goal. I gained a broader agency perspective because

DEBRIS ANALYSIS 101

of all the debris work with multiple centers and was amazed at the amount and level of expertise within [NASA]."

John and the SEI team appreciated Justin's value, which was why he had such a good working relationship with John. What impressed Justin most about John was his work ethic and commitment to getting the job done. John was not going to miss out on anything, even if he had to roll up his sleeves and work with the engineers and analysts on solving a problem. Justin's level of respect increased even more when he discovered John had taken a nap under a conference room table late one night rather than go home in order to support resolving a debris issue on STS-114—the first RTF mission. There were not many engineers chronicled in *Rolling Stone* magazine for technical careers, but John was one of the few. Overall the impact model development effort took Justin's team nearly two years of impact testing before the damage models were fully developed.

All the debris sources and impact models were documented in NSTS 60559, Volume One. This effectively became the program's debris bible. The orbiter impact algorithm determined an impact location and conditions based on the debris transport model output. The conditions of collision, including the angle of incidence and the impact velocity or kinetic energy for a given debris source, were subsequently computed based on the trajectory and orbiter geometry at the impact site. The RCC capability was defined by the Orbiter team in terms of kinetic energy, and the tile and special tile capabilities were defined in terms of velocity. The damage impact model compared the impact conditions defined by the release and transport model against the capability of RCC, tile, and special tile to determine whether a capability had been exceeded. Exceeding a capability was considered a failure and assumed to be a catastrophic event. In several instances the vehicle structure capability was not defined. For example, there was no capability model for the open ET umbilical doors and door mechanism. If debris hit these structures, it was considered a catastrophic failure. Simply reporting the impact probability where every impact resulted in failure did not provide an adequate risk estimate. Instead it represented an unrealistic worst-case risk index that was not representative of flight history. Typically the elements evaluated the DTA on a case-by-case basis for impacts to structures lacking defined capabilities. This analysis would help determine the structure's vulnerability. Another example was the orbiter windows. These did not have a capability curve due to the complex fracture mechanics of fused silica glass. Window impact assessments relied on testing of worst-case conditions ("enveloping") by utilizing similar debris parameters such as impact velocity, impact angle, size, mass, material hardness, and so on. Since the windows had demonstrated acceptable performance for a given set of test conditions, any impact condition below this was considered enveloped and acceptable. If the impact conditions were exceeded, the orbiter was subjected to additional window impact testing to determine if the windows were susceptible to damage under those conditions.

Deterministic Analysis

Deterministic analysis typically utilized worst-case conditions to demonstrate that liberated debris could not lead to catastrophic damage. These worst-case conditions were also referred to as "certification rigor" assumptions. A subclass of deterministic analysis involved using the certification impact tolerance levels against a nominal transport environment. A deterministic analysis was considered acceptable when the vehicle impact capability exceeded the debris environment or impact conditions. This meant there was no risk of catastrophic damage to the orbiter.

Probabilistic analysis was used for debris impact cases that did not clear deterministically. In these cases the probability of exceeding an allowable (such as tile or RCC capability) was greater than zero. Probability in an applied sense is a measure of the likeliness that a (random) event will occur and is measured on a scale between zero and one. The higher the probability of an event, the more likely it will occur—all the way up to a probability of one (a mathematical certainty). Early in the Apollo program a question was allegedly asked about the probability of successfully sending astronauts to the moon and returning them safely to Earth. A risk, or reliability, calculation of some sort was performed, and the result was a very low success probability value. So disappointing was this result that NASA became discouraged from performing further quantitative risk analyses until after the 1986 *Challenger* accident. NASA relied instead on other means such as the failure modes and effects analysis (FMEA) method for system safety assessments.

Putting It All Together

The United Space Alliance and Boeing debris teams were instrumental in the development and integration of the debris release-transport-impact model. John Muratore started working on the new model with these contractors, NASA engineering, and the ET project well before the CAIB recommendation to replace the CRATER model. USA was the prime contractor for the Space Flight Operations Contract (SFOC). This consolidated thirty heritage contracts into one for the day-to-day management of the space shuttle fleet and planning, training, and operations for the space shuttle missions and International Space Station. Lockheed Martin and Boeing formed their partnership in 1996. Since then USA has served as NASA's primary industry partner in human space operations. Chief Engineer Rick Barton led the USA team and ensured all communications and technical decisions relating to the Space Shuttle Program went through him before going to NASA. Although Boeing had built the orbiters, it was considered a subcontractor to USA for all the engineering, subsystem, and technical support needed to operate the shuttle. This unusual arrangement sometimes put strain on the contractor relationship because NASA had a tendency to bypass USA and consult directly with Boeing when dealing with orbiter technical issues.

Rick was a tall, white-haired man in remarkable shape for his age due to his passion for running. He was an old-school engineer from the Apollo days and still wore the standard white shirt, suit, and black tie. He even carried a slide rule because he insisted it was more accurate than a calculator, and he never had to worry about his batteries going bad. I always enjoyed working with Rick because of his great intellect, sense of humor, and ability to sketch out complex problems. He was more like a teacher than a technical manager. In addition to the Boeing staff, Rick had at his disposal a talented group of retired NASA and shuttle engineers that became affectionately known as the "Graybeards." The Graybeards had a long history of spaceflight engineering and operations experience, and they acted as consultants on a multitude of engineering disciplines to double-check Boeing's work or conduct independent analysis. Although the group size varied from ten to fifteen experts, Bass Redd was always the leader. He was a crusty, no-nonsense engineer who had probably forgotten more about spacecraft design than I knew. Bass always had a story to share. Coupled with his dry sense of humor and suspenders, this made him a very colorful character. He was well respected within the human spaceflight community. When Bass spoke people listened because he had lived it.

Darby Cooper led the Boeing debris team. Darby was a self-proclaimed corn-fed Iowa State engineer. He was an excellent communicator and technical wizard. Although, he never admitted it. "After working payloads for a number of years," Darby recalled, "I was interested in working on the A-team on the vehicle side, which was right before the *Columbia* accident. Debris really wasn't an interest, but I was hired into the systems engineering and integration team out of necessity after the accident because I had a pretty diverse background and aerospace engineering degrees. I had not done much work in aerodynamics, so this was a bit of serendipity with the new job."

When Darby worked for me at Boeing, he earned a Silver Snoopy award—one of NASA's highest spaceflight honors—for his debris work during the RTF wind tunnel testing. He also had experience working for John Muratore while working payloads. "John always drove us to rigorous technical solutions and demanded we understood the physics of the problem," Darby said. "It's hard to believe, but John was much more relaxed during Return to Flight than when I worked with him on the payload side before the accident."

Because of his specialized debris knowledge and expertise, Darby became the leader of the talented team I managed when I worked at Boeing. Although he was a midlevel manager within the organization, he was the face of debris for Boeing and one of the people I counted on the most when it came to debris risk.

Aerodynamically Sensitive Transport Time (ASTT)

The term aerodynamically sensitive transport time (ASTT) was defined as a mission time were a specific type and size of debris would not lead to catastrophic damage. It was a deterministic analysis that could also be used to look up the acceptable mass size corresponding to a specific release time and location. If the release location and mass were known, then one could determine the time(s) at which the debris was not a threat. If the release location and time were known, then the largest safe debris mass could be determined. If the release time and mass were known, then the location(s) at which the debris would not be a threat could be determined. Typically the worst segment of ASTTs (when impacts could cause the most damage) was between seventy and one hundred seconds. Nearly 80 percent of the total debris risk exposure from all debris sources and failure modes occurred during this critical window.

Originally the ASTT assessment started with a request from the ET project to define an upper time limit where large foam debris would no longer be a catastrophic threat. The ET project recognized that larger foam releases late in flight were likely, but if they were late enough, they would not pose a threat to the vehicle due to the reduced aerodynamic acceleration potential at high altitudes. Initially an ASTT of 135 seconds was calculated. The team used a specific set of assumptions based on an initial RCC panel capability and the largest foam mass size expected. Many interpreted 135 seconds as the time at which debris was no longer an impact concern. However, an ASTT had not been calculated for tile impacts or ice debris. The ASTT results were highly sensitive to the ground rules defined for the debris environment and capability models. Documentation of the assumptions, therefore, was a critical part of defining this time frame. In reality ASTT was a series of time curves for specific debris types and masses from various debris locations. The forward debris release areas tended to be the most lethal locations on the tank. This allowed only smaller debris sizes to be acceptable. The curves would shift toward larger allowable debris sizes as the release location moved aft on the vehicle. ASTT curves were not intended to replace detailed assessments but to provide a quick, real-time lookup tool that was useful during ascent.

Key Debris Risk Drivers

Of the multiple parameters that affected risk characterization, only a few really drove the impact conditions and ultimately the risk level. The key parameters were mass, release location, time of release, release rate, aerodynamics, and the impact or damage capability. Typically the larger the mass, the higher the risk. The nuance with mass was the relationship between volume and density. Materials with higher density (ice) have a much smaller volume for a given mass compared to low-density materials such as foam. In this case the ice might prove more harmful because the smaller volume would concentrate the impact load over a smaller area compared to foam and make the structure more likely to fail. Release locations farther forward on the vehicle were also generally worse than aft locations for several reasons. First, if the release location was far enough forward, the potential target area was much larger and could include additional targets such as windows, OMS pods, and tail structure. Second, the impact velocities could be much higher because the debris would have a longer transport distance to travel. The ASTT curves with the higher impact velocities (typically occurring seventy to one hundred seconds after liftoff) drove time of release. The early and late releases outside of ASTT were generally benign regardless of debris mass or release location. Another important parameter driving the risk was the aerodynamics of both the transport and debris shape. High-lift debris shapes expanded the possible target area, and high-drag shapes increased the impact velocities. The final risk driver was the impact or damage capability of the structure. Target areas with high impact capability had lower debris risk. Characterization of each parameter was not absolute. There were always conditions and situations during flight where the exact opposite behavior than the model predicted was observed. However, those situations tended to be exceptions rather than the norm.

Debris Classifications

The Space Shuttle Program had two classifications of debris: unexpected and expected. Unexpected debris was defined as those potential debris sources for which design controls were in place to ensure debris would not be liberated. Debris sources that fell into the expected category did not have design controls that could eliminate the possibility of release during ascent. These debris sources were documented in NSTS 60559 and were the focus of the transport analyses performed. Expected debris was further classified in order of risk hierarchy with respect to the disposition and flight rationale that would ultimately be documented in the integrated debris hazard report (IDBR). Liftoff debris was considered any debris source up to the point the vehicle cleared the tower. Ascent debris was any debris source after tower clearance. As the liftoff debris assessment entailed substantially different liberation and transport mechanisms than ascent debris analysis, it was considered outside the scope of debris risk assessment. The remaining rungs in the hierarchy were arranged with the strongest flight rationale toward the top. Debris sources with masses less than 0.0002 pounds were followed by those with no transport to critical structure. The final category was debris sources that cleared deterministically. These could not cause critical damage to the vehicle based on certification rigor capability for worst-on-worst debris environments. The debris assessment process differed dramatically depending on the class of debris. It ranged from the simplest to the most complex assessments.

CHAPTER 10

Debris Risk Assessments

Foam Debris

The foam risk analysis was highly dependent on the foam liberation model. This was different for the various foam liberation failure mechanisms. Before a foam debris probabilistic risk assessment (PRA) could be determined, the failure mechanisms needed to be understood to develop a release model. Analytic models combined with testing results were used to develop a relationship between the specific failure mechanisms and foam loss. As part of the *Columbia* accident investigation and RTF activities, in-depth investigations of the foam failure mechanisms were conducted. Initially four foam failure modes were identified: cohesive strength failures, bond line delamination, bond adhesion failure, and outer fiber cracking. These failure modes were considered "historical" because they were known before the *Columbia* accident. The risk assessment for the first RTF mission (STS-114) considered only these four failure mechanisms. During that flight several foam losses were attributed to other failure modes such as air load failures, thermal cracking or delamination, and crushed foam. These had to be added as follow-up assessments. Secondary impacts and combined failures were possible but not included as part of the foam debris risk assessment. They were ruled out as accepted risks due to their low likelihood, or other failure modes enveloped them.

Void Delta P Foam Failure Mode

There were several variations of the cohesive strength failure mode. These included void delta P[9] losses, "popcorning," and cryoingestion or cryopumping. Foam was applied to the tank either by automated spray or manually. Automated spray was applied in areas of accessibility such as the acreage areas while manual spray was applied around areas of complex geometry such as the feed line attachment brackets. Inherent in the spray process was the development of internal voids caused by the curing process or geometric voids created adjacent to the structural elements. Voids created during the automated spray process were rare, but they were unavoidable during the manual spray process. During ascent the internal pressure began to build as the thermal heating increased. The external dynamic pressure (aerodynamic load) also built as a function of speed. It peaked around sixty to sixty-five seconds. This was known as maximum dynamic pressure, or "max Q." Then it rapidly decreased after that. If the combined pressure stress of the internal void was more than the fracture toughness of the foam, then a void delta P failure (divot) would result. When it failed in this fashion, the foam was called a frustum. It looked like an upside-down diamond with the top chopped off. On every flight there were thousands of half-dollar-sized foam divots referred to as popcorning. These were small void delta P foam losses on the surface created from the internal pressure buildup in the closed-cell foam. They tended to occur during maximum heating between 100 and 120 seconds mission elapsed time (MET) but were typically small enough to clear deterministically.

---

[9] Void delta P failure was the most common foam failure. It occurred if the pressure increase inside existing voids in the foam exceeded the fracture toughness or strength of the foam. The pressure increase inside the voids were due to the aerodynamic heating of the foam surface as the vehicle accelerated. Surface temperatures could reach as high as 600°F.

Cryoingestion and Cryopumping Foam Failure Modes

Cryoingestion and cryopumping were similar to void delta P but could produce additional pressure within an existing crack or foam defect during ascent. Cryoingestion was a concern near the liquid hydrogen tank and intertank flange joint due to the possible leak path of nitrogen from the intertank. Cryoingestion occurred as the pressure in a foam void dropped during tanking due to the temperature decrease from the propellant. The resulting pressure drop drew gaseous nitrogen from the intertank into the void through a leak path. The nitrogen then condensed into a liquid. Rapid heating during ascent caused the condensed nitrogen in the void to either boil or rapidly expand. If the leak path was large enough, the pressure buildup could be relieved. If not the foam would fail.

Cryopumping occurred when pressure in a foam void dropped during tanking, and ambient air (instead of gaseous nitrogen) was drawn into the void through a leak path. As with cryoingestion, rapid heating during ascent caused the condensed air in the void to boil or rapidly expand. It would either vent or cause a foam failure. Both cryoingestion and cryopumping failure modes tended to be explosive. They caused foam debris of irregular shapes and typically much larger sizes than void delta P failures.

Other Foam Debris Failure Modes

Three other foam failure modes—bond line delamination, bond adhesion, and outer fiber cracking—had strenuous process controls that prevented failure. Test and analysis verified these. A bond line delamination resulted when "peel" stresses exceeded the strength of the foam–substrate bond in the presence of a defect or crack. Bond adhesion failures occurred between the foam and metallic substrate and were a subset of the bond line delamination failures. Both failure modes rarely occurred because the bond strength of the foam to metal was so much greater than the foam's fracture toughness. In other words the foam would fail before the bond to the metal tank. Outer fiber cracking was isolated to high-bending regions and stress relief cracks from tanking. This was especially true in areas where the foam was thick. The external tank shrunk several inches when filled with cryogenic propellant and caused the internal stresses on the foam to build. If the internal stresses were higher than the foam strength, a stress relief crack would form. Stress relief cracks were fairly common during tanking, and the external tank project evaluated each on a case-by-case basis to determine whether the crack was within its established launch commit criteria (LCC) and whether the crack was in an area that would harm the vehicle if a failure occurred. If so the mission was scrubbed and the cracked foam repaired.

After the first RTF mission (STS-114), three additional foam debris failure modes were identified. None had been known or accounted for before the flight. Foam losses occurring during conditions of maximum dynamic pressure, which corresponded to the highest aerodynamic loads, were classified as air load failures. Any foam-insulated protuberance such as the liquid hydrogen ice/frost ramps were prone to this failure mode due to exposure to the high-velocity airflow and shock waves formed by supersonic flight. At numerous locations on the external tank, different types (recipes) of foam came in contact with one another. Such was the case for the adjacent acreage around the liquid hydrogen ice/frost ramps. The NCFI acreage foam and the BX-250 ice/frost ramp foam had different coefficients of thermal expansion and were located in areas of high heating. Consequently the different foam expansion rates created areas of localized thermal stress. If large enough it could lead to cracking. This condition would weaken the foam and make it more susceptible to failure from high air loads, void delta P, or cryopumping. The final foam failure mode uncovered during STS-114 was a failure attributed to "crushed" foam. During manufacturing mats and protective gear were placed around the high-traffic areas during the foam application process to allow access to areas where the foam was applied manually. The highest-traffic areas were typically on the liquid hydrogen and intertank acreage areas next to the feed line, ice/frost ramps, and bipod. Crushed foam resulting from a misplaced hand, foot, or tool impacting the unprotected tank would become substantially weakened or develop small cracks. The bigger the imprint and amount of impact weight, the more the damage. Foam could be crushed to about 20 percent of its original thickness without being detected. The weakened foam was then more prone to air load, void delta p, and cryopumping failures. All the process and access control procedures for manually sprayed foam were changed after STS-114 to mitigate the crushed foam risk.

Foam Debris Release Model

For both deterministic and PRA analysis, the key parameters needed from a foam debris release model included the location of the debris release, size or mass of the debris, time of debris release, release rate, pop-off velocity, and release angle. The void delta P and cryopumping or cryoingestion release models were developed first and were based on extensive analysis and empirical test data. The ET project dissected foam applications on flight hardware to characterize the frequency, size, and distribution of manufacturing voids inherent in the foam application process and voids due to the local effects of tank geometry. Statistical models developed from the dissection results provided a defect representation of the manufacturing process. Information regarding the substrate tank temperature and ambient conditions provided the conditions for the heating rates and heat transfer calculation. Thermal vacuum ground testing used artificially machined voids in flat panels of foam and established the fracture strength of the foam when exposed to the heating conditions during ascent. With this empirical input, the release model could predict whether each void in the model would liberate debris based on its position and when. Parameters such as the external heating rate, freestream pressure, and fracture strength were applied to characterize the uncertainty in debris liberation that ground testing alone could not replicate.

While the void delta P failure mode was possible in all foam applications, most either had no dissection data or the data showed manufacturing voids were too infrequent and/or small for a statistical model to accurately characterize them. Therefore, the use of empirical models was restricted to regions where only meaningful dissection data existed. Such was the case for the acreage regions sprayed using an automated, controlled process that rarely produced any defects or voids. In cases such as this, flight history or a model based on engineering first principles was used to construct a release model for a specific debris source such as ice or orbiter debris.

Foam Debris Tables

For external tank foam sources, the preferred release model inputs were the empirically derived foam debris tables. These were mass distributions as a function of time and release rate. Debris tables for the risk assessment of critical foam debris were developed from probabilistic analyses of void delta P and cryopumping foam failure mechanisms. Debris tables showed the predicted debris mass and release frequency per ten thousand missions. They utilized empirically derived divot–no divot curves as well as statistical distributions of defect size and location to generate random defects for each foam source. A divot–no divot curve was a statistical representation of the dissection data and a predictive tool that combined the effects of base temperature of the foam, surface temperature, aeroheating damage, and the structural properties of the foam matrix. The ET project was responsible for developing all release model inputs for the purpose of generating probabilistic risk assessments for the twenty-nine identified foam debris sources. Unfortunately, given the extensive data and complexity needed to generate a representative debris table, only a few of the twenty-nine foam debris sources on the external tank had one. Only the bipod foam debris table was developed prior to STS-114. If a debris table was not available, a physics-based model or mass and release rate distributions derived from flight history or test data were used instead.

John Brekke was a tall, blond, blue-eyed project manager, and his Aerospace Corporation team from El Segundo, California, worked in parallel with the ET project to develop their own linear elastic fracture mechanics (LEFM) release model. The CAIB recommended an independent entity to provide safety assessment. Consequently the Aerospace Corporation contracted with NASA to serve this function for RTF. Shortly after the contract was signed, Brekke was asked to join the Aerospace Corporation team supporting the shuttle RTF missions and the SEI organization. Brekke's SoCal tan and blond hair complemented his relaxed demeanor, and his team played a key role in developing the PRA approach and generating all the foam debris risk results. Although similar to the ET project debris table for approximating the void geometry of the foam application process the Aerospace foam release model had a key difference. The ET project had its statistically based divot–no divot curves and debris tables, while the Aerospace Corporation team used a physics-based fracture mechanics model to determine when foam would liberate. In the Aerospace Corporation LEFM model, if the void pressure exceeded the fracture toughness of the foam, a piece of foam debris would be generated.

One of Brekke's key analysts was Matt Eby. Brekke and I relied on him the most for developing and running the Aerospace Corporation PRA model. Matt was a tall, slim man in his midtwenties with blondish-red hair. He was quiet and an introvert but excelled at calculating PRA and presenting his results. Matt's ability to explain complex concepts to the engineering community always impressed me, and I developed the utmost confidence in his presentation skills. As a novice engineer, Matt had just graduated, and the Aerospace Corporation hired him at the beginning of 2005. It was his first real exposure to the space industry and his first real engineering job. This goal had been sparked when his older sister had brought home a model rocket from a school sponsored Space Camp. Matt's manager asked him to attend an internal Aerospace Corporation shuttle meeting. This was where he first met John Brekke. "John was a charismatic leader and striking in his ability to pull in people to his project. He was assigning work within minutes of the first meeting," Matt said.

Brekke outlined the Aerospace Corporation role with NASA and identified who would work on which pieces of the PRA model to study the foam loss analytically. Much of the focus was centered on how foam was carried downstream from the tank toward the orbiter. Matt was assigned to determine if a calculated foam debris trajectory would strike the orbiter's wing leading edge RCC panels. It was a small task that lasted a day or two. It involved drawing flat, two-dimensional rectangles. Each rectangle faced upstream in front of each panel. This was an approximation of the complicated three-dimensional panels. If the calculated trajectory ran through a rectangle, it was called a strike. Brekke liked this approach so much he quickly identified how to completely rework it with more data sources. Not long after Matt went with Brekke's team to Houston to present their initial modeling approach to the NASA engineers. Bob Ess and Ray Gomez also liked this approach so much they quickly identified how to improve it to make it more realistic and accurate. Ray offered several suggestions about using CFD solutions of the flow field around the shuttle, porting the code over to the NASA supercomputer at Ames, and integrating the 1800 Winds Database. "I didn't have the courage to tell him I was a mechanisms engineer and new to the program. I hadn't a clue what he was talking about but figured I would find out later from my colleagues," Matt recalled. "So I nodded in agreement, and my involvement with the shuttle community was under way."

The small sample sizes used to curve fit the dissection and test data needed to build the models limited both the ET and Aerospace Corporation teams' foam release models. Debris tables were also only generated for select foam debris sources where dissection data were available. Empirical or physics-based models were applied when possible and compared against available flight history for validation. When the physical understanding proved insufficient to provide a complete debris table or LEFM model, flight history was applied as a substitute. In these cases the limited physical understanding was still used to supplement or refine the analysis. Two things usually limited the release models developed from flight history: the unknown time of debris release and applicable failure mode. The flight history sample sizes were also typically small for most foam debris sources. It was, therefore, difficult to assign a debris release with a given failure mechanism. There were also limitations associated with flight history data due to the imagery conditions, resolution, night launches, and limited viewing of all areas on the tank. It was impossible to determine from imagery alone if a release site was from a single release or a series of smaller releases. Consequently all debris releases were assumed to be singular events—even when video footage showed most releases liberated in multiple, smaller pieces. The flight history distributions also did not always take into account vehicle design changes or give credit for improved debris release performance. Marshall Space Flight Center provided all the flight history models (mass distribution plus release frequency). Physical insight provided the engineered time of release distributions. For example, the void delta P losses were linked directly to aerodynamic heating, the air load losses were primarily centered on the maximum dynamic pressure, and the cryopumping losses were tied to propellant levels and the corresponding substrate temperature. The release location, unless otherwise specified, was assumed to be uniformly distributed over the appropriate foam application area. Unlike void delta P and cryopumping, the crushed foam, thermal cracking, delamination, and air load failure modes all relied exclusively on flight history to develop their respective release models.

Risk Assessment Mass

For each of the twenty-nine foam debris sources, the ET project provided the maximum expected debris mass for each ET foam application and documented these values in NSTS 60559, Volume One. These values were derived primarily from the divot potential for void delta P and cryoingestion failure mechanisms (where available) and flight history for other foam debris locations and failure modes. Unless otherwise stated, the other failure mechanisms—adhesive failures, bond line delamination and bond adhesion, crushed foam from collateral damage, air loads, and thermal cracking—were enveloped by the maximum expected debris mass. This mass represented a benchmark used in a risk assessment to determine if a debris-target pair cleared deterministically. If the maximum expected mass cleared deterministically, no further work was needed, and the orbiter was considered certified for that particular debris source. This was the initial debris certification approach taken after the *Columbia* accident. It was later abandoned in favor of probabilistic risk assessments because only a few foam debris sources cleared deterministically. Even so, the risk assessment mass was still useful for conditional probability calculation and relative debris risk rankings and for helping determine if there was a change in risk posture. For example, an observed mass release larger than a risk assessment mass would automatically trigger an integrated IFA and investigation into the cause.

Enveloping Debris Sources

Enveloping was a technique utilized when release model inputs such as mass distribution, time of release, or release rate data were unavailable. A debris source was considered enveloped by another source and its corresponding risk if all the following conditions were met: one, the debris type was the same; two, the enveloped source had a lower conditional probability; three, the enveloped source had a lower mass; four, the enveloped source had consistent failure modes; and five, the enveloped source was located farther aft. These rigorous criteria were established after the first RTF mission because the first definition of enveloping was too loose. (It was anything farther aft of the reference source.) This proved to be a terribly flawed assumption that exposed the orbiter on STS-114 to a much higher debris risk than originally estimated and reported. For example, the forward liquid oxygen ice/frost ramps, which had a very low foam debris failure rate, enveloped the aft liquid oxygen ice/frost ramps. Several liquid hydrogen ice/frost ramp losses were observed during the mission, and a few exceeded the largest mass size predicted for the liquid oxygen IFRs. After the ice/frost ramps were dissected, it was discovered the liquid hydrogen ramps had much bigger and nearly three times more internal defects. This was largely due to the ramp construction. Different type molds were used to manufacture the ramps, and the liquid oxygen ramps were poured when the tank was in a vertical position. The hydrogen ramps were poured while the tank was in a horizontal position. To make matters worse, the flow field around the hydrogen ramps was more severe due to higher debris impact velocities and angles on the orbiter underbelly.

Integrated Hazard Report

Results from the debris tables, LEFM, and flight history release models defined the debris environment and fed the transport model with the key parameters of release location, debris size and shape, release velocity, and liberation time. This allowed computation of the subsequent impact, damage, and hazard analysis for each piece of debris. Each foam release model predicted debris release during ascent. Results of the model were then arranged in a table that represented the total predicted debris environment for a specified number of flight simulations. Release results could be fed into either the Aerospace Corporation or the Boeing PRA model and produce a risk index. The transport model determined the trajectory of the predicted debris based on the time of release and the flow field around the shuttle. The impact algorithm determined which trajectories impacted the orbiter and assessed the amount of damage. All foam risk analysis results and methodology were then documented in the integrated debris hazard report (IDBR). If the ET project debris tables or flight history inputs were unavailable, then the risk assessment mass documented in NSTS 60559, Volume One, was used to generate conditional probabilities as a function of ascent time. These conditional probabilities were used for relative risk comparisons and bounding the worst-case risk during the ascent profile for a particular debris source. Imagery analysis of each subsequent flight was used for model validation and to assess the release inputs—particularly the release location, mass, time of release, and frequency.

I considered Roy Glanville the "safety conscience" of the Space Shuttle Program relating to debris risk and IDBR risk classification. Roy was a member of the Integrated Safety and Engineering Review

Panel (ISERP) and eventually became the chair prior to STS-121. He held this position until the end of the shuttle program. He developed an interest in rocketry in grade school and built and launched many Estes model rockets. In the 1960s chemicals were a lot easier for sixteen-year-old kids to buy, and his parents bought extra homeowner's fire insurance to protect them from his basement bomber experiments. In high school he received the nickname "Rocket Roy" and was made a chemistry lab assistant. He performed several after-hours experiments in the high school lab trying to make nitroglycerin. Fortunately he only succeeded in making a lot of smoke. Roy was the president of the Southwest Texas Aeronautical Research Association and built rockets so big and dangerous they had to be launched from the US Army McGregor Missile Range. Roy enlisted in the US Air Force and received training as a nuclear weapons specialist while earning enough college credits to graduate with honors and a bachelor of science from a local college in physics. Fresh out of college in 1979, he accepted a job with Boeing Aerospace to work on designing and developing spaceflight laboratory equipment for NASA's Space and Life Sciences (SLS) Directorate. Eventually the NASA Reliability Engineering Division hired him to certify the laboratory equipment for the STS-9 Spacelab SLS-1 mission. Promotion opportunities eventually led Roy back into the Space Shuttle Program in 1989, and he worked as a program-level system safety engineer. At that time he became a member of the Integrated Safety and Engineering Review Panel. This phase of his career exposed him to all aspects of shuttle systems and operations, and this included interacting with all project and element offices, mission operations, and launch operations. Following the *Columbia* accident, Roy became involved in ISERP to help meet the STS-114 RTF schedules.

Ice Debris

Development of an ice release model and conducting probabilistic analysis for ice had all the difficulties of foam plus some additional challenges. Ice formation and liberation was much more complex compared to foam because it was an environmental condition—not a manufactured process. Thus there was no control over it. Ice morphology such as density and size was a function of several variables. This included time, location, ambient temperature, relative humidity, wind, solar exposure, rain, airborne contaminants, surface roughness, and the extent of thermal shorts in the foam.[10] Ice growth from the formation of condensation was inevitable due to the cryogenic propellant in the external tank. The biggest questions were how much ice would form and where. Condensation is the forming of water when warm, moisture-laden air comes in contact with a cold surface. Very limited flight or test data to support development of input distributions for ice mass, time of release, and release rate existed. Available data sources included limited ice testing, historical trends, flight observations, and flight reconstructions. Making ice during the ice characterization and ice liberation testing proved difficult. This decreased the number of test runs, and this resulted in limited data of low fidelity. At best it was doubtful that additional testing would significantly improve the understanding of ice formation and liberation. Engineering judgment, therefore, was used to synthesize available data sources and develop mass and release time distributions. Due to uncertainty in the inputs, a set of parametric distributions was used to study and define a parametric risk space. Ultimately engineering judgment would be used to select the most realistic scenario from the parametric set of input distributions.

Ice Locations

Although ice could form anywhere on the tank, the ice debris risk assessment using probabilistic analysis for impacts to the RCC panels, tiles, and special tiles was completed for the external tank feed line brackets and bellows locations.

---

[10] A thermal short is an area with poor or inadequate insulation that would allow the transfer or flow of thermal energy between conductive materials. A thermal short would typically lead to an area of excessive condensation or ice formation on the external tank.

DEBRIS RISK ASSESSMENTS

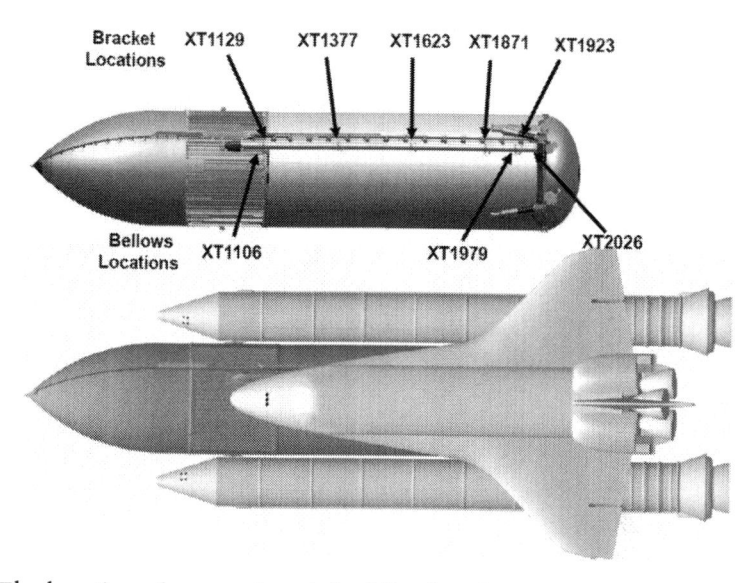

The location of external tank feed line brackets and bellows, which were prone to ice formation. A slight debris impact bias to the orbiter's starboard side was attributed to bellows and bracket ice debris.

The orbiter windows were not included as possible ice targets because the DTA showed no transport from the bracket or bellows locations, and the no-ice zone defined in NSTS 08303 was designed to protect the windows from ice impacts during ascent. It was also impractical to generate a PRA for ice formation on the liquid hydrogen and oxygen umbilical connections to the orbiter due to the lack of test data and the umbilical ice sizes exceeding the capability of the damage map. The umbilical ice flight rationale was based on the far aft location and flight history that indicated the majority of umbilical ice was liberated at space shuttle main engine (SSME) start-up and liftoff when the vibroacoustic environment was the greatest. The aerodynamic transport of umbilical ice was very complex due to separation regions and recirculation (reverse flow) just aft of the external tank crossbeam. This was definitely an area of concern because flight history showed a significant number of small damage hits around the umbilical area. Unfortunately it was difficult to distinguish between Kapton baggie, foam debris, or ice hits. Kapton baggies were used around the umbilical lines as moisture barriers to prevent ice from forming on the seventeen-inch, supercooled metal surfaces of the fuel connectors from the external tank to the orbiter. A nitrogen purge kept the baggie pressurized to further minimize the possibility of moisture reaching the exposed metal. Orbiter wind tunnel test data demonstrated the Kapton baggie separated between Mach 0.25 and 0.35. Flight observations of baggie loss during flight corroborated this. However, the Kapton baggie often flapped around and impacted the tile before tearing away. The transport analytic tools were unable to predict the regions of recirculation, and the large size and unusual shapes of the umbilical ice exceeded the capability of the tile damage model. Major testing and model development were needed to mature the analytic tools, but none were planned due to the enormous complexity and cost. The umbilical lines were capable of having very large ice formations based on observations during previous missions, the multiple tanking cycles, and rain exposure during tanking. Imagery analysis showed most of the umbilical ice liberated in thousands of pieces during liftoff and prior to tower clear. Fortunately there was no transport mechanism to the RCC panels or the windows, and the residual ice that survived liftoff did not have long distances to build up enough velocity to become a serious impact threat. The NSTS 08303 document was also updated to ensure umbilical ice formation prior to launch was within historical limits. The launch would be scrubbed if anything exceeded these limits. It was one of the few debris cases where the program had to rely on flight history as the primary flight rationale.

Bracket and Bellows Ice

Although the driving ice-liberation mechanism for bracket and bellows ice was mechanical articulation and vibroacoustic loads, there were some key differences. Bracket ice formed in multiple pieces in multiple locations around the bracket structure. Bellows ice formed in a single, continuous ring. Tests of bracket ice liberation generally showed the debris was a different shape and smaller size than bellows ice.

Ice around the bracket geometry also tended to adhere to the structure better than the bellows ice. This meant less total ice mass loss. The larger size and curved shape of the bellows ice meant larger lift and drag characteristics, and these produced greater transport range and impact velocities.

Five feed line brackets were located on the external tank and referred to by their position or station numbers. Although the feed line brackets were eventually redesigned with titanium yokes, the parametric process for assessing bracket ice risk was the same. Switching to titanium yokes reduced the amount of ice growth and risk from this debris source. Although each feed line bracket had six subzones defined before the titanium yoke change, seven ice subzones existed after the change. The bracket ice rate of release distributions were based on the engineering understanding of ice liberation observed during liftoff and centered on the highest vibroacoustic environment. This was consistent with ice characterization testing that demonstrated a general trend toward bracket ice liberation during high vibroacoustic conditions. The limited amount of ice characterization testing prevented the building of release rate distributions directly from the test results, but it did provide insight into the environmental conditions most likely to result in ice liberation.

Ice Release Rate Distributions

Different release rate distributions for both bracket and bellows ice were built based on engineering judgments consistent with the test results and imagery observations. Engineers had the most confidence in the "liftoff, max Q" distribution. This distribution predicted the highest rate of ice release at liftoff and later at the maximum dynamic pressure (max Q). The liftoff component was assumed to be 50 percent of the total mass loss which was consistent with image observations and testing and a peak value of 10 percent was assumed at max Q. The remaining 40 percent was spread evenly over the ascent profile from five seconds after liftoff to one hundred twenty seconds where the ice was assumed melted from the high heating. Any remaining ice that survived and liberated past one hundred twenty seconds was not considered in the distribution because it was past ASTT and posed no catastrophic threat.

Ice Mass Distribution

Results from the bracket ice characterization testing provided a total ice loss amount and maximum liberated size. Both were used to construct the mass distributions. The largest piece of ice measured during ice characterization testing was 0.013 pounds—about the size of a golf ball. The 0.013 pounds represented the largest piece of ice that could be liberated during ascent. Due to the limited testing, another mass of 0.026 pounds (arbitrarily doubling 0.013) was utilized to account for the possibility of a larger piece liberating. The limited testing did not allow mass distributions to be developed directly from the data but provided the key insight necessary to build a set of mass distributions as was done for rate of release. Later during the testing, a reduced ice environment showed the maximum ice release to be much less than 0.006 pounds. This test condition was selected to provide a set of test conditions that was more indicative of what might be seen during launch as opposed to the extreme ice conditions imposed during the initial ice characterization testing.

Ice Damage Map

The transport model was developed based on the aerodynamic characteristics of several sizes and shapes of ice. Ice impacts were assessed a modified foam-on-tile damage map—the only alternative prior to STS-114. This was a very conservative damage model implementation because ice (with a smaller volume for a given mass) penetrated deeper into the tile but had a smaller damage area compared to foam debris of the same mass. The foam-on-tile damage map treated the ice impact as a deeper foam impact and thus a higher predicted risk because the impact was more likely to exceed the tile capability.

Ice Risk Assessment Limitations

Overall the ice PRA risk assessments had key "conservatisms" in the model's release, transport, and impact components. These conservatisms (much larger than with the foam models) manifested themselves into higher-than-actual risk numbers for ice. Eventually the conservatisms were reduced to calculate a more representative risk number. The release model utilized the largest total ice mass from the ice liberation tests—ten hours of growth with no solar exposure or wind. The worst-case mass in each bracket sub-location was also used, which was not consistent with flight history observations. The transport model for the bracket ice shapes assumed the same aerodynamics as for bellows ice. This increased the target area and impact velocity. Limitations on the impact model were not addressed until after STS-114 and the development of an ice-on-tile damage map.

Ice-ball Allowable and No-Ice Zones

Once the components of the ice debris transport model were completed, deterministic results defined an ice-ball allowable and a no-ice zone for the orbiter windows and the nose cone—the most vulnerable structures to ice debris hits. Ice-balls could form anywhere on the external tank as a result of stress cracking in the foam during tanking and could grow to be two to three inches in size, depending on the locations and sizes of the cracks. They posed a legitimate ice debris threat and were assessed against an ice-ball allowable table in NSTS 08303. The no-ice zone extended from the top of the external tank downward to the orbiter nose and covered the majority of the liquid oxygen and intertank acreage. This zone was expanded from that used for STS-107 and prior missions because the new DTA indicated the orbiter windows were vulnerable to ice impacts from outside the previous zone. These results were documented in NSTS 08303, which provided the specific prelaunch inspection criteria for ice debris and the external tank thermal protection system defect or damage requirements.

The final inspection team used this document three hours prior to launch to ensure operational controls supported the flight rationale documented in the IDBR. During the launch commit criteria time frame (T minus six hours through launch, scrub, or abort), the KSC ice and debris team used the criteria for assessing observed potential debris and damaged foam. Conditions outside the provided acceptance criteria were reported as an LCC violation. If the violation was not resolved during the remaining countdown, the launch was scrubbed.

Debris Other Than Foam and Ice

The majority of debris sources other than foam and ice were once classified as unexpected debris until repeated releases over two or more flights warranted the classification change to expected debris. Standard program protocol when unexpected debris liberated was to pursue two parallel paths of corrective action. One path involved understanding the failure mechanism and developing mitigation plans or controls designed to prevent the debris from liberating on future flights. The second path involved characterizing the risk. Implementing mitigations or design upgrades that prevented debris from liberating was the preferred path. This represented the strongest form of control and lowest debris risk to the vehicle. Effective controls were also required for unexpected debris. Unfortunately, all the debris sources that would be reclassified as expected had nonexistent or immature probabilistic debris models. This made it difficult to characterize the risk in terms of a PRA risk index number. Compared to the foam and ice debris models, there were limitations in understanding the failure modes and release mechanisms, transport characteristics, and impact tolerance capabilities of critical structures. The debris models created for these debris sources were effective for deterministic transport and the development of vehicle impact conditions. The various element teams would assess the impact conditions to determine if there was a debris threat and, if so, what could be done to mitigate that threat.

CHAPTER 11

Failure to Meet Debris Requirements

Design Certification Review and Return to Flight

After the scathing assessment in the CAIB report, NASA and contractors began to focus all their attention on addressing the CAIB recommendations and returning the shuttle to flight status. The CAIB report identified numerous weaknesses and factors that contributed to the *Columbia* accident other than just a flawed bipod design. There were fifteen recommendations NASA needed to implement before flight operations could resume. The next scheduled flight, STS-114, became known as the Return to Flight mission. NASA responded to the CAIB by making significant organizational changes in terms of people, processes, and procedures. The Space Shuttle Program was reclassified as "experimental" and restricted flights to only ISS assembly missions. The CAIB also wanted NASA to conform to the existing debris prevention requirement documented in NSTS 07700, Volume Ten: "The Space Shuttle System, including the ground systems, shall be designed to preclude the shedding of ice and/or other debris from the Shuttle elements during prelaunch and flight operations that would jeopardize the flight crew, vehicle, mission success, or would adversely impact turnaround operations." Debris was defined as broken and/or scattered remains emanating from the element(s) of any flight or ground systems. The requirement also referenced conformance to the shuttle launch commit criteria. This contained the specific locations where the presence of ice was unacceptable for launch or flight and other locations where limited amounts of ice or debris were acceptable. Ice was defined as frozen water of eighteen pounds per cubic foot or greater density. Frost was acceptable and defined as frozen water of less than eighteen pounds per cubic foot. (For comparison the density of ice cubes in a kitchen freezer range from forty to forty-five pounds per cubic foot.) Ultimately meeting the ice requirements proved the most difficult CAIB recommendation to satisfy.

In addition to all the debris testing already under way, John Muratore wanted all credible debris sources analyzed from a wide range of release locations in order to predict the impact locations and conditions. Results were used to develop critical debris source zones that would provide maximum allowable debris sizes for various locations on the vehicle. The first step in the process involved identifying all the possible debris sources, and the second step involved performing the debris transport analysis (DTA). Debris transport analysis for any debris source included the three main parts: release, transport, and impact. John felt confident the bipod redesign and debris environment could be cleared using the traditional certification approach that had been used on the shuttle since the program's inception. Recovery from the 1986 *Challenger* disaster took less than two years, and John was certain the recovery time from *Columbia* would be the same or less. The shuttle program planned a design certification review for December 2004—twenty-two months after the accident—to cover debris and all vehicle changes. This meeting had to be moved to January 2005. The debris testing, identification of all expected debris, and respective debris analysis pushed the schedule out. Consequently the plan for an RTF launch of STS-114 in 2004 was also pushed back and tentatively scheduled for the end of January 2005 after the design certification review completion.

Problems on the Space Station

John and the shuttle program managers were beginning to feel the pressure of getting the shuttle operational due to the time delays with assessing the debris and a series of critical failures on the ISS. Although the space station had been continuously staffed for more than four years before the *Columbia* accident, station managers reduced the fulltime crew from three to two and suspended all assembly operations after NASA grounded space shuttle flights. After that NASA had to rely on the smaller Russian Progress and Soyuz spacecraft to carry supplies and astronauts to the station respectively. In 2004 the space station experienced a number of malfunctions—problems with the oxygen generator, loss of the control moment gyros (CMGs) used for attitude control, and a food shortage that jeopardized keeping the station crewed. The ISS had four CMGs designed to control the space station's attitude in orbit. CMG-1 failed in 2002 and was slated to be replaced on the next shuttle mission after *Columbia*. On April 21, 2004, CMG-2 was nonoperative when a circuit breaker failed. At least two CMGs were needed to adequately control the station's orientation, and another failure would shift attitude control over to the Russian segment, which used thrusters. Attitude control would be maintained so long as there was enough propellant to fire the Russian thrusters or if an additional CMG was recovered. Eventually power to CMG-2 was restored after a space walk to replace the circuit breaker. On October 6, 2004, ISS crewmembers Padalka and Fincke began troubleshooting the Elektron oxygen generator, which had only been operating intermittently for nearly two weeks. Like a submarine oxygen generator, the system created oxygen from water. It vented the hydrogen overboard in the process. The crew eventually hooked up the system's hydrogen venting line to a different overboard valve in the station's Zvezda module. The valve was normally used as part of an atmospheric contaminant control system, but when hooked up to the Elektron, the system began operating normally. In December 2004, the station was faced with a food shortage. Supplies were so low that if the usually reliable Progress spacecraft missed its delivery, the station would have to be abandoned and order American Leroy Chiao and Russian Salizhan Sharipov back to Earth by mid-January—halfway through the six-month mission. An unoccupied space station and a grounded shuttle would not have reflected well on NASA. The two crewmembers were ordered to eat less for about three weeks until a Russian supply ship would arrive on Christmas Day. To stretch food supplies, Chiao and Sharipov shaved five hundred calories from their typical three thousand-calorie-a-day diets. The restrictions did not include drinking water, but that would also run out by mid-January. The Russians pointed to a range of factors that created the dilemma. This included the decision to postpone regular space shuttle supply missions after the *Columbia* accident nearly two years prior. The problems with the oxygen generator also tangled plans. In October 2004 food was removed from the Soyuz capsule to make room for spare generator parts. It was clear returning the shuttle to flight was critical for space station operations for resupply, assembly, and full utilization.

Debris Classifications

John and Bob had established a set of debris classifications to help evaluate multiple debris sources. These classifications were arranged in a hierarchy and utilized to develop flight rationale. This was documented in the debris hazard report and approved by program management for flight. The first class of debris was defined as anything less than 0.0002 pounds for only nonpropulsive debris sources. This limit was intended to provide a lower limit for debris masses to be assessed. Originally this option assumed a preliminary fifty foot-pounds of kinetic energy impact capability for reinforced carbon-carbon panels along with a relative impact velocity of four thousand feet per second. Given these assumptions the 0.0002-pound criterion was only applied to nonpropulsive items because propulsive debris (char, exhaust particulate from the booster separation motors, etc.) reached considerably higher velocities. The 0.0002 limit was equivalent to a quarter-sized piece of foam a quarter-inch thick and less than the mass of a six-millimeter BB. This classification was important because over 40 percent of the expected debris sources consisted of small masses less than 0.0002 pounds. Most of these releases were due to ablation processes, erosion particulates, paint blistering, or separation debris particles.

The next classification was "no debris transport." This classification eliminated debris–impact pairs where no impact could be determined analytically or by inspection. For example, during ascent ice impacts from outside the analytically derived no-ice zones did not threaten the orbiter windows. A subclassification was also used for debris releases past a defined aftmost limit. This was documented in the integrated hazard report as part of the flight rationale. Any debris liberating aft of this limit could not reach the RCC panels or violate the tile impact tolerance limits.

Deterministic analysis utilized certification rigor assumptions and worst case conditions to demonstrate that liberated debris could not lead to catastrophic damage. In addition debris–impact pairs could use test data to empirically demonstrate the impact could not cause critical damage. Such was the case with the orbiter windows. Impact testing of various debris sources was used to determine if the windows were safe against debris strikes.

A subclass of deterministic analysis involved "enveloping." This technique was applied universally to large sections of the external tank to minimize the number of analytic computations. Prior to STS-114, there were only two criteria for enveloping: the debris had to be located farther aft of the reference source, and the debris had to be the same type. This was subsequently changed after the first RTF mission when it was discovered the enveloping criteria and assumptions were flawed based on different failure modes, transport, and impact conditions. Debris risk from some of the enveloped sources proved a much higher risk than originally expected. The oxygen tank acreage foam was used to envelop the hydrogen tank acreage foam, but the oxygen tank feed line brackets were not because of the different foam material, construction, and debris transport characteristics. Enveloped debris sources were assumed to have the same failure modes and conform to the worst-case conditions such that the certification limits would not be exceeded. Another subclass of deterministic analysis involved using the certification rigor or near certification rigor impact tolerance levels against a nominal damage map and transport environment instead of a worst-case scenario. Results derived in this way were used to establish an ice-ball mass allowable table for ice formed as a result of ET insulation flaws or cracks. The ice-ball mass allowable table (along with other debris allowable conditions) were incorporated into the space shuttle ice and debris inspection criteria document—NSTS 08303. This document provided the KSC final inspection team with the acceptance criteria for observed potential debris conditions for the shuttle flight elements, ground support systems, and external tank thermal protection system. Conditions that did not fall within the LCC contained within this document resulted in an LCC violation and either constrained or scrubbed the launch.

Expected Debris Sources

A list of credible ascent debris sources was compiled for each hardware element—solid rocket booster, reusable solid rocket motor, space shuttle main engine, external tank, orbiter, and the launchpad area around the vehicle at launch. More than two hundred expected debris sources were identified by their locations, sizes, shapes, material properties, and, if applicable, likely times of debris release. The initial set of analyses was completed in 2004, and it culminated in a comprehensive review at the SICB in December 2004. At this review all the space shuttle projects presented their lists of potential expected and unexpected debris sources. With the sources identified, the debris transport analyses could be used to predict impact location and conditions such as velocities and relative impact angles. John had several teams working to improve the end-to-end process for predicting debris release, transport, impacts, resulting damage, and characterization of risk level. Engineering support for this approach crossed multiple organizations within NASA and support contractors. JSC and MSFC engineering provided key algorithm development for liberation mechanisms and debris transport. The ET project and element engineers developed engineering material properties data and test validation data to support foam and ice debris release mechanism predictions. KSC staff and other members from the launch day final inspection team led "Project Ice-ball." This provided basic ice characterization data and a summary of launch day debris observations. Justin Kerr's orbiter debris assessment team developed the impact and damage tolerance algorithms for RCC wing leading edge, and tile. Propulsion element engineering and MSFC propulsion SEI engineers developed empirical data for impact sensitivity to debris on the SSMEs and SRBs. KSC and MSFC engineering covered launch and ascent debris issues along with coordination on launchpad imagery and debris inspection criteria for day of launch. This was such an extensive effort that nearly all NASA research centers participated in providing tools and/or validation data for debris assessment at some point. On the contractor side, the Aerospace Corporation in El Segundo and the USA–Boeing team in Houston did model development and validation as well. Initially the Aerospace Corporation focused on the foam debris, and USA–Boeing focused on ice and other debris. This parallel path development was undertaken to minimize the time it took to develop the models in series. It also acted as a mitigation plan in case either model development effort failed. Eventually both models were checked against each other and could evaluate any type of debris source.

Extensive testing enabled SEI to validate and refine some of the debris transport model assumptions and produce better results. These tests included wind tunnel tests and F-15 flight tests to assess foam flight characteristics. NASA also conducted aerodynamic ice testing that focused on ice flight behavior and breakup characteristics in a flow field. NASA engineers also grew ice of various densities and performed vibration and acoustic testing at MSFC to understand how ice on the feed line brackets and bellows liberated during ascent. These results indicated that, contrary to early assumptions, low-density (soft) ice posed a greater risk than high-density (hard) ice because of the former's tendency to release earlier in the flight and break into multiple pieces. Hard ice tended to stay attached longer. It broke off only after the shuttle had reached an ascent stage where liberation did not pose a large threat to the orbiter. The ice-to-foam bond strength for hard ice was also so strong that Mike "Rudy" Rudolphi, the deputy shuttle program manager at MSFC and director over all the ice testing, had to use a screwdriver to chip the ice away from the foam. This became known as the "Rudy screwdriver test" and was used as an example to illustrate the strong ice-to-foam bond strength. If hard ice did liberate, it was typically due to weak ice morphology or cracks caused by thermal expansion or mechanical articulation. Overall the series of debris tests performed help validate the release and transport models, and they indicated the initial model assumptions were too conservative. NASA also completed a supersonic wind tunnel test at the NASA Ames Research Center. This test validated the debris transport flow fields in the critical range between Mach 1.0 and 2.5. The results showed excellent agreement with the analytically derived flow field predictions. This helped validate the CFD solutions used for debris transport and gave engineers confidence in the debris transport analysis results.

Problems with Debris Certification

During 2004 there was substantial community-wide debris review through a series of four debris summits. The Space Shuttle Program had been working diligently to evaluate test results and determine expected debris liberation scenarios, transport mechanisms, and impact tolerance. Even before all the expected debris sources were identified and brought forward to the SICB in December 2004, preliminary analysis clarified that several foam debris sources, including the bipod foam and all ice debris, would not clear deterministically. The vehicle, therefore, could not be certified against damage from the debris environment. This was a huge issue because it meant NASA could not meet its debris requirement and one of the key CAIB recommendations. The inability of the shuttle program to certify the vehicle against the debris environment also brought about additional scrutiny from outside the agency—especially from the Federal Aviation Administration and the range safety organization at KSC. These and other organizations were concerned for public safety and the impact on expected casualty risk (E_c). Expected casualty was a measure designed to ensure that risks to public safety presented by launch and reentry operations were limited to an acceptable level. The E_c requirements for launch and reentry of a reusable launch vehicle was not allowed to exceed an average number of 0.00003 casualties per mission. It was a miracle no one on the ground was killed or injured when the *Columbia* wreckage crashed over Texas. The shuttle could not meet the E_c requirement for entry before the accident, and now there was even more of a public safety concern given the shuttle program could not certify the vehicle against the debris environment. The only way for the shuttle to meet the E_c requirement during ascent, which followed a trajectory up the Atlantic Coast of the United States, was to perform an evasive maneuver away from land if an abort was declared during this flight phase. It would be a NASA decision and contingency maneuver that decreased public risk at the expense of increasing crew risk and potentially sacrificing their lives. However, the entry E_c requirement could not be met even with the most aggressive evasive maneuver because the trajectory typically stretched across the entire United States or other southern landmasses. An uncertified vehicle against debris damage meant the entry E_c risk was even higher. Consequently the E_c requirements for the Space Shuttle Program were extensively tailored to allow continued flight operations. If an entry took place under off-nominal conditions or when critical crew safety factors (e.g., landing site weather, orbiter consumables, crew health, and more) required the consideration of alternate landing site opportunities, the shuttle program could balance the mitigation of public and crew risk in selecting the entry opportunity and landing site. This provision was necessary due to the space shuttle's established design. Any significant alterations to space shuttle entry operations would have had the potential for negative effects on crew and mission. The shuttle program and NASA headquarters convinced the outside stakeholders that NASA had quantified and thoroughly evaluated the risks associated with this provision, and the public safety risk was considered acceptable for the remaining space shuttle missions. Although the E_c requirement was clear, the entry E_c requirement was essentially waived for the shuttle program.

Dr. Bruce Wendler from the Aerospace Corporation had a PhD in aerodynamics and made sure everyone knew it. He was one of the key people John and the SEI group employed to perform deterministic assessments of debris sources. Bruce utilized a simple figure of merit known as "C/E." This compared the structural impact capability (C) against the debris environment (E). E comprised the debris impact parameters such as velocity and impact angle that were derived from the debris transport analysis. If C/E was greater than one, the debris strike was cleared deterministically. The structure could perform its function without issue. If C/E was less than one, the structure was suspect and treated as a failure. In reality, though, the structure might have survived the impact. This was because the capability computation was based on the minimum impact tolerance, and that was based on the worst-case input values such as worst-case reentry temperatures, end-of-life degradation, and maximum factors of safety. Although the impact capability model had negligible uncertainty by taking the most conservative value, it was more likely to have a C/E less than one. In reality both RCC panels and tiles were more damage tolerant than assumed and had the ability to complete a mission even after debris impacts altered them.

John's debris certification plan was falling apart, and there were no alternatives or mitigations for the foam to bring the C/E to greater than one. NASA clearly would not meet the NSTS 07700 debris requirement, and it was late in the game to discover such a showstopper. This left John and NASA in a precarious predicament and scrambling to search for a solution. Numerous foam replacement options (alternative insulators, net coverings, heaters, and the like) were considered but dismissed for either failing to meet the insulation requirements and causing more ice formation or creating debris even more hazardous than foam. The worst-on-worst (WOW) analysis was the most conservative certification approach because it used the minimum impact tolerance, and in many cases that was reduced even further to cover all uncertainties. For instance, the orbiter project reduced the minimum impact tolerance for the RCC panels from approximately 1,500 pounds (nominal case) to 630 pounds to cover all uncertainties. Consequently this reduction in impact tolerance drove C/E to less than one. Since the worst-on-worst methodology no longer supported certification, John decided to alter the methodology, relax the minimum impact tolerance approach, and base the C/E values on a best estimate of capability and environment. John rationalized this approach was more indicative of the real RCC and tile damage tolerance, but the analysis was characterized as risk assessment instead of certification. Risk assessments were performed using the best estimate of actual conditions (not worst-on-worst) to avoid overstating the risk. Another approach John chose to pursue, more out of desperation than anything, was a probabilistic risk assessment.

Probabilistic Risk Assessment Proposed

John Brekke was working with Bruce Wendler to help NASA with debris analysis. Brekke sketched out the PRA approach for John Muratore and Bob Ess on a napkin at a dinner social following a debris meeting in December 2004. Brekke recalled the meeting. "While attending the debris meeting at MAF in New Orleans, I was talking with a colleague at dinner and used a napkin to outline an end-to-end PRA process. I even wrote down the names of people at the Aerospace Corporation who could help implement the process. After returning to California, I presented my plan and worked with folks at Aerospace before proposing the PRA methodology to Bob Ess in November 2004. Bob was receptive to the idea and wanted it worked in parallel to provide some insurance if the C/E process failed." The PRA was a minimal investment for NASA, and if Brekke was wrong about the process, it could be easily discarded. However, if others came to a similar negative conclusion about C/E, the PRA process would allow NASA to continue with development of flight rationale without losing a step toward a launch date.

John Muratore had initially been exposed to the Aerospace Corporation working with Bruce and trusted him implicitly—almost to a fault. Bruce was an outstanding engineer but tended to do everything himself rather than coordinating with experts. John and Bruce worked out the C/E approach without much involvement with the debris community. When Brekke joined the Aerospace Corporation team and was first exposed to the C/E approach, he was concerned about whether it would do the job. The C/E seemed arbitrary and had no physical basis for what the numbers meant. This drove Brekke to approach Bob about the PRA—an alternative physics-based approach that covered how the debris liberated, was transported, collided, and caused damage. Although Bob liked the concept, he was still apprehensive because he was trying to support the C/E methodology John and Bruce were working.

After running several thousand cases over a three-month period, the proof of concept for the PRA process was presented at a foam debris meeting in February 2005. At the time NASA showed little interest because the program was still trying to make the C/E methodology work. Although Bob didn't give the PRA methodology much of a chance, he did like it as a backup plan and allowed Brekke's PRA team to attend the various debris meetings and learn more about the integrated end-to-end debris problem. Brekke built a strong working relationship with Bob—despite some disagreements. He recalled this fondly. "Although I did not agree with everything he did, I tried to put myself in his place as a customer. From this vantage point, I understood the different pressures and stresses that Bob and NASA were under with respect to solving the debris problems and getting the shuttle back to flight status. Muratore was even more persuasive when it came to implementing his ideas, and even though Bob seemed somehow caught in the middle, I was glad Bob had the confidence in me and the Aerospace Corporation team to work the PRA model."

John Muratore was very skeptical of the PRA approach because developing flight rationale based on PRA was not an approach the shuttle program had ever utilized. The NASA engineering and safety communities did not even accept it. John was already being accused of manipulating numbers by using the best-estimate approach to meet his certification objections, and he would have faced even more skeptics by adopting a PRA approach. Although the PRA methodology produced a risk number, the difficulty had always been trying to get agreement on an acceptable risk level. NASA had no defined risk thresholds or requirements. There were only opinions, and those varied widely within the agency. John was also concerned that springing a new PRA approach on the shuttle program using complicated algorithms and probabilistic calculations would quickly be rejected. Despite the PRA shortcomings, though, John needed a backup plan in the event the best-estimate methodology floundered. He approved Brekke's plan to move forward with his PRA approach. "Characterizing the risk using the PRA methodology and developing flight rationale based on the results would be the biggest debris challenge for NASA," Brekke said. "It was impossible to eliminate foam from coming off the tank. As a result NASA would be unable to certify the tank."

Darby Cooper and the Boeing team also recognized it would be nearly impossible to certify the vehicle against ice debris, and the PRA alternative seemed reasonable to them. "I considered moving beyond the deterministic analysis and adopting the PRA methodology one of the biggest debris challenges," stated Darby. "Many folks in the debris community were kind of stuck in the deterministic mind-set. There were lots of discussions on how to certify the vehicle against the existing debris environment, and for some items that was possible. But for foam and ice, with the large number of variables and the variability of the process, accepted risk was really the only way to get to flight rationale."

One concern was how much consideration one put into the debris risk (PRA) numbers when determining if the tank was ready to fly. The other concern with PRA was how the results were interpreted. Helen McConnaughey best described this concern. Her CFD background taught her that computational tools needed to be benchmarked in order to produce confidence in the numbers they generated. This meant comparing the computed values with the results from a problem with a known solution to gauge the calculation's accuracy. The foam debris risk methodology (newly developed after *Columbia*) had test results to validate only parts of the methodology, but the only benchmark for the integrated PRA approach was the bipod loss from STS-107. "The PRA values were useful to gauge relative risk, but I never had confidence in the absolute values of the PRA numbers," stated Helen. "When we calculated a PRA value that was in the ballpark of the STS-107 risk value, I became very nervous. I became increasingly more alarmed as I saw the decision-makers relying on the PRA numbers as if they reflected actual risk."

Although John approved Brekke's PRA plan, he knew it would be a monumental task to convince the Space Shuttle Program to fly under an "accept risk" posture rather than certification.

Design Verification Reviews Proposed

With the RTF launch date set for the end of March 2005, John needed time to implement the best-estimate methodology and allow the Aerospace Corporation and Brekke time to build the PRA model for a few select foam debris sources. Instead of presenting debris certification results at the vehicle design certification review meeting in January 2005 as originally scheduled, John proposed his new debris plan. Instead of certification, SEI would hold a debris verification review (DVR) to present validation data for the debris transport, C/E analysis for debris cases that cleared deterministically, and best-estimate results for debris cases where the C/E was less than one. If the best-estimate C/E results were less than one, then probabilistic risk analysis would also be performed on a few select cases. Flight history would be used for flight rationale as a last resort if insufficient information was available to perform the best estimate or PRA analysis. The flight rationale for all expected debris sources including the integrated hazard analysis would be presented as well. A DVR board consisting of senior shuttle program managers would be assigned to review the results and approve the flight rationale. At the end of the DVR, a board recommendation and an independent NESC evaluation

FAILURE TO MEET DEBRIS REQUIREMENTS

would be made to the shuttle program during the STS-114 flight readiness review.

Jeff Pilet and the ET project became heavily involved in all the DVRs. Jeff and his team were constantly trying to develop an approach to certify the tank for RTF, but they also recognized they could not protect the vehicle from all possible foam losses. Even a small to medium-sized foam loss released at the wrong time could lead to catastrophic damage. The "eureka" moment for Jeff's team was when they related foam loss size to voids within the foam. From this they developed what became known as the divot–no divot curves that predicted when foam would fail. Although these curves required complicated algorithms and statistical data to develop, they were easy to understand and became symbolic of the ET project's RTF plan. Essentially the amount of foam loss could be predicted based on the size and location of a void within the foam. For a given foam application, the curves could show the maximum expected foam loss. These data were then used in the C/E calculations and were eventually modified for the PRA computations.

The PRA and debris verification approach did not gain a lot of momentum until the second DVR. During this meeting the new NASA administrator, Mike Griffin, introduced himself to the debris community and suggested using a PRA-based methodology to characterize and explain the integrated debris risk. "I can still remember him standing up at the whiteboard, drawing two normal distributions, and saying, 'This is how you do it!'" recalled Jeff. "Mike provided much-needed leadership and really helped get us moving in the PRA direction. From that point forward, we developed the 'debris clouds'[11] used for the PRA. What really made me confident the tanks were ready to fly was that we had done everything we could to minimize debris, and we could now communicate the risk through the PRA."

Dr. Michael Griffin was an engineer at heart, and despite his relaxed appearance of khaki pants, polo shirt, and loafers, which

---

[11] A debris cloud is a mathematically derived table that characterizes foam debris size as a function of time for a particular failure mode. The debris cloud was used to analytically predict the risk level for that particular failure mode.

he wore even for the highest-level meetings, he was well qualified to lead the agency. Prior to being nominated as NASA administrator, Griffin headed the space department at Johns Hopkins University's applied physics laboratory in Maryland. He was previously president and chief operating officer of In-Q-Tel, Inc., and he also served in several positions within Orbital Sciences Corporation. This included as chief executive officer of Orbital's Magellan Systems division and general manager of the Space Systems Group. Earlier in his career, Griffin served as chief engineer and associate administrator for exploration at NASA. He also authored a textbook, *Space Vehicle Design*, and taught several courses regarding spacecraft design, computational fluid dynamics, and introductory aerospace engineering as an adjunct professor at the University of Maryland, Johns Hopkins University, and George Washington University. Dr. Griffin received a bachelor's degree in physics from Johns Hopkins and went on to earn five master's degrees and a PhD in aerospace engineering from the University of Maryland. He was a recipient of the NASA Exceptional Achievement Medal, the American Institute of Aeronautics and Astronautics Space Systems Medal, and the Department of Defense Distinguished Public Service Medal—the highest award given to a nongovernment employee. Even with all the accolades, it was not the pay, power, or prestige that motivated him when he took over as the NASA administrator. It was simply his love for spaceflight and the agency. His major concern when he took over was that the Space Shuttle Program appeared to be approaching RTF on a tactical basis. "We were always three months from launch," said Mike. "We were not looking at the larger picture. What would be required to step back and really figure out what was going on with foam? What other systems needed attention that they hadn't received? What would it take to fix it *all*? We had our heads down and were working hard, but the strategic plan for fixing the shuttle was not clear."

Although reasonable, heavy criticism and a great deal of skepticism met John's new debris verification plan. Many felt the best-estimate methodology was too subjective. They felt the numbers could be easily manipulated to get the desired result. There was even stronger resistance to the PRA approach, and no one liked using flight history to generate flight rationale. Relying on flight history to justify flying the next mission was a contributing factor in the *Columbia* accident and a CAIB-identified weakness. The program, therefore, was unwilling to make the same mistake no matter how compelling the flight history argument. John responded to the critics by stating that the majority of expected debris sources had already been cleared. These debris sources were either less than 0.0002 pounds, had no transport, or their C/E values were greater than one. John added that extensive hardware designs and mitigations had been implemented to eliminate or reduce debris risk, and PRA and flight history would only be used as a last resort. Furthermore, operational changes such as contingency shuttle crew support, launch on need, and on-orbit repair capability were not accounted for during the debris risk assessment but were measures intended to ensure flight safety—even if a debris event occurred. Despite the resistance John managed to convince the program managers to set a mid-March DVR. This was a precursor to the RTF launch date set for the end of March. The only thing left for John and the SEI team to do now was deliver the PRA results and flight rationale—a task that proved much easier said than done.

CHAPTER 12

Dream Job

Season of Debris Verification Reviews

The mid-March DVR was a three-day marathon of analysis results that did not go as planned. I had recently made the move from Boeing to NASA, and for the first time in my career, I was now acting as the customer. I had led the team that compiled most of the Boeing presentations on debris classifications, transport, and impact capabilities. To a certain extent I felt as if I was grading my own homework. John announced at the beginning of the meeting that, despite redesigning the bipod and implementing major design and process changes to eliminate or mitigate debris, NASA was still unable to certify the vehicle for all debris cases. He went on to describe the best-estimate technique and how probabilistic analysis might be used as a last resort in a few foam debris cases. If the program needed to perform probabilistic analysis, it would be forced to accept some small amount of risk. The handpicked DVR board members included shuttle program leaders from each of the project elements, program offices, Astronaut Office, the Safety Office and NESC representatives. Board members shifted in their seats when John made this announcement. The DVR board collectively represented several hundred years of human spaceflight experience, and most had worked on the shuttle program in some capacity all the way back to the early design phases of the 1981 maiden flight.

The board members were not accustomed to probabilistic analysis or the idea of accepting residual risk from an uncertified vehicle. After trudging through hundreds of PowerPoint slides full of test and analytic data, it became clear there were more questions than answers as the number of action items continued to build. The board saw enough after Dr. Wendler from the Aerospace Corporation presented the C/E results. They showed several values less than one, and many felt best-estimate results had no basis other than an engineer's guess made to get the desired result. John Brekke and his team were still developing the probabilistic model and were only able to present their methodology. John and the board decided the RTF mission would have to wait another two weeks to work out all the debris issues. Another follow-up DVR was scheduled for the first week of April, and the launch date was set for the end of April. Over the next two weeks, the one-day meeting grew into another three-day event. Helen, John's SEI counterpart at MSFC, knew the community was striving to understand the debris risk. "A tremendous amount of work had been done to identify, analyze, and characterize sources of debris, debris transport, and damage tolerance," Helen said. "Although lengthy, the DVRs provided a forum for Elements, Engineering, and Safety to see what had been done across the board to provide a basis for assessing the debris risks. SEI's job was really to characterize the debris risk from each source because the tanks had been built years earlier, and the ET project was really limited in what they could do."

Although great progress had been made, John and the DVR board decided again that more work needed to be done. The flight was delayed to the end of May.

Round Three of DVRs

On April 27, 2005, NASA was in its third round of debris verification reviews with a May launch three weeks away. The review contained the usual cast of debris characters from throughout the agency, the DVR board, and John—the chair. The expectation heading into the third review was to close out action items from the previous DVR reviews, finalize debris assessments and risk numbers, and approve the vehicle safe for flight. Mike Griffin was attending and looking for the DVR board's recommendation to proceed with the May launch. "Personally Steve Poulos and others thought having the NASA administrator in the early meetings was at too low a level for his direct involvement. However, each leader must decide to what level he or she wants to engage any activity."

In this case Dr. Griffin wanted the shuttle operational and knew solving the debris issues was the critical challenge.

A debris concern slated to be addressed was the ice risk. This had essentially been ignored because of the higher visibility of the foam risk. The main issue at hand was determining the ice risk from the feed line bellows and brackets that attached the seventeen-inch stainless steel pipe used to transfer the −423°F liquid oxygen from the forward section of the external tank to the orbiter and the shuttle main engines. A recent audit of historical impact data on the solid rocket boosters, which were recovered and inspected after each flight, revealed a number of unexplained impacts that left long, deep indentations in the corklike material surrounding the field joints. There had been two significant impacts that gouged out large chunks of cork over the 113 flights and many smaller impacts. The cork surrounded the field joints at three axial locations and wrapped around the full circumference of the SRB. It protected the retainer band and pins from environmental and thermal effects and provided insulation over the joint heater for preflight temperature conditioning. These joints were redesigned after the *Challenger* accident and had had considerable attention paid to their well-being and performance. If the cork was compromised, the seals would be exposed to high temperatures during ascent, and this could lead to another seal failure. Although unlikely, no one in the program was willing to take that chance. The SRB project team had done the usual forensic detective work and could not find any evidence of residual material from a foam impact. The impacts were characteristic of ice, but the challenge was determining where the ice had originated and whether it could have hit the orbiter. Extensive debris transport analysis showed possible impacts from both the forward bellows and first two feed line brackets. Ice from the brackets was quickly ruled out, though, because a much larger piece of ice than the brackets could produce caused the amount of historical impact damage to the SRB field joints. The trajectory analysis showed the odds of a field joint impact from ice release from the bellows was one in fifty-five. That was consistent with the flight history damage. If ice could damage the field joint, the odds were even higher it could damage the tiles and RCC panels. Given the parametric set of assumptions used in the analysis, the RCC impact risk could have been as high as one in fifteen, depending on the timing and mass of the released ice. Although the analysis results just represented an impact risk, no one wanted a high-density material such as ice striking the RCC panels or tiles. Since the problem had to be corrected, it meant yet another launch delay, and it was my job to deliver this recommendation.

Bipod Foam Debris Still an Issue

The board and participants were testy from their third set of DVR meetings in two months, and the mood was tense due to the presence of the new administrator, who was eager to proceed with the RTF mission. The emotional scars left from the *Columbia* accident clearly affected the audience and induced heated debates over debris issues and shuttle risks. The last two years coupled with the grind of three lengthy debris verification reviews was beginning to wear on people's patience. The media were also aiming their criticism at NASA's inability to quickly solve the debris problems that had doomed *Columbia* and get the shuttle flying again. The first day focused on the foam debris risk and the second day on ice debris. Most people felt comfortable with the external tank design modifications intended to eliminate and minimize the foam debris risk. The bipod ramp (the culprit in the *Columbia* disaster) had been removed, and the foam closeouts around the area had been completely redesigned. As the first day wore on, Lockheed Martin and the ET project presented evidence that geometric voids or defects adjacent to metal structures and attachment fittings around the bipod could lead to a foam debris release during launch but of a much smaller size. Unfortunately the debris transport analysis showed that at certain flight times (most notably between seventy and one hundred seconds), foam debris released from this location could exceed a tile's impact capability limits. There was, therefore, still a residual risk that foam releases from the bipod area could lead to catastrophic damage. However, a debris release from the newly designed bipod was considered unlikely, and a large piece of debris would need to impact a tile at a critical time to be catastrophic. No one knew the risk for this possibility. John was growing increasingly agitated as the bipod story unfolded, and he called for a short break. John wasted little time signaling Bob Ess and me over to his side. He stated that the story on the bipod foam was "unacceptable." Bob broke the uncomfortable silence and explained to John that he had John Brekke's Aerospace Corporation team working on a linear elastic fracture mechanics model to produce a risk number for the bipod foam debris. The residual bipod debris risk was expected to be low due to the small debris size and unlikely release probability. John did not seem appeased and asked how the ice story was shaping up. Bob replied that Boeing was still completing the debris transport analysis but should be finished later that day. He added that preliminary results showed ice transport to the vehicle (including the SRB impact locations), but the update was needed to determine the impact percentage. John took a deep breath, walked away, and shook his head. He paused and fired back that he wanted to be the first to hear when the ice results were complete—regardless of the time. Bob immediately turned his attention to me and asked that I do the ice presentation. "I've got to get the Aerospace guys working on the foam story," Bob said. "Besides, you are the most familiar with the ice assessments, and this would be a good time to get your face in front of the debris crowd."

Bob was right. I had been working on the presentation throughout the morning and was most familiar with the ice risk assessments. After the dismal morning session on the bipod foam risk, I was all too eager to let Bob handle the foam work.

Assessing the Ice Risk

The first day came to an exhausting close around six in the evening, and I headed over to the Boeing building (about three miles away) to work with Darby and his team of engineers performing the ice analysis. I planned to stay and work the remaining details of the ice risk presentation while waiting for the final results. They came in around midnight. I had completed the majority of the presentation anticipating the results would show minimal concern, and I even wrote the conclusions to reflect this presumed result. That all changed after I reviewed the Boeing results. They showed that ice released from the forward bellows had a high probability of impact to the SRB field joints and an even higher impact probability to the tile. The only good news was the low impact probabilities from the first two feed line brackets. These lower impact probabilities were driven by the smaller sizes of ice compared to the bellows and the relatively benign aerodynamic transport. The impact probabilities fell off significantly for the middle and aft bracket locations on the tank, and the overall impact percentage to the SRB field joints was much lower for the brackets than the bellows. The smaller ice particles tended to follow a ballistic (straight-line) trajectory when released due to their inability to produce much aerodynamic lift. As the size of any debris particle decreased, it became more likely to behave like a cube or even a sphere that would generate zero lift and thus be limited in its range of travel.

Because of the significant ice risk from the forward bellows, the presentation package had to be changed—especially the conclusions. Analysis results showed an undeniable correlation between ice debris releases from the forward bellows and the impact damage seen on the SRBs. The ice impact risk to the tile from the forward bellows was much too high to ignore. The ET project had discussed the possibility of installing heaters on the forward bellows as a precautionary measure but had opted to forgo the installation due to time constraints and cost. They wanted to focus on higher-priority foam debris sources. The program priority was fixing the foam debris problems, but now it was unclear how much risk there was from ice debris. Recommending a heater installation for the forward bellows would cause another launch delay. It was one o'clock in the morning when I notified John and Bob about the results. I told them the only logical conclusion was to install heaters on the forward bellows. John knew the news was bad, but he concurred with the launch delay after I briefed him on the analysis results. I managed to get a couple hours of rest before the second day of the DVR meeting began, and I felt surprisingly alert due to the adrenaline surge in anticipation of my presentation. The room was still buzzing from the bipod foam discussion the previous day, and John had alerted the board members that the ice risk was most likely going to cause another launch delay. A few presentations were before mine, and I tried to patiently review the key points and rehearse what I would say. Instead I found myself reminiscing about the progression of events that had brought me to this point in my career—finally reaching my childhood dream of working for NASA.

My First Job on the Space Program

After moving to Huntsville in 1991 to begin work for Boeing on Space Station *Freedom*, I immediately enrolled in the engineering PhD program at the University of Alabama. I set myself a goal of applying for an astronaut position as a mission specialist during the next selection cycle. Over the next couple years, the move to Huntsville started taking a toll on my marriage. My wife and I were farther from our families and friends in Ohio, and there were limited social options compared to Lexington, Kentucky. It was challenging to meet new friends. In the fall of 2004, our relationship was deteriorating, and after our third visit to a marriage counselor, the divorce recommendation shocked me. I thought things were bad but not that bad. Clearly something was happening I did not know about. I was stunned when Jennifer decided to drive back to Ohio alone to spend Christmas with her family. With our marriage falling apart, I decided to make a last-ditch effort to save it. I drove to Ohio in a snowstorm to try to reconcile, but she was as cold as the weather. Her parents were usually enthusiastic to see me, but they hardly spoke a word. She refused to join me and my family on Christmas Eve at my sister's home. I could hear the howling wind outside whipping the snow and off in the distance a train whistle. I thought to myself this was the loneliest I had ever felt. Even today hearing a train whistle takes me back to that point.

Jennifer returned after the holidays to sign divorce papers and pack up her things. As she drove off, I realized our lives were now on separate and divergent paths. As I struggled to recover from my failed marriage, I experienced a number of other setbacks. Although work and school offered a temporary reprieve from the ordeal, I did not pass my doctoral board. I did not fail either, but I had to repeat the process. While flying alone at night in mid-December to maintain my pilot currency, I had to make an emergency landing in the Tennessee River just a few miles south of the Huntsville Airport. The engine failed—most likely due to carburetor icing from a warm, moist air mass I flew into while crossing the river. I tried in vain to restart the engine, and I activated the emergency beacon while communicating my position to the control tower before splashing the plane into the icy water. The impact jammed the cabin door, but I finally broke it open to free myself from the craft as it sank behind me. Adrenaline was surging through my body. Despite severe neck pain and an inability to use my left arm, I made my way through the ice patches toward the shoreline. It was make shore or die, and this was a powerful motivator to make the swim to land. The plane's taillight was still flashing red under the surface. The moonlight reflected on the calm river and brightly illuminated the water. Not knowing the extent of my neck injury and concerned I would make it worse if I moved, I lay still on the bank. The rescue team located me there about forty-five minutes after the crash. I was shivering from hypothermia.

Move to Houston, Texas—Center of Human Spaceflight

After recovering from the airplane accident and passing my PhD board, I decided a trip out west to California was what I needed to take my mind off the last year's series of tumultuous events. I was on a leave of absence from Boeing and did not have to report back to work until late October. My plan was to take a much-needed vacation, work on my dissertation, and work out a transfer to Houston, Texas, in the fall after I returned from California. Moving to Houston would place me at the center of human spaceflight and provide a good change of scenery from Alabama. I was relieved to move out of the house and leave all the bad experiences in Huntsville behind. I was tan and sporting shoulder-length blond hair when I reported to work in Houston. Anticipation was high that NASA would be launching the first space station elements soon. My first assignment was working space station payloads and operations for Jim Buchli. He was a former astronaut, US Marine Corps pilot, and wrestler from the Naval Academy. He provided significant assistance with my astronaut application and often teased me about my bad hair days.

The transfer to Houston was an excellent move. It felt like a new beginning. Texan hospitality was more to my liking than the judgmental attitudes I often encountered in Alabama. There were a lot more things to do in the nation's fourth-largest city, and I quickly took advantage of them. In the summer of 1996, a group of coworkers and I decided to attend a Houston Astros baseball game against the Atlanta Braves on a Friday night after work. Then we planned to hit a downtown nightclub afterward. By the sixth inning, the Astros were losing by eight runs, and most fans had already filed out of the Astrodome. I had seen enough and coaxed a few hard-partying coworkers to head out early to the Roxy—a local hotspot. If we left immediately, we could get there in time for happy hour. It had to be better than drinking expensive beers and watching the Astros get clobbered. They were persuaded, and we left the game for the Roxy. I arrived at the club ahead of my friends and staked out a table before it became too crowded. My attention quickly focused on the main bar where a very attractive woman was repeatedly declining the advances of interested males. She was tall and classically dressed with a long black skirt and white sweater. She looked like a blond Julia Roberts. She had the bearing of an ice queen with all the rejections she had doled out. Seeing a challenge and growing impatient with my now-very-late friends, I walked over and introduced myself. I was not going to take no for an answer—no matter what her rebuttal. As it turned out, she was waiting for late friends as well, and we sat together comfortably and conversed for the next hour until our friends showed up. Her name was Brenda, and she was a third-generation Texan. She was born and raised in Bellaire—the prominent suburb of Houston known for its championship high school baseball teams. Brenda was a successful bond broker, and this matched her conservative analytic personality. We began to see each other more and more, but we were cautious with our newfound friendship. We were both on the rebound from recently failed first marriages. The more time we spent together waterskiing, working out, and traveling, the more our friendship and trust grew. After a year or so of dating, we took our first skiing trip together in Colorado. I knew Brenda would someday become my wife, but it took me a couple more years to propose.

Changes to the Space Station Program

The Space Station *Freedom* program changed significantly during President Clinton's administration and evolved into Space Station *Alpha* as a cost-cutting effort designed to save it from the congressional budget ax. In 2004 Congress passed the Space Station appropriation, which was part of NASA's budget, by a single vote. This signaled to NASA that major changes were needed to garner continued funding support. NASA scrambled to overhaul the Space Station program. For the first time NASA brought on international partners that would contribute various modules and hardware as part of the construction plan to share costs, risks, and technical expertise. NASA and contractors set aside their normally confrontational differences and worked shoulder to shoulder for nearly six months to save the program. The Russians were the biggest contributing partner in terms of hardware and launch vehicle support. They used their Soyuz for crew transport and Progress module spacecraft for cargo. Staunch space rivals since the late 1950s, the only collaborative experience of Americans and Russians was the Apollo-Soyuz mission in 1975 and nothing until the International Space Station (ISS).

The new space station plan was threefold. During phase one the Americans would conduct a series of visits to the Russian space station *Mir* to learn how to work together in space. Phase two involved assembling ISS to the point of initial operating capability. That meant crewing the station. The third and final phase was to complete the assembly. That included all the remaining international partner modules from Russia, Japan, and the European Space Agency. With international partners involved, it was difficult for Congress to recant support because it would appear NASA and the United States were backing out of international commitments. I had missed most of the restructuring and long hours in the bunkers—little more than large open rooms with minimal office necessities such as phones and computers. Those who lived through the experience shared countless stories of the stressful working conditions, intense negotiations, and arduous milestones.

The mastermind behind the international partner plan was George Abbey—the outgoing and controversial director of NASA's Johnson Space Center. Abbey had the appearance of a basset hound but the demeanor of a pit bull. Former astronauts and NASA managers who worked closely with him offered either glowing praise or scathing criticism—sometimes both. Largely unknown to the public, Abbey wielded so much power within NASA that many observers ranked him second only to the agency's administrator. Abbey grasped the political forces affecting NASA and its future. A former USAF fighter pilot, Abbey came to NASA in 1967 and rose to prominence at JSC in the late 1970s. He had an encyclopedic knowledge of events and an uncanny grasp of human behavior. Abbey distinguished himself at the center and was promoted to increasingly responsible positions. He eventually served in the coveted position of director of flight operations. According to some astronauts Abbey had a dark side. During the space shuttle era, the man nicknamed "Darth Vader" held absolute power over the astronaut corps. He dictated who flew in space and who did not, and all astronauts had to jump to his tune. Many described

Abbey's management style as Machiavellian, and he made enemies both within and outside NASA. Many saw his departure from JSC in 1988 for a series of posts at NASA headquarters and on space advisory groups as an effort by higher-ups to remove the controversial director from power. Abbey, however, displayed a talent for political longevity, and during his tenure in Washington, he became a close ally of NASA's administrator, Daniel Goldin. Goldin ultimately restored him as the JSC center director in 1995.

Many credit Abbey as the person who saved the Space Station Program from a decade of delays by championing Russia as a full partner. This move helped rescue the program from cancellation by the White House and Congress. He also served as a driving force behind the visits of US astronauts to Russia's *Mir* orbital outpost to give NASA badly needed experience in space station operations.

While Abbey was busy saving the space station, Boeing shocked the aerospace and defense industry in December 1996 by its acquisition of Rockwell's aerospace and defense units and by initiating a merger with the McDonnell Douglas Corporation. The Federal Trade Commission approved the merger in July 1997. It made the Boeing Company the largest defense contractor in the nation. This series of events allowed the merger of the three largest space station work packages into one and made Boeing the prime contractor for the ISS. In 1998 ISS hardware started showing up at the Cape. This included the first US element—Node 1. Much of the integrated testing had been scrubbed from the program during the ISS redesign as a cost-cutting measure. This was primarily because the ISS was being touted as a paperless program. All the end-to-end integration testing and element-to-element fit checks were done on the computer. This quickly became a flawed strategy when preliminary fit checks unveiled a number of integration issues.

Launching the International Space Station

The first two ISS elements were launched from the Cosmodrome launch facilities in Baikonur. Node 1 from KSC followed. This served as the intersection between the Russian and US elements. Node 1 was attached to the Russian Functional Cargo Block (FGB) by a portable mating adapter (PMA). After arriving at the Cape, some simple interface checks between Node 1 and the PMA revealed many connection problems. Connectors were incorrectly clocked, wire and fluid line lengths were too short or improperly positioned between elements, alignment critical for leakproof docking was off, and there was inadequate spacing for an astronaut with a gloved hand or tool to work. Connection problems on-orbit could have disastrous consequences. Problems could jeopardize mission success and the well-being of the hardware and potentially the astronauts. Rich Clifford, the integration manager for ISS and a former astronaut and West Point graduate, asked me to lead the element-to-element integration at KSC and ensure the astronauts could connect all the hardware on-orbit. The work involved developing launch-to-activation procedures, verifying those procedures in NASA's 6-million-gallon Neutral Buoyancy Lab, conducting all the hardware fit checks at KSC, and completing all the crew element integration testing (CEIT) for all the US pressurized modules. Rich knew I was close to receiving an astronaut interview during the 1998 selections and believed that working with the primary and backup crews for the next seven ISS missions would provide networking opportunities with the crew office that might give me the extra boost needed to make the 2000 astronaut class. In addition to the ISS assignment, Embry-Riddle University asked me to teach graduate courses at their Houston extension campus. This had just opened in 1998. I developed their graduate-level space operations curriculum, and I soon realized I had a passion for teaching. I enjoyed the classroom and learned a great deal from the students. They were generally older and very knowledgeable in their areas of technical expertise. This broad experience enriched my knowledge of the space industry, space operations, and NASA as a whole. Rich supported my newfound teaching passion and thought it too would be an advantage on my astronaut application.

 The first three launches of ISS hardware went well. That included Node 1, which was launched in December 1998. However, it would be another two years before the next ISS element (the Russian service module) was launched. This was due to numerous delays stemming from technical issues and unstable funding that caused many Russian engineers to go without pay for months on end. Since the collapse of the former Soviet Union in 1991, the once-great space power struggled to keep its space program viable. The Russians were finding it increasingly difficult to operate their *Mir* space station and left one cosmonaut stranded indefinitely until enough funding could be scraped together to fly another Soyuz mission to change out the crew and bring him home. He ended up setting the mission endurance record for time in space at well over a year. The Russians' spaceflight infrastructure was crumbling from years of neglect and lack of maintenance. This carried over into the space shuttle–*Mir* missions and eventual delay of the service module. One near disaster after another marred these missions as the aging *Mir* space station encountered fire, collision and subsequent depressurization, toxic gas, leaky thermal systems, and countless other less life-threatening malfunctions. These events raised serious concerns at NASA over the well-being of the astronauts. It put the continuation of the space shuttle–*Mir* phase one missions in doubt. It also raised questions about whether the Russians could deliver reliable and safe hardware to the ISS. The delivery delays of the service module only exacerbated the problem and created tensions within NASA, which was already under stress because of an increasingly impatient Congress. American-built hardware was piling up at the Cape and eventually filled the space station assembly building. The truss segments had to be housed in the old operations and checkout building where the Apollo moon rockets had been assembled in the 1960s and early 1970s.

The service module delay was not only causing strained relations with Russian partners but also jeopardized the very survivability of the fledgling space station. The FGB had a limited battery life, and this led to taking aggressive measures to conserve power and prolong its endurance. After the service module was finally launched in September 1999, the launch of the US elements began in earnest. Eight shuttle assembly missions were flown in 2000 and ten missions in

2001. No major space policy changes occurred when President Bush took office in 2000, and ISS was first crewed in October 2001. ISS was well on its way to assembly completion as planned.

2001 Astronaut Selections

The large number of astronauts selected in 1996 and 1998 and the rigorous pace of shuttle launches and ISS assembly missions caused the 2000 astronaut selections to be delayed one year. I was excited I had been selected for an interview as a mission specialist. As such I would be required to undergo the grueling, weeklong physical and psychological testing that culminated with a one-hour interview in front of a twenty-member astronaut selection board. Every selection cycle is a little different, but typically the thousands of applicants are screened and downselected to four to six groups of twenty highly qualified candidates to be scrutinized during the weeklong assessment. The selection board then had the challenge of naming a predetermined number to that astronaut class. It was rumored that eighteen to twenty-four people would be chosen for the 2001 class from the six groups of twenty. I was in the fifth group and went through the interview process in February 2001. The experience was simultaneously exciting and nerve-racking. The amount of psychological evaluation surprised me as well as the great extent of physical examination. The military pilots were even warned that results from the physical exams could potentially disqualify them from flight status if something was uncovered outside their standard military screenings.

The process for choosing those who would be interviewed and ultimately selected remains a mystery to me. When the selection committee asked me why they should select me, I told them the following story. "While growing up, I was the youngest and smallest in my class. This typically caused me to be picked last when teams were chosen for sporting activities. Although I hated this, it motivated me to try harder to avoid being picked last the next time. When I later became interested in wrestling, the multiple weight classes evened the playing field. Winning was now based on performance rather than size, and that was my biggest attraction to the sport. I was excited when I finally made Navy's nationally ranked wrestling team my sophomore year just in time for a match against our rival, Army. I was the only Navy wrestler to lose, and it was the first time in the individual sport of wrestling that I felt worse for my teammates than myself because they had wanted a shutout. After that match I made a vow to work even harder at whatever I did to avoid letting down my teammates. This vow ultimately provided the motivation that propelled me to earn All-American honors as a wrestler. I pledge to do the same if selected as a mission specialist, and if this is the type of person the selection committee wants on the astronaut team, I am your man."

When the announcement was made, I was disappointed to be passed over for the 2001 class but not discouraged. I later submitted my application again with what I felt were even better qualifications such as the publication of a graduate-level textbook titled *Spacecraft Systems Design and Operations*, promotion to full professor at Embry-Riddle University, leadership of the debris modifications and risk assessment after the *Columbia* accident, and the backing of the NASA administrator. Even with these additional accomplishments and recommendation, I was not invited back for another interview. When interested parties ask me about how to become a NASA astronaut, I tell them I do not know. However, one could do it the old-fashioned way and make enough money to pay the Russians several million dollars to fly aboard the Soyuz. Odds are much better for making a fortune and flying with the Russians than being selected as a NASA astronaut. The price could become even more affordable when global entrepreneurs such as Burt Rutan of Virgin Galactic or Elon Musk of SpaceX develop commercially crewed rockets for space tourism.

Transition to the Space Shuttle Program

In July 2001 I was relieved to see the air lock (the last of the space station's pressurized modules) finally launched. The pace over the last two years had been blistering, and I was growing tired from all the travel back and forth to the Cape. Also Brenda and I were married on the beach in Destin, Florida, in March 2001—one of our favorite vacation spots—and we moved into the new home we had built on a private lake. Brenda and I had the opportunity to work together on Boeing's Transportation Security Administration program federalizing the nation's airports after the 9/11 terrorist attacks, but I wanted to get back to my passion for human spaceflight. Therefore, when the opportunity became available to work on the Space Shuttle Program as manager of systems engineering and integration, I took it. Unfortunately the shuttle SEI job after the *Columbia* accident instantly changed from flight operations to redesign and fixing the debris issues.

It was about nine months after the *Columbia* accident when Bob Ess, my NASA counterpart, approached me at a debris review about working for NASA. Bob worked in the rejuvenated SEI group for NASA that John Muratore now headed. John had already established a reputation as an uncompromising manager and astute engineer with a low tolerance for substandard work and attitude. He was notorious for working long hours and demanding strict attention to detail. He would not hesitate to dismiss even the most senior engineers out of the room for mistakes, missing data, or inadequate analysis. As in my military days on the submarine, here was a no-compromise standard of excellence that everyone was expected to follow. If one's work was exceptional, John was quick to provide praise and acknowledgment. Recognition was a scarcity at Boeing, and this acknowledgment was something that appealed to me most about working for John. Bob was John's protégé was just as relentless as John. He was looking for someone to step in and share the enormous debris workload and RTF activities. It was normal protocol for NASA to advertise internally across the agency for open positions, and I was curious why NASA was looking externally. I asked Bob. He stated that the position had been open for six months, and no one wanted the job. I asked why he was interested in me. He replied they needed someone with my expertise and experience. I agreed to the interview and met with John in his office the next day.

My Interview with John Muratore

Based on what I knew of John's personality and reputation, I was expecting an Admiral Rickover-style interview. The fact I had survived an interview with Rickover comforted me, and I asked myself how bad it could be. John's office was astounding and definitely reflected his passion for space. Spacecraft and aircraft models of all types, pictures, books, figurines, coins, pins, and mission badges all the way back to Mercury occupied every available inch. Miscellaneous hardware from previous projects along with other collectibles and action figures such GI Joe Astronaut littered the room. I made my way to a small, circular meeting table piled with countless PowerPoint printouts, shuttle documents, foam samples, and engineering drawings of the external tank and orbiter. Bob cleared off some chairs, and John closed his office door to provide some momentary privacy from the group already lined up outside and waiting to see him for work related things. We discussed my background and résumé, and then he asked why I wanted to work for NASA. I told him I had always wanted to work for NASA—ever since visiting the opening of the Neil Armstrong museum near my home years before. I was looking for a larger leadership role than Boeing had been able to provide. I wanted to work for NASA for the same reasons I went to the Naval Academy and served on the USS *Hyman G. Rickover*. I admitted the challenge and high expectations seduced me. Knowing John was a retired air force officer, I also added that I thought NASA could use the help of an ex-navy "nuke" to get the shuttle flying again. John chuckled and appreciated the reference to interservice rivalry. John asked what I thought the biggest difference was between NASA and a company such as Boeing. I rattled off a few things relating to culture, organizational structure, and contractor and customer relations. John interrupted me as my answer began to run out of substance. He concluded I was not zeroing in on the answer he had in mind. Boeing, he said, was in the space business to make profit. They had shareholders and board members to satisfy on top of the customer base. NASA was in the space business to launch humans into space safely. He said my job would be simple—work with Bob to fix the debris problems and get the shuttle flying again. He then stated I was perfect for the job and asked how soon I could start. Although salary was not discussed, I did not care. I said I would be ready to change badges in two weeks.

Transition to NASA

My Boeing colleagues thought I was crazy to make such a move, and some former NASA employees who had been purged from NASA after the *Columbia* accident and were now working for Boeing said I should seriously reconsider. My boss asked why a person like me would like to work for a person like John after thirteen exceptional years at Boeing. Not usually caught off guard or at a loss for words, I jokingly replied I liked my plebe year at the academy and needed another. He laughed and wished me well. I reported to work in February 2005 on a cold, rainy day. Thus I started my dream job at NASA. I found out later from Bob I was not the first choice for the job but the best fit. John's first choice was Darby Cooper. Darby had worked on the shuttle program before the *Columbia* accident and had established a stellar reputation for his knowledge of aerodynamics and expertise using the analytic models. Darby had been an exceptional member of my systems engineering team at Boeing, and his NASA customers (John and Bob included) held him in high regard. In true systems engineering fashion, Bob felt the shuttle program would be better served with me at NASA instead of Darby—even though I lacked Darby's technical expertise in debris analysis. Bob was right, and it did not take me long to master the debris discipline. The hiring situation reminded me of making the wrestling team at the Naval Academy and the national wrestling tournament. Persistence and preparation had again paid off, and all I needed now was a chance to make the most of my opportunity at NASA.

Debris Verification Review—April 27, 2005

I was happy with my decision to work for NASA, and this thought raced through my mind as I awaited my ice debris presentation at the April

2005 DVR. The forward bellows ice risk was too high, and corrective action needed to be taken to reduce the risk. Installing a heater would solve the problem and eliminate the bellows ice risk but would involve moving the shuttle back to the Vehicle Assembly Building (VAB) for the installation. That meant another launch delay.

"Dr. Peters, you are up," boomed John.

It was a long journey to get to this point, but it was my time to take center stage and shine. After I completed the presentation, the DVR board immediately implemented the recommendation to push the vehicle back into the VAB and install a heater on the forward feed line bellows to prevent ice formation.

Another memorable highlight from the DVR event was when Mike Griffin, the NASA administrator, approached a whiteboard during a C/E methodology presentation and stated, "We need to look at the problem differently. From a Monte Carlo [PRA] approach." Griffin confirmed Brekke's apprehension with the C/E methodology and provided validation of the PRA approach that he had proposed and developed. "This made me feel really good because the NASA administrator was saying what I was saying. That meant I must be right," stated Brekke.

This review and support from Administrator Griffin was a turning point for PRA acceptance. Mike thought the DVR and ice presentation were well done and very informative, and he said so. He also thought it was obvious we needed a heater on the forward bellows to prevent ice accretion and subsequent liberation in flight. With all the DVR data as background, Mike had a private chat with Bill Parsons—the shuttle program manager. Mike told him that if he chose to delay the launch to June, he would back him on that decision. From Mike's perspective, no decision to launch the shuttle could be routine. Flying the shuttle brought great risk—risk that could not be eliminated. Sending brave men and women into space was never without risk, but NASA would do everything possible to reduce those risks. If it took more time to achieve that goal, then so be it. "Our intent with this effort is to make certain we are as safe as we know how to be before we launch the shuttle and its crew," stated Mike. "We will return to flight. Not rush to flight." At a post-DVR press conference, Mike publicly announced the launch delay. "As a result of these reviews, our team has come to the conclusion it is prudent to have additional verification and validation of our extensive engineering work to ensure a safe flight for *Discovery* and the STS-114 crew. Therefore, we have moved our target launch window for the shuttle's mission to the July opportunity."

CHAPTER 13

STS-114 Flight Preparations

Schedule Pressure Builds

Schedule pressure caused by the numerous launch delays and problems on the space station was now at a maximum. Many on the program were growing concerned the pressure was causing the workforce to make mistakes that led to mishaps and required rework. For instance, worker fatigue was a contributing factor behind the death of a worker who fell from the VAB while repairing damage from Hurricane Frances, which pounded the Florida coastline in early September 2004. Both the VAB and orbiter processing facility were damaged in the storm and needed timely repair to support shuttle assembly operations. The long hours put in by John and everyone involved with debris kept getting longer. There was an enormous amount of work that needed to be finished. It reminded me of days at the Naval Academy as a midshipman and on the USS *Rickover* as a young submarine officer. The familiarity of the pressure actually brought a certain amount of comfort, and I was confident about getting the work completed. Shortly after the third DVR in April, I held a technical exchange meeting in early May to address the feed line bellows and bracket ice risk, the no-ice zones, and the ice-ball allowable table in NSTS 08303. The ice characterization and liberation testing were late getting started and were still in progress. There were no ice release models yet, and ice analysis was limited to only a few debris transport cases. Everyone knew ice was a significant debris risk, but there was no way to quantify the risk or generate flight rationale for the hazard report. With a tanking test two weeks away and the flight readiness review five weeks away, my job was to make sure the ice risk assessment would be ready.

Debris Risk Briefing with the STS-114 Crew

Shortly after the vehicle was pushed back into the VAB, John agreed to meet with the STS-114 crew for a face-to-face overview of the debris risk. Unlike other debris meetings, this one was small. It involved only six of the seven STS-114 crewmembers, John, Bob, and me. John started the meeting by giving an overview of all the debris work that had taken place over the last two years, and he explained why the launch had been delayed several times. Bob followed with a discussion of the process for developing flight rationale for the expected debris sources and foam debris risk PRA. I finished with an overview of the ice debris risk assessment and our mission operations plan. This included the multiple imagery tasks the crew was to perform during various mission phases. The crewmembers had been following the closure of the CAIB recommendations, progress of the debris risk assessment, and flight rationale development through the multitude of debris meetings. They had many pointed questions and a few critical comments during the discussion. At times it became fairly contentious. Their concerns centered on the full understanding of the debris environment (particularly ice debris) and whether the program had taken sufficient steps to ensure the crew safety. The shuttle program had not spent much effort hardening the orbiter against debris strikes because the goal was to eliminate the debris threat altogether. The orbiter also lacked a certified approach for debris damage repair. Ground testing showed most repair techniques had failed in some fashion either during the application process in a vacuum or when exposed to the simulated high-temperature reentry conditions in the arc jet test facility at JSC. They were also very skeptical of the PRA approach. John answered most of their questions and reiterated all the positive steps taken. This included CSCS to ensure their safety. He concluded the nearly three-hour meeting by telling the crew STS-114 would be the safest shuttle flight ever flown.

Finalizing the Debris Risk and Flight Rationale

The totality of all the debris work was staggering. In addition to the foam and ice risk assessments, Bob and I still needed to generate the flight rationale for all other debris sources, document the rationale in the integrated debris hazard report, obtain ISERP concurrence, complete the updates to the ice-ball allowable in NSTS 08303, and prepare the mission operations plan. This plan included integrating all the imagery and radar assets and establishing the day of launch and on-orbit support roles and responsibilities. Bob and I agreed we would work the flight rationale separately. He would handle the foam risk and flight readiness review presentation, and I would be responsible for the rest. Resolving the debris issues spanned all the shuttle project offices—not just the ET project. It involved assistance from all ten NASA centers for resources, test facilities, and expertise in a multitude of technical disciplines. Except for initial development in the late 1970s and early 1980s, the amount of testing was unprecedented. All the element teams worked intensely to characterize the debris environment and classify debris as either expected or unexpected. A good example of the latter was the thermal tile and insulation blankets. They were designed to stay intact through reentry and landing. Examples of expected debris included numerous foam, ablator, and ice debris that could liberate during flight as well as dozens of exhaust constituents from the booster separation motors. These fired about 120 seconds into flight to separate the spent SRBs from the vehicle. More than 250 expected debris sources on the vehicle had to be analyzed to ensure they would not lead to catastrophic damage if liberated. Flight rationale for each of these debris sources was generated to ensure the vehicle was safe and ready for flight. The mountains of test data and analytic results were all reviewed during the four sets of marathon debris verification reviews that started in March, continued into the summer, and ran all the way until the STS-114 flight readiness review.

There always seemed to be more questions than answers and very few reliable facts. Redesign of the bipod area (the culprit in the *Columbia* accident) was one of the easier debris sources to remedy. The large foam ramps that covered the bipod fitting were removed, and heaters were installed to prevent ice formation. After aerodynamic and thermal testing of the new configuration, the design was certified for flight. The challenge for the program was how to address the vast number of expected debris sources and more importantly how to show the vehicle was safe from those debris threats. Since the debris models were just being developed and validated, it was difficult to determine which debris sources posed the biggest threat. Each debris source required its own set of release, transport, and impact inputs into the debris model. This made it nearly impossible to prioritize debris threats. The one known fact was that foam removed from the tank was foam that no longer posed a threat to the vehicle. Unfortunately removing or shaving foam off the tank increased the likelihood of ice formation and liberation. The windows, tiles, and RCC panels were even less tolerant of ice impacts. Encapsulating the external tank in a cover such as a net or applying a coating designed to prevent foam from liberating was considered but quickly dismissed due to feasibility issues. There was always a weight penalty associated with any proposed solution, and finding a material that could withstand the aerodynamic forces and searing surface temperatures that would reach 600°F within the first one hundred seconds of flight was unachievable. If the material failed, it would become a debris source that was potentially more deadly than the foam.

The Tanking Test Debate

After the first debris verification review ended in a no-go recommendation for launch, the program determined it would be prudent to conduct a tanking test on April 14, 2005—a full dress rehearsal for the actual launch. John Muratore pushed for the test because it would determine the operational readiness of the external tank and exercise the new debris modifications, procedures, and debris detection assets. It had also been more than two years since the launch support teams had been on-console, and the tanking test would serve as refresher training to sharpen their skills for the upcoming flight.

During the test the launch team ran into a number of issues. The external tank contracted when loaded with the supercooled liquid oxygen and liquid hydrogen propellants. The diameter and length were measured before and after loading the propellants, and the tank shrunk about four inches in diameter and nearly a foot in length. The structural contraction imparted internal stress to the foam that led to some cracking. Cracking was undesirable for a number of reasons. Cracks could lead to foam liberation—most likely from the force of air loads trying to peel the foam away from the tank. Depending on the location of the crack and its severity, cryopumping could occur. Imagery taken during testing and on subsequent flights after STS-114 showed the cryopumping foam releases occurred explosively. Cracks could also lead to thermal shorts that enhanced the likelihood of ice formation (ice-balls).

The size and location of an ice-ball that formed during the tanking test was such that a launch scrub might have been ordered had it been an actual flight. Even though it was a hot, sunny day at the Cape, the ice-ball continued to maintain its size. Ice formation on the external tank was very difficult to predict due to the complex set of causal variables such as temperature, humidity, wind, solar angle, location on the vehicle, the amount of condensation or rain on the tank, and the extent of thermal cracking in the foam. There were also different varieties of ice such as sheet ice that formed on the surface, ice-balls that formed in thermal shorts and in cracked foam, frost, and either hard or soft ice (based on density). Hard ice was high density and clear like the ice cubes frozen in a freezer. Soft ice had a lower density and cloudy white appearance similar to slush. The ideal condition to form surface ice was during the winter months with low temperature conditions. All other ice formations occurred best under warm (85°F) temperatures and high (about 80 percent) humidity. The rain and amount of condensation on the tank could either melt or promote ice growth, depending on the amount and liquid temperature. A large amount of warm rain typically melted most ice formations, but a cold drizzle only added to the growth rate. The foam cracks tended to be concerns if they were located far forward on the tank and/or had an elevated edge or surface area for the air loads to assert a force or allow cryopumping. Cracks that formed during the tanking test would not pose a launch constraint because there was no elevated edge.

Along with the ice-ball and foam debris concerns, the tank suffered a failure with two of the four liquid hydrogen depletion sensors known as engine cutout (ECO) sensors. The ECO sensors were designed to secure the shuttle main engines before the tank ran out of either liquid oxygen or liquid hydrogen propellant. If the engines were allowed to run until the tanks ran dry, the large turbopumps that feed the propellants into the combustion chamber would rev to the point of catastrophic destruction. It would be similar to popping an automobile clutch after revving the engine and causing damage by exceeding the maximum revolutions per minute. In addition to the failed ECO sensors, a pressure relief valve designed to help maintain hydrogen tank pressurization in the final two minutes of the countdown cycled more than normal. The new tanks modified for STS-114 and subsequent flights featured a newer-style dual-screen diffuser. This injected a jet of helium gas into the hydrogen tank to help keep the super-cold fuel circulating at the proper temperature. It also provided the pressurization needed after the tank was isolated from ground systems two minutes prior to launch. Engineers believed the unusual valve cycling during the April 14 tanking test was due to the newer-style dual-screen diffuser. The valve cycled thirteen times, and that was above the allowable and normal performance of eight or nine cycles. After some extensive troubleshooting of the ECO sensors, the tanking test was secured, and explosive propellants were drained from the tank. It was not a stellar performance. This prompted a question whether to conduct another tanking test, which would mean another

STS 114 FLIGHT PREPARATIONS

launch delay of the RTF mission. To John RTF preparations took a dramatic turn after the tanking test. "Just as we were starting the tanking, Helen showed me an internal external tank [team] presentation about the change to the diffuser. This was the cause of the pressurization problem, which was the first of the really contentious battles. Up until the tanking test, everyone had been working together pretty well, but the tanking test started off some really ugly stuff, and of course the first ECO sensor anomaly happened here."

NASA had hoped to launch *Discovery* in May, but problems cropped up during the April 14 tanking test and contributed to the decision to conduct another tanking test after engineers evaluated the ECO sensor failure data and delayed the STS-114 flight until June. "What struck me most about the ECO sensor troubleshooting and anomaly resolution effort," Helen stated, "was the dissension and posturing that occurred. I naively thought everyone would work together to try to understand what went wrong so we could fix it and move on, and it made sense to me that John and SEI work the problem since it was obviously a system issue. Getting to the root of the problem was very difficult. The orbiter and ET projects were both trying to prove the problem wasn't in their element, resulting in a real lack of cooperation in the overall effort."

Decision to Swap the External Tank

Two weeks after the first tanking test, there were many questions about the potentially dangerous buildups of ice around the forward liquid oxygen feed line bellows assembly that could shake loose during launch and damage the orbiter's fragile heat shield tile or wing leading edge RCC panels. The program had accepted the recommendation to install a heater around the bellows assembly to eliminate the ice debris threat but had to delay the launch yet again until mid-July. To minimize the schedule impact, the decision was made to install the bellows heater on the next tank slated to fly (ET-122) and swap it out with the current tank (ET-120). After the tank swap decision was made, there was intense discussion about which tank to use for the second tanking test. A number of engineers recommended the second tanking test be performed with ET-122. If ET-120 were used for the second test, they argued, then a third tanking test with ET-122 was recommended. The third test had the potential to push the launch date even further into the summer. (External tank designations are sometimes a source of confusion because they typically do not match the flight number. A match was actually more the exception than the norm.) After a few weeks of heated debate, the program managers decided to conduct a second tanking test, which was set for May 20. They would forgo a third tanking test after swapping tanks, and this decision avoided the possibility of a launch delay into August. The rationale behind the decision was that after the tank was swapped, why not launch if it was ready to go? If the launch was scrubbed, it would be the equivalent of a third tanking test. John had been a proponent of dissecting production tanks ever since he took over as the SEI manager shortly after the *Columbia* accident. He was overruled, however, due to the cost of doing the dissections and refurbishing the tank back to flight readiness status. It was also unclear how much could be learned beyond all the foam debris testing that had already been planned and was being implemented. John finally got his wish after the tank exchange decision was made and preparations were made to ship ET-120 back by barge to the Michaud Assembly Facility after the second tanking test. It would later prove one of the best decisions the program made. It provided an opportunity to uncover the secrets of additional failure modes through the knowledge gained from the dissection work. The engineering and safety teams were still concerned about the tanking test performance and were the most vocal about running a third tanking test. John reassured them their concerns would be addressed during the debris verification reviews and worked to their satisfaction before declaring the vehicle ready for flight.

The Second Tanking Test

On May 20 the second tanking test with ET-120 was performed and produced much better results. The ECO sensors worked properly, and this led NASA managers to believe the intermittent operation on April 14 was most likely due to a wiring issue that troubleshooting resolved. This gave the shuttle team confidence the sensors in *Discovery*'s new tank, ET-122, would behave normally as well. The only glitch during the second test was excessive cycling of the pressure relief valve. Once again it cycled thirteen times. Since ET-122 was being retrofitted with the bellows heater, the program managers decided to replace the dual-screen diffuser with the original design. By switching back to the older single-screen diffuser, engineers were confident the excessive cycling problem would not recur during the STS-114 launch countdown. Based on the performance of the second tanking test and progress made on the bellows heater installation, a new July 13 launch date was set.

Stafford–Covey Assessment of Flight Readiness

On June 8, 2005, the Stafford–Covey Return to Flight Task Group announced that NASA had closed out all but three of the fifteen recommendations needed to complete before it could launch *Discovery* in July. The task force, led by veteran astronauts Thomas Stafford and Richard Covey, was established in the wake of the *Columbia* accident to monitor NASA's compliance with the CAIB's recommendations. Covey and other members of the task force said, "We saw no showstoppers that would interfere with Return to Flight."

The group had nearly all the data in hand it needed to reach a judgment and expected to be able to do so in time for the STS-114 flight readiness review. Some additional analysis was needed to close out two of the three open recommendations. That left only the item related to the orbiter inspection and repair capability. The CAIB said NASA must have a "practicable" solution for repairing "the widest possible range of damage." The ambiguity of the recommendation's wording had been a source of ongoing debate among Stafford–Covey task force members. Jim Adamson, chair of the group's operations panel, was of the view that, "Until that debate was resolved, the group would not be able to determine whether or not NASA was in full compliance with that particular CAIB recommendation."

Either way the Shuttle Program managers made clear they intended to resume shuttle flights with the repair capabilities it had in hand without knowing for sure whether they would work in an emergency. Ralph Roe and the NESC worked with the orbiter and made significant contributions toward developing an on-orbit repair capability. After STS-114 Dr. Charles Camarda joined NESC and utilized his astronaut and materials engineering expertise in the final development of the repair techniques. "We understood that the repair was not certified, but while Charlie was the NESC representative at JSC, he pushed very hard to do as much ground testing of the material as possible. He was instrumental in getting on-orbit data of the material during the various missions. We had gained significant confidence in the repair capability but understood the risk of trying to implement it on-orbit," said Ralph.

For small RCC holes and tile damage, Justin Kerr and the orbiter team had high confidence in the various repair techniques. This was especially true toward the end of the program, as the techniques were refined and certified. Probably more than anyone in the program Steve Poulos had great confidence in the repair capabilities. "The development teams had done all the necessary and appropriate analyses and tests to demonstrate the performance of the repair capabilities in relevant environments. There were certainly some locations where repair was not possible. However, for those areas where we had the data to show a repair could successfully be performed, I personally would have been ready to implement it," stated Steve.

However, Justin and others on the orbiter team knew the repair techniques were limited for large RCC breaches and tile damage. In these cases the DAT would most likely recommend the rescue mission. "Although we wanted to minimize debris damage to the orbiter, we were always looking for an opportunity to try some of the repair techniques on select damage sites to assess the performance," noted Justin. "As it turned out, the program never did get the opportunity to perform a repair, but it was comforting to know we had this capability at our disposal in case we needed it."

Pressure to Produce a Foam Debris Risk Number

A number of challenges to Bob and the Aerospace Corporation team's ability to run the risk model and generate a valid risk number existed. There were several model inputs such as potential mass release as a function of time, release rate, release location, release angle, and pop-off velocity that had to be established and backed by engineering and the debris community as legitimate. The most difficult challenge was determining the relationship between the foam fracture toughness and temperature—a risk driver in the critical parameters of debris mass, release time, and liberation frequency. The work went on around the clock and seven days a week. Midnight teleconferences with Brekke and the Aerospace Corporation team in California to discuss test or analysis results were common. The frantic pace was even worse for those working the ice debris risk assessment. Test results from the ET project's ice characterization and separation testing were coming in late and had to be reviewed. I was working with Darby and the Boeing team to develop a probabilistic risk model in parallel with the Aerospace Corporation, but my primary focus was on characterizing the ice risk. It was nearly impossible to gain much insight from the ice testing results. The ice characterization testing only evaluated the worst-case ice-growing conditions, and these were rarely seen. The tests only provided ice morphology for the largest size and highest density. The ice separation testing was designed to determine how and when the ice would liberate along with a distribution of mass sizes and number of released pieces.

Unfortunately the ET project was only able to provide a single test run of data before running into test facility problems and subsequent delays. The test teams quickly recognized the difficulty in both growing ice and simulating the launch environment. A single test run was hardly worthy of any statistical significance or confidence, but it was the only thing available. Time was running out. It was only five days until the debris verification and flight readiness reviews. So much emphasis had been placed on the foam debris there was precious little ice impact data for the thermal tile and RCC panels. Consequently the orbiter team was only able to provide a foam impact model for tile. They did, however, make some adjustments to the foam-on-tile impact model based on initial ice impact testing and recommended we use that for the ice risk assessment. I knew this was very conservative because a piece of ice of the same mass would produce a much smaller impact cavity on the tile. This smaller damage cavity intuitively meant a lower risk, but we did not have the data to quantify how much lower. The initial ice probabilistic runs for the feed line bracket and bellows ice were producing risk numbers as bad as one in five. I knew these results were flawed because they were inconsistent with flight history. They were a manifestation of the conservative worst-case ice testing inputs and application of the foam impact damage model. The other challenge facing the ice team was the amount of run time required to complete a probabilistic computation. The analysis was being performed on NASA's Columbia supercomputer and was supplemented with whatever computational assets Boeing could provide. The run time had been improved substantially from forty hours to ten but could easily take longer. It depended on the complexity of the inputs. Despite this improvement Boeing was working frantically to automate and streamline the computational process to minimize the run time.

It was late Friday evening (five days before the flight readiness review), and the ice debris team was desperate. I called a triage meeting at Boeing to discuss options and determine if we would be ready for the upcoming DVR. I was not about to give up hope and certainly did not want to tell Bob and John we would not be ready. Delaying the DVR would mean another launch delay. We knew it would be a long weekend but brainstormed some feasible ideas. Bob eventually joined us at the Boeing building and decided to remain and work on the DVR presentation.

The Aerospace Corporation team had completed its risk assessment of the bipod foam. It had a very low risk as expected. Bob had finished with the DVR package with the exception of the final ice risk numbers. It was a race to the end, and we had every intention of crossing the finish line. It was well past midnight when the team finally finished their work. Most of the team would be flying to Orlando in another eight hours to support Bob's presentation at the DVR at Cape Canaveral. I collapsed into bed after setting the alarm. I was too tired to change my clothes, and was glad I would get at least two hours of uninterrupted sleep on the flight from Houston to Orlando.

My wife woke me as the alarm beeped in the background. Darby and Bob had left several messages on my BlackBerry. The messages were discussing the worst-case ice risk number—one in one hundred. More importantly they were asking what I thought we should report to the program. I called Bob and Darby for a quick conference call, and I recommended going with the results and emphasizing the conservatisms—especially from using the foam impact model for ice. I had been working with Boeing and the final inspection team (FIT) at KSC to complete an ice inspection document—NSTS 08303—that would offer protection from ice debris by establishing ice growth limits. These limits were consistent with our risk assessment for critical areas of the tank. Three hours before each launch, the FIT inspected the vehicle and launchpad in search of anomalies, ice growth, and general debris-related issues. They were the shuttle program's last line of defense against debris and would utilize NSTS 08303 as a guideline to determine if there was any ice growth threatening the vehicle. The problem was the document was not ready. We agreed to press ahead anyway and simply assure the program it would be ready for flight, there was solid flight rationale from the foam debris risk story, and NSTS 08303 offered protection against ice. The worst thing that could happen was another launch delay until we could refine the ice models.

The Final STS-114 Debris Verification Review

On June 25, 2005, the long series of STS-114 DVRs finally came to a conclusion with a one-day review at Cape Canaveral. The final DVR was scheduled along with the flight readiness review. This was to allow time to close out the remaining actions from the previous DVR—most notably generation of the residual bipod foam debris risk and ice debris risk numbers. Bob was working with the debris team from the

Aerospace Corporation that just a few weeks earlier had proposed utilizing their newly developed linear elastic fracture mechanics model to generate a risk number from the small amount of foam that could still potentially release from the bipod. I provided Bob with all the ice and nonfoam debris risk data, and Brekke and his team provided most of his foam debris data and presentation materials going into the review. Since the first foam PRA presentation in February, and through all the debris verification reviews in the spring of 2005, Brekke and his team were frantically building the Aerospace Corporation PRA model. Building the foam PRA model involved a lot of logistics and data collection. This included obtaining and integrating transport and CFD solutions from NASA and Ray Gomez, impact allowable tables for tile and RCC panels from the orbiter project and Justin, and foam defect data from the ET project. Going into the final DVR, Brekke felt the Aerospace Corporation team had made the best attempt possible to characterize the risk using the PRA tools they developed over so short a time. The Aerospace Corporation results were a key part of the foam debris flight rationale—especially for the bipod area—and Brekke felt confident we had done the right things in the time available. "It was not perfect, but NASA had an informed idea about the foam risk. The PRA results gave the key decision-makers the opportunity to assess the risk and determine whether it was acceptable or not," asserted Brekke.

I had provided Bob with most of the ice risk materials he presented. This included analysis results Darby and his Boeing team had produced over the last couple days leading up to the DVR. Even though the ice bracket risk was one in one hundred eighty and the bellows risk one in one hundred, the results represented worst-case conditions, and the real attention was on the foam risk. The DVR board recognized the ice risk numbers were very conservative, and the members were comfortable knowing the highest ice risk had been eliminated with the installation of the heater on the forward bellows. Darby and the Boeing team thought the flight rationale for the ice risk was sound based on the historical comparisons with the PRA tools and the flight history evidence. "I also thought our damage history was telling us that ice was a manageable problem," recalled Darby. "Since we did not see a right side damage bias, where all the ice sources were located, I felt the risks were conservative."

According to Ralph and the NESC, the series of debris verification reviews was a learning process for everyone. "When Mike Griffin joined NASA as the administrator, I think we developed a methodical approach for characterizing the debris risk," stated Ralph.

With Mike's leadership and the probabilistic approach, the emphasis changed from fixing just the proximate cause of the STS-107 bipod foam loss to fixing the highest risk areas on the ET. Ralph viewed this as a critical and necessary philosophical change. At the time of the STS-114 FRR, the NESC team believed more work needed to be done on ice debris and worked with the program on a set of recommendations. This included better characterization and controls for the ice debris risk, and that eventually helped make NESC comfortable we were ready to fly STS-114. For Darby the "go for launch" decision made at the June STS-114 DVR turned out to be his most memorable debris event. This was especially real after seeing the headlines in *Florida Today* the next morning and knowing we had beaten the monster that had kept us grounded for two and a half years.

In preparation for the June DVR, NASA executed and analyzed over 1 billion debris transport cases. These cases included various debris types, locations, sizes, and release conditions. This work provided NASA with unprecedented insight into the ascent debris environment to which the space shuttle was subjected. In feasible cases debris sources that could cause significant damage were redesigned. Examples of vehicle redesigns driven by debris risk include the bipod closeout to eliminate the debris source, intertank thrust panel venting to minimize mass size and rate of release, new foam application process for the hydrogen IT flange installation, installation of forward bellows heater to prevent ice formation, and installation of bellows drip-lip to reduce ice formation. The foam ramps that protected ET bipod fittings from ice buildup were replaced with a new electrically heated joint. In the redesign the insulation was removed and the fitting mounted across the top of a copper plate that contained electric heaters to prevent ice formation. Liquid nitrogen was used to purge the intertank connection of any potentially explosive hydrogen gas. However, liquid nitrogen could freeze around the bolts in that area and cause foam insulation to liberate. Therefore, the bolts in that area were redesigned to prevent liquid nitrogen leaks. The foam ramps that protected the liquid oxygen feed line bellows were angled and could permit ice buildup. A design called a drip-lip that prevented ice buildup replaced these. The bellows, which compensated for expansions and contractions when the liquid hydrogen tank was filled and emptied, was modified. Previously the foam insulation overlaying the bellows was angled. This allowed water vapor to condense and run between the foam insulation, and this sometimes caused the foam to crack and liberate. To correct this problem, the foam skirt of this joint was extended over the insulation such that water could not run between the foam. Explosive bolts that released the SRBs from the external tank at SRB separation could fragment and damage the shuttle. Therefore, engineers redesigned the bolt catcher to prevent the bolts from damaging the external tank or hitting the orbiter. Finally Tyvek covers were installed on the orbiter's forward reaction control system thrusters to prevent water from collecting and forming ice that could easily impact the orbiter's windows. The Tyvek covers were designed to release within the first fifteen seconds of flight. They replaced butcher paper, which had an unpredictable release time.

    The DVRs provided a good forum to disseminate PRA numerical predictions for debris sources to help in the identification of leading redesign candidates. For Roy Glanville and the safety community, the DVRs communicated which organizations were responsible for risk hazard control of both expected and unexpected debris. Maintaining hazard controls for unexpected debris during liftoff and ascent was the responsibility of the element project teams and was achieved through improved design, processes, and procedures. John Muratore and the SEI team were responsible for the integration risk assessment for the expected debris. The debris hazard report (IDBR) ranked the debris source–target pairs considering probabilistic and deterministic risk ranking but also the likelihood of cause manifestation considering the strength design, processing, and procedural controls as demonstrated by ground tests and flight history.

The June 2005 DVR accomplished two major objectives to support the rationale for RTF. First, NASA developed an end-to-end estimate of the orbiter's capability to withstand damage relative to the ascent debris environment. This was done by using the ET project's best estimate of the foam and/or ice debris that might be liberated and the worst-case assumptions about the potential of that debris for transport to the orbiter. Second, for those cases where the initial assessment indicated the orbiter could not withstand the potential impact, NASA performed a probabilistic risk assessment to determine the likelihood of a critical debris impact. NASA assessed four foam and two ice debris cases. These cases represented the worst potential impacts of a general category and were used to envelope similar but less severe transport cases from the same areas.

The DVR board members were rather sedate compared to previous debris meetings and asked very few questions during Bob's daylong debris presentation. NESC concurred with Bob's debris risk assessment and flight rationale, and when John polled the board, it was a unanimous decision to go fly. We had crossed the finish line and would begin final launch preparations for the first shuttle flight since *Columbia*. STS-114 was finally a go for launch.

CHAPTER 14

STS-114 Go for Launch

L Minus Two Meeting

The shuttle team knew it was getting close to launch because a number of bizarre things began to happen. The launch countdown clock actually starts at the L-2 (launch minus two days) meeting. July is right in the middle of hurricane season, and as luck would have it, one was brewing in the Atlantic and slowly creeping west. Fortunately for the Cape, Hurricane Emily was tracking well south of Cuba and would not pose a threat to the flight—provided it did not change course and head north toward Florida. The more intriguing issue facing the shuttle team was a massive sandstorm in the Sahara desert. Strong winds whipped fine grains of sand high into the atmosphere, and the particulates were slowly drifting across the Atlantic straight toward Florida and the Cape. John Muratore ordered the debris team to perform a full assessment to determine if the fine grains of sand would pose a debris threat to the shuttle. This was a highly unusual scenario that required assistance from outside the agency. We needed data on the composition of the sand—its size, density, and concentration—as well as atmospheric conditions. NASA coordinated with the National Oceanic and Atmospheric Administration and had the Hurricane Chaser Squadron from Keesler Air Force Base in Biloxi, Mississippi, send a C-130 (normally used for hurricane evaluation) to retrieve a sample. Although it was a mad scramble for the debris team, and there was a genuine concern about potential window damage and cracking, John was enjoying all the activity. The program managers and media were following the story, and John was using the opportunity to showcase NASA's debris team and their ability to respond to any debris issue. John was even more jovial when the debris team learned the sand particles were too small to pose a threat. They were also at a low enough altitude that there would have been minimal impact energy even if the particles were larger. The winds were even due to shift away from the Cape and eliminate the problem altogether. Bob Ess and I had made the final updates to NSTS 08303 and reported a go for launch at the L minus two meeting. After that meeting John wanted the flight support team well rested and alert during launch. He requested they stand down for the next day. He told those who had to work to limit themselves to a "normal" day. I joked that a "normal" day for the debris team was fourteen to sixteen hours. He laughed and said he wanted us out of the office by noon.

Final Launch Preparations

Twelve hours prior to launch, the pad was cleared. A launch perimeter extending out a mile from the center of the launchpad (called the "launch danger area") was cleared as well. There was a difference between the actual time and official countdown clock due to planned holds. These were injected for key system checks and go-no-go decision points. The countdown entered a two-hour hold at the T minus six hour mark to allow time for the MMT to review the weather and vehicle system status before giving the "go for tanking" recommendation. At about eight hours prior to flight, the external tank fueling began, and the countdown resumed at the T minus six hour mark. This also marked the time when the crew was awakened, fed breakfast, and given their final medical exams before suiting up for the flight. It took about two hours to fill the external tank and ensure the ECO sensors indicated "wet." Six and a half hours prior to launch, the countdown entered a two-and-a-half-hour hold at the T minus three hour mark. During this hold the external tank was placed in a stable replenish mode to maintain fuel levels as some of the cryogenic propellant boiled off. The final inspection team was sent to the launchpad for the walk-down inspection. Armando Oliu, a mountain of a man who looked like a defensive end in the National Football League, led the team. He was the person on the STS-107 postlaunch assessment team who had reported the foam strike after analyzing the seventeen frames of video that captured the bipod release and impact to the port wing. Armando recognized the impact's severity and often wonders today whether he did enough to communicate his concerns to the MMT. The video still haunts him because additional imagery requests were denied, and no further action was taken during the fateful mission. His job now was to ensure all launchpad systems were ready and free of debris threats to the vehicle per the NSTS 08303 document. During the final inspection, the ascent team reported to their consoles in the mission control room, and the SEI personnel reported to their stations in the various mission and engineering support rooms. John Muratore supported the launch from KSC as part of the MMT. The rest of the debris team was scattered among KSC, JSC, MSFC, NASA Ames, and contractor sites at Michoud, Houston, and Huntington Beach, California.

Debris Team Launch Support

During missions the overall debris team was made up of other various teams. This included Integrated Imagery (I^2), NASA debris radar (NDR), orbiter damage assessment, ET thermal protection system, and debris transport analysis. The debris team's goal prior to launch was to determine if there was an NSTS 08303 threat from ice or ice-ball debris. If there was, the debris team would assess the situation and give a recommendation for either a waiver and go for launch or a launch scrub. During the mission the team's goal was to determine whether the orbiter had sustained a debris impact. If so they had to determine whether it was safe for reentry or whether a repair or rescue mission was warranted. Imagery and other debris sensor sources were critical for mission support and performed real-time debris assessment by providing the following critical information: identification of debris sources during ascent or liftoff, determination of orbiter impacts and damage characteristics, comparison of debris data with aerodynamically sensitive transport time (ASTT) to assess impact threats, assessment of the vehicle debris environment against predictions, correlation with other debris-tracking assets, and determination of a need for on-orbit, focused inspections. All reports and findings collected during the detection and inspection process were provided to the DAT. They analyzed the data, ran damage models, and requested additional information such as detailed inspections (if necessary). The DAT was responsible for informing the orbiter project manager whether damage was minor and acceptable for nominal reentry, a repair was required (and what type of repair), or the damage was significant enough that a rescue mission be launched. Imagery was also critical for postflight assessment and was used for IIFA determination, debris reconstructions, postflight debris performance evaluation, and debris model validation.

Debris Imagery Assets

Numerous ground cameras were utilized to track the performance of the launch vehicle during prelaunch, liftoff, ascent, and landing. Ground imagery data provided an indication of vehicle health and served to focus attention for on-orbit inspections. The enhanced launch vehicle imaging system (ELVIS) was a suite of cameras primarily mounted to the outside of the launch vehicle, but it also included some crew handheld components. ELVIS was an RTF upgrade intended to provide imagery for observing general launch vehicle and orbiter conditions, specific launch vehicle debris performance, and debris tracking and characterization during liftoff and ascent.

The separation camera was designed to capture launch vehicle performance and the associated debris environment primarily at the

RETURN TO FLIGHT

SRB and ET separation time frames. After MECO and ET separation, the orbiter performed a pitch maneuver if there was sufficient lighting. This allowed the crew to obtain still and video imagery of the ET from the overhead windows. Out of all the imagery, the crew handheld pictures proved the most valuable for pinpointing locations of foam debris releases on the tank. On-orbit sensor systems were primarily used to determine the integrity of the orbiter TPS prior to clearing the vehicle for reentry. Airborne assets such as the WB-57 were used at the discretion of shuttle program managers in support of ascent or entry imagery operations. The WB-57 (eventually phased out) used its WB-57 ascent video experiment (WAVE), cast glance (CG), and high-altitude observatory (HALO) remote-sensing assets to secure high-altitude imagery data outside the range of the ground cameras and closer to the shuttle flight path.

In orbit a rendezvous pitch maneuver (RPM) was a 360-degree maneuver performed to capture orbiter imagery data from the ISS prior to docking operations. It was used to determine areas of debris impact and damage. Detailed on-orbit imagery inspection of the orbiter TPS was largely done using the shuttle remote manipulator system and the orbiter boom sensor system (OBSS). Other sensors included: payload bay cameras, ISS cameras, extravehicular mobility unit wireless video system cameras, crew verbal reports, vehicle telemetry, and non-NASA assets. All reports and findings collected during the detection and inspection process were provided to the debris team and orbiter damage assessment team. The DAT was ultimately responsible for recommending to the orbiter project manager whether the vehicle was safe for reentry.

Integrated Imagery Team

The JSC Integrated Imagery (I^2) team was responsible for reviewing liftoff, ascent, on-orbit, and landing imagery as well as screening and analyzing all on-orbit imagery. From the mission evaluation room, I^2 supported the mission beginning at L minus two hours through the end. I² was responsible for consolidating and posting all imagery, radar, and wing leading edge data into the launch and landing imagery management system database. All the on-orbit inspection imagery and observations were documented in the TPS inspection imagery management system database. Typically there were several hundred and sometimes more than a thousand debris sightings every mission. The vast majority of the debris either missed the orbiter or was not considered a threat due to the release time, liberation location, or small size (e.g., popcorn foam). To narrow the total sightings to debris that could pose a threat, a smaller database called the NASA imagery reporting database (NIRD) was created. This database contained all the "reportable" debris sources that struck the orbiter, were large and within ASTT, exceeded a risk assessment mass size, or were new or unusual. Typically three to four dozen entries per mission were considered candidates for the integrated issues tracking matrix that was reported to the MMT each day. The JSC integrated imagery team was also responsible for all flight day products reported during the daily MMT meetings. The first four days were the most intense. Flight day one products included data from the digital ground cameras and videos, external tank oxygen feed line video, umbilical well digital stills, crew handheld video and digital stills, wing leading edge sensor data, and the radar quick look report. Flight day two products included ground camera film drops, wing leading edge sensor report, OBSS video and still images, window four minicam, and crew handheld digital stills. If space station rendezvous was conducted on flight day three, then eight hundred millimeter and four hundred millimeter revolutions per minute images were available. If not they would be available on flight day four along with the radar final report and the SRB ELVIS videos.

KSC was the lead for screening and analyzing prelaunch, liftoff, launchpad, and landing imagery. A subset of the KSC imagery team also executed pad walkdowns and final inspections on launch day. The KSC imagery team had dozens of people and representation from all the element projects, SEI, Office of Safety and Mission Assurance, ground operations, and the KSC visual services team. MSFC had an imagery team responsible for reviewing liftoff and ascent imagery as well, and it was the lead for screening and analyzing propulsion system performance. This team supplemented the JSC imagery databases with its own photographic analysis database, which contained historical debris losses from all the shuttle flights. MSFC personnel used the data contained in this database to assess foam debris mass loss and develop all the flight history foam debris distributions used in the PRA. The DAT was responsible for reviewing all on-orbit inspection data and imagery data showing ascent debris or other events related to or potentially affecting the orbiter. Justin typically led the DAT. It was made up mostly of Boeing thermal protection system engineers responsible for providing assessments on NIRD-reportable events that had or potentially had impacted the orbiter. The DAT was expected to provide closure rationale. This included on-orbit and postlanding evidence for the presence or absence of orbiter damage for all applicable NIRD-reportable events.

The foam observed loss database (FOLD) was an independent and historical ET foam loss database maintained by MSFC engineering and supported by MSFC's imagery team. It contained ET foam debris dimensions and masses as determined by the ET postflight assessment team. The mass estimates and debris dimensions for every mission were presented to the ET Chief Engineer Review Board to support the ET project's postflight assessment report. NASA also utilized radar assets to detect debris releases. NASA's debris radar (NDR) team was a cooperative effort with the US Navy. It was responsible for managing the ground- and sea-based radars used to track launch vehicle performance and help characterize the ascent debris environment.

The NDR team was responsible for providing I^2 with the flight day one and final written radar reports. The orbiter project also had a wing leading edge integrated debris sensor (WLEIDS) team responsible for managing the system used to detect RCC panel impacts or other structural g-loads along the forward wing spars. The accelerometer-type sensors were used during ascent, and a subset of these were used on-orbit for indications of potential micrometeoroid and orbital debris impacts. During prelaunch the Mission Operations Directorate (MOD) was responsible for providing lighting assessments for launch and ET separation. On-orbit MOD was responsible for downlinking all the crew's handheld stills and video images and the planning, execution, and downlinking of all the on-orbit inspection imagery. Finally SRB engineering was responsible for downloading and processing the SRB camera imagery following SRB retrieval. All these teams (along with JSC and MSFC engineering, USA–Boeing, Aerospace Corporation, and Lockheed) supported debris integration, which was Bob Ess and me. Debris integration synthesized all the data and provided debris-related reports to John Muratore and the flight control team on launch day and to the MMT during the mission.

Final Countdown Sequence

About four hours prior to launch, the countdown resumed at the T minus three hour mark. The crew departed the operations and checkout building for the launchpad, and the MMT reported on-console. Three hours prior to launch, the crew arrived at the pad to begin the boarding process. This took about an hour. After the crew loaded, the hatch was closed and latched for launch. About an hour before launch, the countdown entered a ten-minute hold at the T minus twenty minute mark. During this hold the firing room computer programs were verified, the abort landing sites were checked, inertial measurement unit alignment was verified, launch director weather briefings were given, and preparations were made for computer software transition to terminal count. This was also the point when the final inspection team provided its briefing and summary of any launch commit violations. The countdown resumed at the T minus twenty minute mark and entered an approximate forty-five-minute hold at the T minus nine minute mark. Since this was the final hold in the countdown, the flight controller and support teams completed all their final system checks, and the MMT conducted the final go-no-go poll for launch. When the countdown resumed at the T minus nine minute mark, there were nine minutes left to conduct dozens of steps in the launch procedure before engine ignition. Some major steps included retracting the orbiter access arm, starting the orbiter's auxiliary power units, pressurizing the external tank, retracting the vent hood, and switching to internal power. At T minus thirty-one seconds, the orbiter's onboard computers took control of the countdown, and at T minus 6.6 seconds, the SSMEs were started, and SRB ignition and liftoff followed.

Nothing matches the excitement of an impending flight. It had taken such a herculean effort to get STS-114 to this point that the excitement was even more heightened. Even though it was early morning when I drove through the Johnson Space Center main gates, the press trucks, reporters, and a few die-hard spectators were already forming a small crowd. I made my way through the security guard checkpoint and security doors and up the three flight of stairs to the SEI support room. I always took the stairs because I would sometimes forget which elevator button to push (Two A or Two B) to get to the desired floor.

Another ECO Sensor Failure

The launch day excitement for the first RTF attempt was short-lived because a launch scrub had been called before I arrived on-console. This was due to another ECO sensor failure. Most members of the debris team were scheduled to report to their stations five hours before launch—about three hours after the final weather briefing and the start of tank fueling. The ECO sensor failure perplexed John and the rest of the launch team. They thought this issue had been resolved. The last tanking test had been flawless, and the technicians had performed the same level of inspection and troubleshooting on the replacement tank as they had on ET-120. Although the launch was scrubbed, further troubleshooting was performed before ordering propellant detanking. During the ensuing MMT meeting, the status of the ECO sensors was discussed. John believed the additional troubleshooting had pinpointed the problem with a point sensor box in the orbiter. That was a fancy way of describing a junction box that served as an interface between the external tank sensors and the orbiter flight computers.

John requested a few more days to evaluate all the ECO sensor data to validate his hypothesis, and technicians at the Cape made preparations to replace the faulty component. A new launch date of July 26 was set. (At least NASA no longer had to worry about the Hurricane Emily threat or the Saharan sandstorm).

The next launch attempt had its share of debris issues. This included a mysterious grease ball that appeared on the tank, bird droppings on one of the left wing RCC panels, a six-inch gecko that perched on the tank surface, and more cracked foam. The grease was wiped clean, the bird droppings were left as is, and the foam cracks were deemed a go for launch. It was assumed the gecko would shake loose at liftoff, but it eventually disappeared. The last five minutes on the communication loops were unusually quiet. It seemed surreal as the countdown clock hit zero and the vehicle blasted off the pad. Just as it was clearing the launch tower, one of the many ground cameras recorded footage of a large turkey buzzard colliding with the top of the tank. The fate of the bird, how much damage resulted from the strike, and if it would cause a foam failure as the aerodynamic loads peaked were unclear. It was an ominous beginning to the flight.

CHAPTER 15

NASA Dodges a Bullet

Postlaunch Meeting with SEI at Johnson Space Center

"We just dodged a bullet. No. I take that back. We just dodged a cannonball," roared John Muratore to his SEI mission support team gathered for the first RTF postlaunch meeting on July 26, 2005.

John was more animated than usual as he paced around Conference Room 272 located in the fabled Building 30, MCC, at Johnson Space Center. Building 30 was home of the flight controllers and the nerve center for human spaceflight since the *Gemini IV* mission in 1965. Small beads of sweat started to form on John's forehead and his red face grew strained as he searched for answers. Three hours earlier *Discovery* (STS-114), the first space shuttle mission since the *Columbia* disaster, had launched at 10:39 a.m. (EDT) from Pad 39B located at the Kennedy Space Center. It was on its way to the International Space Station for an assembly mission. It had taken over two tumultuous years to resume shuttle operations since the impact of a large piece of foam compromised the left wing of *Columbia* and caused it to disintegrate upon reentry over Texas. Several new and modified cameras known as ELVIS were flying on *Discovery*'s Return to Flight mission. At 129 seconds into flight, an ELVIS camera peering aft on the vehicle observed an enormous piece of foam release from somewhere on the external tank near the large seventeen-inch oxygen feed line, which

NASA DODGES A BULLET

was used to transport liquid oxygen to the shuttle main engines. The large piece of foam was clearly visible during launch on footage from the newly installed cameras designed specifically to monitor debris events and subsequent collision with the shuttle vehicle—most notably the orbiter heat shield. John had had three hours to think about the debris observation and its consequences as he traveled from KSC, where he supported launch operations from the launch control complex, back to JSC. What was perplexing John the most was what went wrong. *Goddamnit,* he thought, *the debris problems that led to the demise of Columbia and loss of seven crewmembers were supposed to have been solved.*

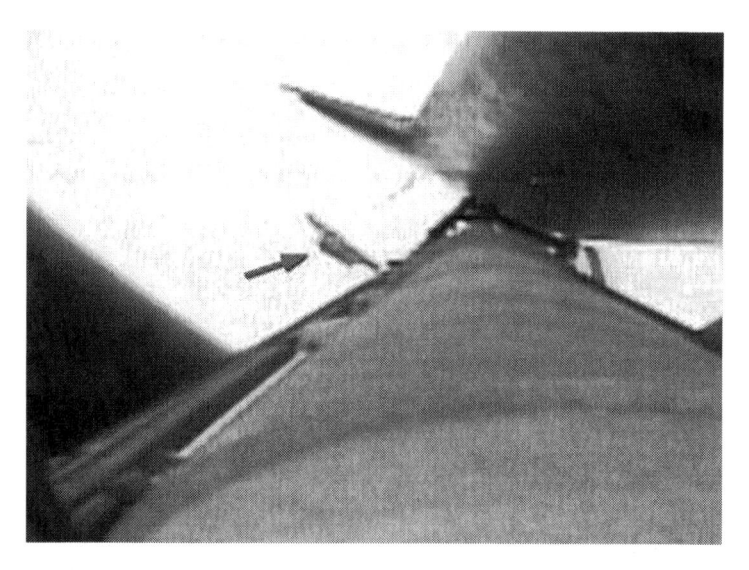

A grainy view of the large hat-shaped piece of foam from the hydrogen tank protuberance air load (PAL) ramp. It was shown releasing at 129 seconds into flight. Although the foam debris did not hit the orbiter, reconstructions later showed the catastrophic risk of this release to be as high as one in five. (Courtesy of NASA.)

It was now 12:30 p.m. local time in Houston, and John had requested a quick SEI team meeting before the daily MMT meeting at 1:00. John had been awake all night starting with the STS-114 weather

RETURN TO FLIGHT

briefing at midnight and the beginning of tanking eight hours before liftoff. He was accustomed to long hours and was not required to report to station until much later in the countdown, but he was not going to miss a thing—regardless of when he had to report to his post. Fatigue exacerbated by launch day excitement and the large foam debris event was starting to show. It was normal procedure for the SEI team to meet two hours before the daily MMT to review the status of each technical discipline and the health of the vehicle and to prepare a report for the MMT. However, time constraints due to the launch and travel time from KSC allowed just a brief meeting. John was only able to give a verbal report to the MMT that contained all the key shuttle managers and board members.

"How could this have happened?" John offered to an already tense and quiet team.

John was clearly focusing the energy of his question by his stare directly at Bob Ess and me. Bob, a rising star at NASA, finally broke the silence. He summarized the facts around the large foam loss and described the detailed data that would be obtained over the next two days as key imagery was downlinked from the orbiting shuttle back to the MCC. Bob added that the foam appeared to release at 129 seconds into flight from the starboard side of the external tank along the oxygen feed line, traveled outside the camera view, and did not appear to impact the vehicle.

There were three main sections of the external tank: the forward oxygen tank, aft hydrogen tank, and intertank section that served as transitional acreage between the two. The oxygen feed line transported −423°F liquid oxygen from the forward oxygen tank through a seventeen-inch stainless steel pipe that ran along the intertank and entire length of the liquid hydrogen tank to the shuttle main engines. Initial size estimates of the foam loss were eighteen inches long by six inches wide. Depending on the release location, it could vary from one to three inches thick. These dimensions equated to nearly a pound of foam. That was close in weight to the 1.6 pounds of foam shed during *Columbia*'s flight. By now everyone working the mission and the media were aware of the foam loss, and even those outside the debris community recognized the event would have been a significant threat if the foam had hit the orbiter. Terri Murphy, a professional and direct NASA veteran in charge of all the imagery and debris detection assets, spoke up next. Head of the I^2 team, she was the first person John sought out after making his way back to JSC. Terri's team's job was to review the thousands of pictures, miles of video, mountains of radar data, and wing leading edge sensor information and consolidate the data into a debris observation report to be presented to the daily MMT in a special imagery briefing. Terri was tough enough to put John in his place and had done so on several occasions during previous mission simulations. She had worked for John since her days as an intern in mission operations. After a successful career in mission operations, she left NASA because of her daughter's health, and John had hired her back into NASA during RTF to lead the imagery effort. Her team was located in the mission evaluation room (MER). This provided the mission flight controllers the additional engineering support and technical expertise needed for problem resolution and mission execution.

The MER was a large room within mission control. It was about the size of a basketball gymnasium and had small pods of workstations. Teams were organized into technical discipline and major spacecraft subsystems. Located in the back center of the room were the desks and workstations for the MER managers. They integrated the various subsystem teams and provided constant updates on vehicle status. They rotated in eight-hour shifts and provided twenty-four/seven coverage. During missions the MER was a beehive of activity with small pockets of intense activity depending on the mission timeline and on-orbit activities. I loved going into this room because there was always something happening. Since 1965 mission control had been at the helm of America's human spaceflights. The flight controllers and engineering support teams working in Building 30 had been vital to every US-crewed mission since *Gemini IV*. That included the Apollo missions that took humans to the moon and 135 total space shuttle flights since 1981. Revamped in the mid-1990s, the newest generation of control rooms was designed with modern computer workstations similar to those used in many high-tech offices. It was the centerpiece of space operations at JSC.

A number of other debris observations were of concern, but the PAL ramp was clearly the five hundred-pound gorilla in the room. John scanned the room as he stood to leave for the MMT meeting. Ensuring he had made eye contact with everyone before he left, he said the PAL ramp loss could end the shuttle program. The words shocked many in the room—me included. The loss alarmed Helen and her team at MSFC, but they were thankful and relieved no apparent damage had been done. "It raised a general sense of unease in me about the possible threat of foam elsewhere on the tank. In a way it made all foam suspect and in need of additional scrutiny to ensure we wouldn't have another catastrophic loss," added Helen.

It had taken NASA more than two years and a huge effort to resume flight operations, and now the program was at risk of being shut down altogether. Clearly John had higher-level insight about the program status and logically concluded another significant foam failure would not be tolerated.

STS-114 Ascent Debris Performance

Discovery's climb to orbit was extensively documented through a system of new and upgraded ground-based cameras, radar systems, and airborne cameras aboard high-altitude aircraft. The imagery captured of *Discovery*'s launch and additional imagery from laser systems on *Discovery*'s new orbiter boom sensor system laser-scanner and data from sensors embedded in the shuttle's wings helped mission managers determine the health of *Discovery*'s thermal protection system. When *Discovery* neared the International Space Station early on July 28, 2005, Commander Sergei Krikalev and Flight Engineer John Phillips used the space station's digital cameras and high-powered eight hundred millimeter and four hundred millimeter lenses to photograph *Discovery*'s thermal protective tile and key areas around its main and nose landing gear doors. All imagery was downlinked for analysis.

Before docking with the space station, Commander Eileen Collins performed the first rendezvous pitch maneuver about six hundred feet below the station. The motion flipped the shuttle end over end at about one degree per second. This allowed ISS Expedition 11 crewmembers to photograph the underside of *Discovery* and its heat-resistant tiles in detail. Additional time was spent transferring equipment and supplies to the station as well as removing and stowing the same on the Mini Pressurized Logistic Module named *Raffaello* for return to Earth.

John's words about possible cancellation of the Space Shuttle Program were resonating in my mind as the mission progressed. The debris events that eventually unfolded during the course of the STS-114 mission made John's comment of even more concern. Debris data from the multitude of cameras, sensors, and radar were pouring in to the I^2 console, and people packed the back corner of the MER. They loitered around the team and tried to decipher the data and prepare for the twelve- and twenty-four-hour debris reports. Imagery assessment teams located at Kennedy, Marshall, and Johnson Space Centers were all independently assessing the huge amount of imagery data. Although each center reviewed all the data, they each had a unique area of focus. Kennedy focused on the ground cameras, Marshall on the in-flight video, and Johnson on the in-flight photos and video from the WB-57 aircraft that had been deployed from Ellington Field in Houston to record high-altitude in-flight video. The orbiter team fed I^2 impact data from the newly installed wing leading edge sensors. Tony Griffin ("Radar Tony") was responsible for providing radar data from the radar installations that were new to the RTF mission. In less than two years and starting with nothing but a concept, Radar Tony had done a remarkable job building an integrated radar system capable of detecting minute debris releases. John, Bob, and I were nestled around the I^2 console in the middle of the action. We were eager to review the data as they became available. Our biggest concern was determining if any debris had struck and damaged the orbiter. We were also interested in how the vehicle and debris detection assets performed. On flight day two, the orbiter boom sensor system was deployed to inspect the RCC panels on the wings and thermal tile on the orbiter's belly.

NASA-Only Meeting in John's Office

As the day wore on, John became increasingly irate over the performance of the external tank. He was particularly upset with the PAL ramp foam loss and other losses around the bipod—areas that had presumably been fixed. There was also a huge foam loss near a hydrogen tank ice/frost ramp that concerned John. Before the flight he was confident the debris problems on the vehicle had been addressed and were no longer a safety of flight issue. This was not the case, and now his and NASA's reputations were at stake. John was in constant communication with the ET project. He tried to get answers for the failures while directly supporting the mission and working with the debris team to assess the numerous debris issues. It was a twenty-four-hour-a-day operation due to the safety of flight concerns associated with the STS-114 debris events. After the MMT meeting on flight day three, John called a NASA-only meeting in his office. It involved just those civil servants working on debris issues. John closed the door behind him, and each person made his or her way to a seat. A closed-door, NASA-only meeting meant something serious was happening. John was visibly agitated. "The reason I called you all here," he stated, "is because you are the only people I can trust."

In addition to the exhaustion and frustration of the last two years of work, John revealed a deeper set of concerns. Before every launch a tanking meeting was held at KSC, and the mission manager would poll every element project about whether it was go for tanking. At the midnight meeting for STS-114, no issues were raised. The meeting ended with the direction for the KSC operations team to load the tank for flight. After the meeting John noticed the external tank team in a cluster having an animated discussion. When John went over and inquired about the commotion, Ron Wetmore, the Lockheed Martin external tank program manager, handed John a bolt. When John examined it, he saw the bolt was necked down—reduced in diameter near its top. This was an indication it had been severely overloaded. "That doesn't look right," John said. He inquired where the bolt had come from. Ron explained it had come from ET-120 in the VAB. ET-120, the tank originally slated for the STS-114 mission, had been replaced after a tanking test revealed failed ECO sensors. Had the tank not been replaced, it would have been on the pad and ready for flight with the damaged bolt. The bolt was one of several holding the orbiter to the external tank. John asked why this hadn't been a topic in the tanking meeting, and Ron explained they hadn't raised the issue because the bolt was from ET-120 and was not being flown. John asked how the ET program knew the tank on the pad didn't have the same problem, and Ron provided some rationale about ET-122's bolts being from a different lot than those used on ET-120. John didn't buy the story and was livid. He rushed down to the console in the Kennedy control room, and a gigantic fight ensued via the voice communication loops between Wayne Hale and the ET project over support for a go for flight decision. Helen agreed with John's concern over the bolt and that the flight rationale was not right. Right before launch Wayne overruled John, and the flight proceeded.

At the time of the NASA-only meeting in John's office, postflight work had already revealed the ET rationale was flawed. All the attachment bolts were from the same lot. After all the foam issues, the problems with the fuel sensors in the tank, and the bolt, John had clearly lost confidence in the ET project. It was clear John didn't trust the ET project any further than he could throw the tank. As the meeting continued, John gave a summary of things he had discovered over the last two days during his own informal investigation of the PAL ramp. The most alarming revelation was a potential foam failure mode that had not made it to John's SEI board or been brought to his attention. The potential failure mode was kept within the ET project. John vowed to uncover the guilty parties and hold them accountable. SEI would lead the investigation effort for the shuttle program, and he described the role each person would play to determine not only the problem but also the fix. Although not yet official, the shuttle program would stand down until the investigation ran its course and the vehicle was safe to fly again. Everyone was stunned. John was clearly on the warpath and would take no prisoners. John wrapped up the meeting by emphasizing the need to focus on the remaining STS-114 mission and ensure its safe return. A number of debris issues remained that jeopardized the mission and needed to be resolved—soon.

First Formal Debris Report to the Mission Management Team

John's first debris report to the MMT was dismal. Several pieces of debris had struck the orbiter. The most notable was at sixty-six seconds. The nose landing gear door had been struck near the seam between the door and the adjacent tile. If the seals located in the seam were damaged, it could allow high reentry temperatures to damage the structure, landing gear, or front tire. By now most everyone had seen video of or heard about the large loss of foam from the PAL ramp on the liquid hydrogen tank. Video footage also showed two separate foam losses from the port bipod at 138 and 148 seconds. The dynamic pressure was low enough during this time frame that minimal damage was sustained to the orbiter at the impact sites. However, the fact there were any losses from the bipod area was deemed unacceptable. Radar had reported thirty contacts with a few deflections that indicated potential impacts. The wing leading edge sensors detected four possible collision events above their threshold or noise level.

The debris reports only got worse as additional data became available. During launch the debris team had access to all the ground cameras and live feeds from the feed line video camera. The high-resolution ground camera could track the vehicle downrange for sixty to seventy miles and even farther under ideal conditions. However, most eyes were focused squarely on ELVIS—the external tank feed line camera. This provided an excellent view of the ET from which foam or ice debris was most likely to liberate and impact the orbiter's bottom tiles. If any foam or debris releases did occur, this camera was best situated to provide impact location and help determine the extent of damage. ELVIS started recording when the SSME fired, and it continued to operate even after the tank separated just a little over eight and a half minutes into flight. Video images from this camera were the ones most commonly aired on the Internet and television networks covering the launch. Other than John, no one was more disappointed with the foam debris performance and PAL ramp loss than the ET project. "We thought we had good flight rationale coupled with excellent flight history performance of the PAL ramp until STS-114. What we failed to understand was how cryopumping from the adjacent ice/frost ramp could impact the PAL ramp," recalled Jeff Pilet. "Although we told folks that foam loss was possible (and expected), we needed to do a better job of setting expectations because it was apparent many struggled seeing foam loss from the bipod and liquid hydrogen flange areas. From that point we really improved our communication of why foam loss occurred and when it was expected."

On flight day two, the OBSS and RPM footage confirmed the impact observations seen during launch and identified additional lower surface damage not observed during liftoff. The RPM footage also enabled the debris team to identify several protruding gap fillers on the bottom TPS tiles, a damaged blanket near the window on the port side, a protruding insulation blanket on the vertical tail, and a damaged insulation blanket on the starboard orbital maneuvering system pod located on the vehicle's aft end. Gap fillers were thin pieces of coated ceramic material wedged between the thermal tiles to prevent hot gases from penetrating the tile gaps, and they served as a soft protective interface between tiles to prevent tile-on-tile rubbing that could lead to cracking or even dislodgement. Excessive reentry heating from a tripped boundary layer was the primary concern regarding the protruding gap fillers. This was especially true for those located forward on the vehicle.

During reentry the shuttle always rode a smooth "bow wave" of hot gases (plasma) that formed a thick boundary layer over the tile surface. This minimized the extent of aerodynamic heating. If the smooth boundary layer were prematurely disturbed ("tripped"), the surface temperatures could increase substantially and potentially exceed the tile capability. All the protruding gap fillers would have to be assessed to determine if they could trip the boundary layer and thus be a reentry debris threat. Reentry debris threats were not considered during the RTF analysis, but threats from the gap fillers and damaged blankets near the orbiter port window, vertical tail, and orbital maneuvering system pod now faced the team. Bob and I were both up all night reviewing the imagery and working with the debris teams to reconstruct the debris losses observed during launch to characterize the impact conditions (debris type, impact velocity, impact angle, and debris size) for Justin Kerr and his orbiter DAT to evaluate.

Along with the PAL ramp loss, there were other significant debris events that concerned John and the debris team. One of the Tyvek covers on the forward reaction control system engines released much later than designed. Most reaction control system engines were located forward of the orbiter flight deck windows. Therefore, they posed a transport and impact threat if ice were to form or if the Tyvek hit a window with high velocity. They were designed to release during ascent before the vehicle reached 170 miles per hour (roughly fourteen seconds into flight). The other debris release that caught the team's attention was a large foam release from a liquid hydrogen tank ice/frost ramp (LH_2 IFR) at 155 seconds into flight. When the external tank separated from the orbiter after eight and a half minutes into flight, the umbilical well camera and crew handheld images detected an enormous loss far exceeding the maximum size predicted to release. The liquid hydrogen tank ice/frost ramp foam was nearly two and a half times larger than the maximum size predicted. A number of other foam losses were detected on the tank, but all were within expected size and predicted performance. Because the ground cameras, ELVIS, and the ET separation images did not reveal any orbiter impacts from the Tyvek covers or LH_2 IFR foam, these failures would be dealt with postlanding. The reentry debris transport analysis was the highest priority for the debris team. It had taken more than four months to mature the ice and foam debris risk models to the point of making risk predictions. Now the team had fewer than four days to determine whether these new debris sources posed a threat during reentry. John was deeply troubled over the vehicle's debris performance and especially the reentry threat that now loomed over the mission. Until the debris team could get a better handle on the extent of the threat,

John recommended to the MMT that they consider making launch-on-need preparations. Launch-on-need (LON) was essentially the rescue vehicle that was a last resort to return the crew from the ISS in the event the orbiter was deemed unsafe for return due to debris damage or other vehicle failure. John placed Bob and me in charge of assessing the reentry debris threats and made it clear we had every NASA asset at our disposal. We would need it.

Debris Team Call to Action

Justin was concerned about the reentry debris damage threat from both the blankets and the gap filler, but he was even more worried about a thermal boundary layer threat from the protruding gap filler. I recognized a big gap in our existing debris assessment because reentry debris threats had not really been considered. Both these debris sources had been considered unexpected debris. Although orbiter debris was discussed during the season of debris verification reviews and RTF preparations, the program was so focused on fixing all the foam and ice problems that potential orbiter sources did not get much attention. Justin had had so much work to do fixing the foam problems that the thought of taking on assessing all the potential orbiter debris sources was too much. According to Justin, "The blankets and gap fillers could cause extensive and even critical damage to the orbiter—especially the control systems—if the blankets ripped away at the wrong time and impacted these surfaces during reentry. In addition the protruding gap fillers could trip the boundary layer and elevate the surface temperatures dramatically, and this could jeopardize the orbiter's structural integrity."

Bob and I decided to split the tasks. He would focus on the blankets, and I would lead the protruding gap filler effort. We needed debris release, transport, and impact data for these two debris sources, and we needed it fast. Bob and I called the debris team, engineering, DAT, and support contractors together to develop a plan. It felt like scenes out of *Apollo 13* or bygone missions where the ground support teams rallied together to create solutions to difficult and unexpected problems on the fly. Bob proposed we generate debris transport solutions assuming the blankets could release at any time during reentry. He wanted Ray Gomez to determine where the blankets would travel and what would be the impact conditions (velocity, impact angle, and kinetic energy) if the blankets hit the vehicle. He also wanted to run a wind tunnel test to determine if the blankets would release under reentry conditions and have the orbiter team conduct impact testing once the impact locations and conditions were determined from the DTA. The final piece of datum for the orbiter team to determine was whether the exposed structure would be at risk from reentry heating if the blankets did cause impact damage. The plan was much the same for the gap filler. However, conducting a wind tunnel test to determine if gap fillers would release was deemed unfeasible given time constraints. The bigger concern related to the gap filler was to determine whether the boundary layer would be tripped and how high the temperatures would subsequently rise. If the gap fillers did release, the team needed to know if there was transport to the RCC panels and what type of damage would be caused to the tile if the gap filler skipped along the orbiter's belly.

    I suspected the impact results for the forward window blankets would be bad even before the Columbia supercomputer at Ames Research Center completed the analysis. The most likely time for the blankets to release was under the same conditions that produced the greatest impact velocity. It was unclear before the analysis what type of damage the aft blankets and protruding gap fillers could cause. The aft location of the blankets limited the impact area and transport time needed to build a high impact velocity. Shallow impact angles to the tile were expected if the gap fillers released, but the index card shape and flexible material made the aerodynamic characteristics difficult to predict. They could pose a serious threat to the RCC panels if transport to those locations on the wing leading edge were possible. In addition to these debris concerns, the orbiter team and DAT still needed to assess the thermal dynamic and structural integrity of the vehicle from the potential boundary layer trips and exposed surfaces where the blankets were missing or damaged. The Columbia supercomputer was busy crunching away the numbers on flight day three. This gave the team a temporary reprieve from working reentry DTA. However, John also had the debris team involved in the PAL ramp loss investigation that now expanded to cover the large loss from the LH_2 IFR. Columbia completed the DTA runs early in the morning on flight day four, and as expected the forward blanket could hit the OMS pods and vertical tail with enough force to cause severe damage. If the testing showed the forward blanket would release, then the MMT would be forced to perform an extravehicular activity (EVA) to fix the damage or accept the reentry debris risk. The DTA also showed the impact risk was high if the blanket released. Unfortunately an overall risk number could not be determined because there was no window impact model. However, it was clear there were impact conditions that exceeded the window's testing limits. The gap filler DTA showed high impact probability to the tile but no transport to the RCC. The orbiter team had conducted a few impact test runs on tile the day before that showed some substantial damage, but that was under conditions much worse than the DTA predicted for every case except the forward two locations. The MMT was not prepared to make a decision regarding the forward blanket or the gap filler until it had the results of the blanket wind tunnel test being conducted at Ames and the boundary layer thermal analysis. The only bright news for the debris team was the solid rocket booster cameras showed no debris issues, and the team could finally get a little rest.

Gap Filler Removal Recommendation to the MMT

The boundary layer thermal analysis was presented to the MMT on flight day five along with the recommendation to remove the forward two gap fillers. Focused inspections were also ordered to get a closer look at the forward blankets, gap fillers, and several impact damage sites on the tile. The combined threat of reentry debris damage and elevated temperatures from tripping the boundary layer led to the removal recommendation. The problem was determining how and when to conduct an EVA (or space walk) to remove the threatening gap fillers. The ISS robotic arm was too short to reach, and it would be the first time an astronaut would perform an EVA on the orbiter's underside. Other issues included determining which tools would be needed for the removal process, when to perform the task, and how long it would take. Astronauts had trained for the first time to make tile repairs heading into STS-114 but had never trained on this type of task. The EVA team needed to work out the logistics and determine if it was even feasible. Two space station assembly EVAs had already been performed as part of the normal mission timeline, and one scheduled EVA remained. If the gap filler removal task were added to EVA3, and it took too long, another EVA would have to be performed. This would extend the mission timeline and begin to consume the orbiter's resource reserves (oxygen and power) for reentry. If EVA3 was dedicated entirely to the gap filler removal, this would jeopardize completion of the assembly mission and require an operational workaround until the next shuttle mission or additional EVAs conducted by the space station crew. The orbiter's mission time was also quickly becoming constrained due to the added focused inspection time and additional time needed to fix the forward blanket. On flight day six, the EVA team reported they needed more time to simulate the gap filler removal task in the neutral buoyancy lab but were leaning toward adding the removal task at the end of EVA3.

On flight day seven, the EVA team brought forward their recommendation to remove the forward two gap fillers at the end of EVA3. They had rehearsed the task in the neutral buoyancy tank the night before and determined it would take thirty to forty-five minutes for the removal process. If the first portion of EVA3 took longer than expected, the ISS assembly task would be halted in favor of removing the gap fillers. The remaining ISS task (along with a possible blanket repair) could be performed on an additional EVA4 space walk. This would extend the shuttle mission, but ISS assembly had gone so well there was plenty of time to execute an additional space walk before exhausting the shuttle's oxygen and power reserves. The MMT concurred with the approach and asked the EVA team to finalize the procedures and bring back the execution plan the next day. I was feeling good about where we stood on the gap filler problem, but the blanket testing at Ames was a different story.

Wind Tunnel Testing of the Blanket at NASA Ames

To execute a wind tunnel test in such a short period of time would be nothing short of miraculous. Bob had worked directly with the United Space Alliance structures and orbiter teams to build the test article at the Cape out of spare tiles and blankets and fly it cross-country on a NASA T-43 jet to Ames for testing. The ad hoc test team at Ames consisted of Ames and JSC engineering personnel. They were busy preparing the wind tunnel, and I was working directly with Ray Gomez, the meticulous aerospace engineer who had worked on the shuttle program since 1985, to establish the worst-case test conditions. Bob had pulled together the test team and test plan. Both Bob and Ray were now in full triage mode. Despite working around the clock, there was nothing to report to the MMT on flight day seven other than a status report. The one key piece of data the blanket test team needed was the configuration of the damaged blanket. That would come from the focused inspection the next day. Other than that testing would occur the next day if everything went according to plan. This gave the EVA team and MMT time to consider their options. Even though the DTA showed there was no transport to the RCC panels, and the likelihood of tile damage was low due to the shallow impact angle, the MMT decided to conduct an extra EVA to remove the gap filler. This was more to eliminate the possibility of tripping the reentry boundary layer (and thus preventing orbiter exposure to elevated reentry temperatures) than out of concern from a debris threat standpoint. If the gap filler did come loose, it would just skip along the tile surface. The MMT would have to wait another day before Bob and I could determine if repairing the blanket would be part of the EVA as well.

PAL Ramp Failure

In addition to the real-time reentry debris issues being worked, the key question still being asked was why the PAL ramp failed. An initial Aerospace Corporation reconstruction showed the debris risk for the PAL ramp loss could be as high as one in five if the debris had released in the sixty to seventy second time frame. The PAL ramp prevented direct cross-flow air loading on the pressurization line and cable tray support structure, and it had an excellent flight history. Early tanks had super lightweight ablator panels under the foam in high heat areas, but these were eventually phased out. There had been only two observed occurrences of PAL ramp foam loss and no observations of PAL ramp loss since eliminating the ablator panels. In the summer of 2004, the ET project, SEI, Office of Safety and Mission Assurance, MSFC engineering, and the Astronaut Office reviewed all the PAL ramp debris data and decided to fly it as is. They rated it as a "remote catastrophic" risk. The PRA before STS-114 for void delta P losses was determined to be less than one in ten thousand. This was consistent with the risk ranking. Clearly some other failure mode had not been considered, and a meeting was held at MAF in the middle of the mission to start the investigation into the cause and develop a resolution plan to fix the problem. Rod Wallace, the SEI chief engineer, was tasked to lead the meeting and develop an action plan that would minimize the time the shuttle fleet would be grounded. ET-120 had just arrived by barge from KSC, and the PAL ramps were the first foam location the ET project dissected to gain insight into the failure mode.

Focused Inspections on the Orbiter

Later in the evening of flight day seven, astronauts used the robotic arm to conduct focused inspections of the blanket damage and several debris impact sites. A closer look at the blanket damage site revealed more damage than expected. There was a nearly eight-inch tear along the twelve-inch seam that joined one blanket to another. The protruding edge was torn and exposed a small area of the underlying structure. The test team would have to shred the blanket seam even more before placing the test article in the tunnel, and this made it even more likely to fail. Another issue the orbiter team was brooding over in the middle of the night was a concern about the amount of potential tile damage during the EVA. Any type of tile collision with the EVA crewmember or robotic arm could gouge the tiles or even cause them to dislodge. Early in the morning, the Ames test team ran into difficulty with one of the pumps necessary to run the test facility. If the problem could not be resolved quickly, the backup plan of performing a manual "strength" test would be done and compared to the amount of aerodynamic force predicted to act on the blanket. If the pull force was greater than the aerodynamic force, the blanket would hold. Otherwise it would fail and have to be repaired. It was not an ideal comparison but better than nothing.

Delay of the Ames Wind Tunnel Testing

During the flight day eight MMT, many members (including Chairperson Wayne Hale) were growing increasingly anxious over the EVA3 concerns and the blanket test delay. Everything was dependent on the Ames wind tunnel test and whether or not the blanket would tear free. Of large concern to the shuttle management team was the knowledge that if the test showed the blanket would fail, the orbiter would be unsafe for a return home. It would have to be jettisoned from the space station. This would leave *Discovery*'s crew trapped on the station. Wayne Hale and Bill Parsons were particularly concerned the conditions of the test were too strenuous and would cause an unrealistic blanket failure. At that point Bill called John out of the MMT meeting. The frustration of the last two years built to a crescendo as Bill stretched to his full six-foot frame and employed his salty US Marine Corps vocabulary. He told John to go back to the MER, where he had been working with JSC's engineering group, to redefine the test conditions and ensure they were not too extreme. John was also at the height of frustration, and the two managers had a heated exchange. Their deep friendship kept any disagreement within bounds, though, and the two men both calmed down. John headed to the MER, and Bill returned to the MMT meeting. John worked with the engineering team to make the test conditions more realistic. John scrambled back to the meeting, and the revised test conditions met with grudging approval. The MMT also made a decision to proceed with EVA4 preparations being conducted on flight day ten, unless the wind tunnel test showed the blanket would stay put, which would lead to cancellation of the blanket repair. The Ames test team would have less than twenty-four hours to get results. Executing another EVA meant the mission would be extended by at least a day. Consequently the MMT decided to forgo the scheduled transfer of oxygen reserves from the shuttle to the ISS—a normal procedure for a departing orbiter. The amount of oxygen the orbiter could carry drove the amount of time the spacecraft could remain in orbit. Oxygen didn't just keep the astronauts alive. It also helped provide electrical power to the vehicle by combining with hydrogen in one of its three fuel cells.

Emergency MMT Meeting

On flight day ten, an emergency MMT meeting was called to determine whether a blanket repair would be added to the EVA. An urgent decision had to be made to fit within the mission timeline. The Ames test showed that the blankets did not release when exposed to worst-case reentry conditions. "Working the blanket issue on STS-114 proved to be one of my most memorable debris events because there was so much happening during this mission," Justin said. "It was one of the few times in my career that I worked one-on-one with one of the astronauts while determining how to transport the test article from KSC to the Ames wind tunnel facility. [He] had pulled out a drawing of the T-38 cargo pod, and we worked together to make sure the test article would fit into the cargo hold."

During the MMT Bob Ess presented conditional probabilities of multilayer insulation (MLI) blanket impacts during reentry. The release probability and damage tolerance capability were not known.

The good news was that the probability of impact to the OMS pod and tail or rudder was less than 1.5 percent and lower than initially reported the day before. Considering the blanket release mechanism and the structural and impact response, the catastrophic risk associated with this threat was even lower than the impact probability. If the blanket did not release (what the initial wind tunnel test results indicated), there was no debris threat to the orbiter.

Justin presented the orbiter assessment. If the blanket struck the orbiter, the windows and rudder actuator would not be critically damaged. Even though rudder damage was possible, control authority would still be maintained. The rudder would survive impacts by blanket fragments as large as 0.013 pounds (one-quarter the mass of the largest expected fragment). OMS pod impacts were only possible between Mach 1 and 2.6 and they could survive impacts by blanket fragments as large as 0.053 pounds (one-tenth the mass of the largest expected fragment). The orbiter team's recommendation was to return the shuttle with the blanket as is. No damage avoidance actions were recommended. The MMT concurred, and the EVA was limited to removing the gap filler. This reduced both the mission timeline and risk of any incidental damage.

The discussion about potential tile damage during EVA3 continued between the orbiter, DAT, debris, and EVA teams as the EVA3 astronauts floated out into space to begin their work. A large crowd huddled around the EVA station in the MER as EVA3 was being performed. Fortunately removal of the gap fillers on flight day twelve proved an easy task. Astronaut Robinson did nothing more than pluck them free with his hand. The reentry debris threats had been addressed, and it was time for the shuttle crew to say its good-byes to the space station and bring the orbiter home. In the MER today, there is a simple case mounted on the wall. It contains two thermal blankets. The first is the blanket tested at the Ames wind tunnel; the second is the blanket removed from *Discovery* after the flight. The blankets look identical. They serve as silent testimony to the team that executed the test and under incredible pressure found the right test conditions that gave assurance of a safe return.

STS-114 Landing

Even the STS-114 landing had its share of drama. Kennedy Space Center was beset with weather issues starting August 8—the original landing date. Two landing opportunities at Kennedy were waved off on the eighth and two more again on the ninth. Edwards Air Force Base in California was chosen as the preferred landing site following the second wave-off at KSC. This marked the sixth night landing at Edwards and the fiftieth shuttle landing in California. STS-114 completed its thirteen-day mission on August 9, 2005, with the MLI blanket intact. As soon as the shuttle landed, Mike Griffin grounded the fleet indefinitely until the debris issues were resolved. Those who were preoccupied with supporting STS-114 had little time to dwell on the impending uncertainty associated with the grounding. It was clear all the possible foam failures had not been accounted for, and the debris risk on STS-114 had been a lot higher than predicted. Although PRA reconstructions of the large LH_2 IFR loss peaked at about one in fifty, they were not as alarming as the PAL ramp. "I was frankly surprised to learn during the postflight analysis on the PAL ramp that a major foam repair had been performed in this area, yet it had not been reported or discussed at the flight readiness review," Mike said. "I considered this to be a major oversight."

Mike was also concerned that, more than two years after the *Columbia* accident, NASA still didn't appear to have a solid understanding of "root cause"—the underlying physical mechanisms behind the various types of debris generation. By this time Mike had made the decision to instate Bill Gerstenmaier (then the ISS program manager) as associate administrator for the Space Operations Mission Directorate. He asked Gerstenmaier "Gerst" to put together a new team that would pursue the underlying physics of foam loss. This was eventually the group that got us where we needed to be with regard to debris mitigation.

CHAPTER 16

Saving the Shuttle Program

Orbiter Damage Inspection

Postlanding debris inspection of the *Discovery* was conducted at Edwards Air Force Base from August 9 to 11, 2005. The orbiter TPS sustained a total of 176 impacts. Twenty-nine had a dimension of one inch or larger. These numbers were within the shuttle flight experience but greater than the averages of previous flights. The orbiter lower surface sustained 152 total hits. Twenty-one had a dimension of one inch or larger. Approximately twenty-six damage sites were located in the area forward of the right-hand main landing gear. Some of this damage might have been attributed to the foam loss during the ascent phase. The largest damage site measuring 2.25 inches by 0.9 inches and 0.5 inches deep. It was most likely ice/frost ramp debris loss from station Xt-1525. Numerous damage sites around the LH_2 and LO_2 umbilical area represented the largest concentration of hits to the lower surface. Tile damage in this region was typical for every flight and was most likely caused by pieces of umbilical purge barrier (baggie material) flailing in the airstream and contacting tile before being blown downstream.

Foam Failure Investigation

From a debris performance standpoint, the first RTF mission was a disaster. Numerous media reports stated that NASA had failed again, and countless advisory groups such as Stafford–Covey voiced criticisms about NASA's inability to solve the debris problem and exposing astronauts to unnecessary risk. Several foam debris sources were outside the risk assessment contained in the debris hazard report the program accepted prior to flight. The program managers accepted twenty-five integrated in-flight anomalies (IIFAs) for debris alone. This was well above the three to four IIFAs the program historically averaged.

Two days after the mission was over, John, Bob, and I were headed on a plane to New Orleans to meet with ET team representatives at MAF to establish investigation teams to determine the root cause of all foam loss events. The meeting at MAF became the start of the RTF Part Two activities. Also attending the meeting were representatives from NASA headquarters. Most notable was Bryan O'Connor, NASA's outspoken chief of the Office of Safety and Mission Assurance. Bryan was a former US Marine Corps fighter, test pilot, and astronaut who had flown as a pilot on STS-61B and commander on STS-40 missions. He was a grizzled veteran who had taken a variety of safety assignments following the 1986 *Challenger* accident. The discussions over the next two days were tense and heated. NASA had worked nearly two and a half years trying to solve the debris problems only to find out they were not fixed. Although the NASA administrator and shuttle program managers never stated it, everyone knew the program would remain grounded altogether unless the debris issues were resolved. In addition to the PAL ramp, the LH_2 ice/frost ramp, acreage adjacent to the LH_2 ice/frost ramp, and the LH_2 intertank flange were elevated in risk posture from "remote catastrophic" to "infrequent catastrophic." Perhaps the most disappointing event was the bipod foam loss in the redesigned area. This was elevated in risk two levels from "improbable catastrophic" to "infrequent catastrophic." Tyvek covers were also elevated in risk level. The new Tyvek covers that were designed to eliminate tearing and incorporated a parachute pocket to promote release prior to T plus eleven seconds of flight did not release in time. Ground film observations confirmed the late release for two of the covers. One occurred at T plus fourteen seconds and the other at T plus twenty-one. Gap fillers were considered an unexpected debris source prior to the flight but were now classified as expected. The debris team would have to start from scratch to build a release-transport-impact model to assess the gap filler risk.

Fixing the PAL Ramp Problem

John had already appointed Rod Wallace to lead a feasibility study to determine the best course of action for the PAL ramp and the ET project to dissect ET-120 to gather critical defect data for risk assessment. Bryan also assigned Richard Gilbrech, who later became the associate administrator for exploration systems, as the leader of a PAL ramp removal tiger team. It was common for NASA to pursue parallel paths to solve complex problems but not in this case. Richard's team was to report back to headquarters in thirty days with his team's recommendations. John thought this was premature given that the dissection work was just getting started. It would take several weeks just to do the load assessment on the cable tray and pressurization lines if the recommendation was to remove the PAL ramp. Nevertheless the meeting ended with Richard's tiger team assigned the task of bringing forward a PAL ramp recommendation to headquarters by mid-September. John and the ET project still had to complete a PAL ramp feasibility study and resolve all other foam IIFAs. In an attempt to explain some of the foam debris, the ET project introduced three additional failure mechanisms that had not been accounted for in the debris risk assessment for STS-114. These new failure mechanisms included crushed foam on the acreage areas, air load failures on the LH_2 ice/frost ramps, and thermal cracking and delamination to account for the large acreage foam loss adjacent to the LH_2 ice/frost ramp. The ET project had considered these as possible failure modes but dismissed them at their Level-3 Engineering Review Board as unrealistic. John felt blindsided and was furious these failure modes had not been brought forward to his Level-2 Systems Engineering Control Board to give the program an opportunity to account for the additional risk. He made several accusatory remarks and challenged the ethical behavior of the ET project leadership for failing to communicate all possible failure modes to him and the program. Wanda and John, the ET project managers, did well to maintain their composure during Muratore's tirade but definitely took exception to his accusations. Despite the friction both sides recognized there was a lot of work to be done before the shuttle would fly again. It was clear when the meeting ended that RTF II was going to require an even greater effort and longer hours than STS-114.

Hurricane Katrina

On August 28, 2005, a week after the RTF II meeting at MAF, Mother Nature added to the ET project's misery by unleashing Hurricane Katrina on Louisiana and the city of New Orleans. At least 1,836 people lost their lives in the storm and subsequent flooding, making it the deadliest US hurricane since the 1928 Okeechobee hurricane. The most severe loss of life occurred in New Orleans. The city flooded as the levee system failed catastrophically. In many cases this was only hours after the storm had moved inland. Eventually 80 percent of the city and large tracts of neighboring parishes became flooded, and the floodwater lingered for weeks. Hurricane Katrina was extremely difficult on the ET team working at MAF because most had to be evacuated to different parts of the country. Their homes were damaged or destroyed, work on the tanks had to stop, and it took more than a week before communications were reestablished as all the area's cell phone towers had been knocked out. Jeff Pilet rode out the storm in Atlanta and was not able to return home for about two weeks. "While in Atlanta my neighbor managed to call and let me know my house survived, and I started watching video footage of the damage. The extent of damage really hit home when I saw the Twin Span Bridge destroyed. This bridge connected the north shore of Lake Pontchartrain to New Orleans and was what most people employed at MAF used every day traveling to and from work. When I saw that, I wondered about the future of the shuttle program and the external tank project and whether MAF had even survived or how people would get to work. Hurricane Katrina coupled with removing the PAL ramp turned out to be my biggest debris challenge."

Except for a few emergency personnel who were part of a "ride-out" crew, all of MAF and its workforce had been evacuated. The MAF rideout crew did a heroic job of keeping the pumps running and the facility dry. They saved MAF and the shuttle. If the facility had flooded, the ET project and the Space Shuttle Program itself probably would have been finished—or at least substantially delayed. Fortunately for NASA the damage to MAF and the external tank manufacturing facilities was moderate, and there were no storm-related injuries or widespread flooding. Eight external tanks remained protected and sustained no damage. However, the roof of the main building was breached, and debris damaged one of the fuel tanks stored inside. All shifts were initially canceled until the end of September, but on September 16, 2005, NASA announced the repairs were progressing faster than anticipated, and NASA would continue to use Michoud for external tank work. On October 3 the facility officially reopened for essential personnel. Though, some key individuals had returned earlier. It was not until October 31 that the facility reopened to the rest of the workforce. In the meantime a core group of engineers relocated to Huntsville, Alabama, and continued the postflight analysis, investigation, and corrective action plans. Through the dedication of these people, the ET project was able to develop the necessary technical data to show the PAL ramps could be safely eliminated for the next shuttle flight.

Wayne Hale Becomes the New Shuttle Program Manager

A week after Hurricane Katrina, Bill Parsons took over as the center director of Stennis Space Center. This was primarily to provide the leadership needed for the recovery effort at Stennis in Mississippi and for many ET project employees who were taking refuge there until MAF reopened. Bill turned over the reins of the shuttle program to his deputy, Wayne Hale. Wayne's wisdom matched his grandfather-like appearance. (He always wore tiny, round wire-rimmed glasses.) He started working for NASA in 1978 and developed a reputation as a no-nonsense leader who worked his way up the NASA ladder. Between March 1988 and January 2003, Wayne served as a flight director in mission control for forty-one space shuttle missions. Then he moved on to become the deputy chief of the Flight Director Office for Shuttle Operations from 2001 to January 2003. As a flight director in charge of reentry, Wayne gave the go call twenty-eight times. "Every time was the toughest thing I had ever done," proclaimed Wayne. "I was never ever one hundred percent certain. It was always a gray area. Never a sure thing. But the team needed to have confidence that the decision was good. It was almost a requirement to speak the words much bolder than you felt...like it was an easy call. Then you prayed you were right."

Wayne relocated to Kennedy Space Center to become the launch integration manager of the shuttle program on February 1, 2003. He returned back to JSC five months later to be the shuttle program deputy manager. He was proud of putting four kids through college and driving the same beat-up, two-tone blue Chevy Suburban for years. He parked his car in the same spot every day at the farthest edge of the parking lot and thus fended off complaints from others about the crowded parking conditions and long walk into the office. It became rather common for Wayne, a compulsive e-mail writer, to compose lyrical reflections on life at NASA and the interior lives of behind-the-scenes workers. I always enjoyed reading these.

From Wayne's earliest days, he had wanted to work in the space program. "I can remember watching Alan Shepard's launch in 1961,

John Glenn in 1962, and on and on," recalled Wayne. "My parents told me that, at three years old, I was very interested in the Sputnik launch, but I don't remember it." Wayne picked Rice University as his first choice for college because it was close to Johnson Space Center and had ties with NASA. "I always aimed to work at NASA," confirmed Wayne. "I was very fortunate to get a job in the mission control team when the shuttle was starting up because the shuttle, of course, was the only game in town for many years."

As a rookie at JSC, Wayne always believed he would work on the shuttle program for a couple years, build the space station, and move on to setting up bases on the moon and expeditions to Mars. Things did not unfold quite as he would have wished, but he had become the shuttle program manager tasked to keep the shuttle flying and to complete construction of the space station. In the wake of STS-114, Wayne had a lot of challenges heading into the job. He rated debris generation as the highest priority, but there were a myriad of other problems to solve. With over 2 million moving parts, the shuttle was not a simple system. This meant he could not concentrate on one problem to the exclusion of all else. Clearly fixing the PAL ramp was the first order of business. Wayne remembered driving home from the first day as part of the MMT for STS-114. He received a phone call from John Muratore, who was reviewing the downlinked video from the ET camera. John reported a large debris event possibly striking the left wing. "It was the worst day of my life," Wayne said.

Wayne immediately made an illegal U-turn and went straight to the video lab to meet with John and assess the situation. The good news the next day was that the remote manipulator system and boom surveys showed no damage. "Although we knew we had more work ahead of us for the next flight, at that point, we had no idea how much work that would be," stated Wayne.

The next priority for Wayne was dealing with the aftermath of Hurricane Katrina and its effects on the shuttle program. Bill Parsons was from the Mississippi area that was affected even more than New Orleans because the hurricane directly hit there. He was called to lead the NASA recovery efforts at both Michoud and Stennis. Very shortly after this, Bill was named as the center director of Stennis. "What can I say?" said Wayne. "It was a devastating blow to New Orleans and the ET community at Michoud." Many workers were in terrible temporary living conditions, and it was months before the site was really open for work. "I was not surprised when they promoted me to program manager but felt a great weight of responsibility for the future," recalled Wayne.

PAL Ramp Removal Recommendation

On September 15, 2005, Richard Gilbrech released his PAL removal final report. This included a number of recommendations for short-term and long-term actions to minimize the possibility of foam loss from future external tanks.

These recommendations were predicated on the assumption that ongoing efforts by the Space Shuttle Program to identify the root cause of debris liberation (including ongoing tests and analyses) would be taken to fruition. Key short-term recommendations included the following: remove and replace of the entire length of the LO_2 and LH_2 PAL ramps using improved application processes, implement modifications required to prevent cryopumping through the bipod heater wiring, investigate the possibility of venting ice/frost ramp "fingers" to minimize air load losses, and improve ET hardware protection provisions to minimize the potential for collateral hardware damage during processing. The key long-term recommendations included: eliminate the PAL ramp at the earliest possible opportunity coincidently with rigorous aerodynamic tests and analyses, develop hard covers for ice/frost ramps and implement in conjunction with PAL ramp elimination, minimize the number of technicians walking on the tank to the extent possible, and implement a no-touch processing policy. NESC and Dr. Charlie Harris, who led the NESC team during the investigation, agreed. Ralph Roe and Charlie came to the same conclusion the PAL ramps needed to be removed before the next flight. "The PAL ramp loss investigation improved our understanding of the additional failure modes that could lead to foam loss. As a whole the team was too focused on void delta P failure mode prior to STS-114," said Ralph.

With the ET project left recovering from the hurricane, John took a more proactive approach and demanded the PAL ramps be removed before the next flight (STS-121). He knew any foam removed from the tank represented the elimination of that debris risk, and he did not want to rely on the ET project to design a replacement. Given Katrina nearly destroyed the MAF, and its workforce was scattered, it was difficult to determine when the ET team would be able to start. This meant further delaying the next shuttle flight. John was also worried about whether a replacement design would even reduce the huge risk to an acceptable level. After all the ET project had claimed the bipod was fixed, but it still liberated foam debris. Removing the PAL ramp meant the aerodynamic loads on the external tank cable tray and pressurization lines would have to be checked to ensure they did not fail during flight and become an even deadlier debris source. This meant running a wind tunnel test to assess the air loads without the PAL ramps. John was confident SEI could run the test in four to five months and be ready to fly again by early spring. The program would not have to wait for the dissection data results or rely on an ET project redesign. He was able to persuade Wayne that his plan was the safest and quickest way to fly again, and that was the direction the program headed.

The Space Shuttle Program was pressed very hard to fly again as soon as possible—just like after the *Columbia* accident. However, Wayne still wanted to understand why the PAL ramp failed and to know whether there were other debris threats still uncovered. The answer became clear in December 2005 when ET-120 was shipped back to MAF. This was the first time a tank that cryogenic propellant had once filled had ever been returned to the factory. The program managers knew there were problems in the ECO sensor system, but the real focus was on the PAL ramp. Sure enough X-rays showed there were cracks in the PAL ramp that had not been there before. This discovery caused the debris community to go back and look again at all the work that had been done on foam from *Columbia* to STS-114. It was clear the foam-on-foam failure mode was missed, and the different thermal expansion characteristics between two types of foam were the key driver causing cracking. John and the ET project started the process to remove as many of those applications as possible. "We had already kicked off the work to verify that removing the PAL ramp would not cause aerodynamic flutter problems for the pressurization lines on the outside of the hydrogen tank, but this was not a certainty," recalled Wayne. "We were at risk the entire time until those results came back—just in time for the STS-121 flight readiness review."

Wind Tunnel Testing to Certify PAL Ramp Removal

John had placed me in charge of all the wind tunnel testing to allow Bob time to work the integrated in-flight anomaly investigations and closures. The testing had grown in scope substantially and was the critical path item for the RTF II schedule. NASA needed to prove the cable tray and pressurization lines beneath the PAL ramp would sustain the aerodynamic loads after it was removed. At the September 20 SICB, John was anxious to get started and wanted test plans for assessing the cable tray and pressurization line loads and crushed foam brought to the SICB as soon as possible. The normally crowded SICB meeting was sparsely attended due to another approaching hurricane that had formed in the Gulf. Rita (a Category Five hurricane) was heading straight for Houston, and landfall was due on September 24. It was less than a month after Hurricane Katrina. The test plans had to wait another week as the cities of Galveston and Houston heeded the evacuation order. Fortunately for Houston and JSC, Rita turned north at the last minute and caused only minimal wind damage and some localized flooding. JSC recovered quickly, and I quickly organized a test team with NASA engineering, USA, and Boeing. Mike Oelke, a former employee of mine when I worked at Boeing, was assigned as the Boeing test lead. He had a somewhat tarnished reputation, and this made me wonder whether Boeing was trying to retaliate for my switch to NASA the year before. I knew from personal experience that once someone was slammed to the mat at Boeing, it was hard to get up. I was, therefore, eager to give Mike a chance. This turned out to be an excellent decision because Mike more than delivered.

Putting together a major wind tunnel test was a huge effort that required a facility, test article, instrumentation, data acquisition system, and a test team. These multiple entities all had to converge in order to run the test. Any one missing piece could derail a test schedule and escalate cost. Testing of this magnitude usually required at least a year to plan and execute and another three to six months to analyze the data. That was under best-case conditions. Our team had five months, and there was only one available wind tunnel at Glenn Research Center (GRC) in Cleveland, Ohio. This facility could also only fit a small test article. Though not a showstopper, it would require scaling the results. The other alternative was the Arnold Engineering Research Center (AEDC) in Tullahoma, Tennessee. It would hold a full-scale test article, but it would be ten to twelve months before we could use the AEDC facility.

There was a similar dilemma with the crushed foam testing to determine under what aerothermal conditions crushed foam would liberate. John was insistent on flying crushed foam samples on the F-15s at Dryden. This was similar to what was done for determining the aerodynamic lift and drag characteristics of foam in flight. This approach was expensive and had some major limitations in terms of maximum heating and dynamic pressure. Brekke and his team at the Aerospace Corporation proposed an alternative combined-environment test. I thought this would do the job quicker, better, and at a fraction of the cost. The problem was convincing John and the SICB this was a better alternative.

First Meeting with Gerst

Not long after returning from the Hurricane Rita evacuation, John called me into his office for a special meeting with "Gerst". He had just been promoted on August 12 to the space operations mission director

(SOMD)—arguably the number three position at NASA. As mission director Gerst had programmatic oversight for the International Space Station, space shuttle, space communications, and space launch vehicles. "As associate administrator I'm supposed to be in a position at the top level looking over things, but I like to operate a little differently," he once put it. "I like to understand the real detail behind what is going on by talking with the experts."

He did not like big fancy graph presentations but instead preferred the face-to-face discussion to understand from the experts their real concerns. Gerst knew John well and wanted to meet with him to scope out the problem from the trenches and understand his concerns, where he was pushing to solve the debris problems, how the debris PRA model worked and what were the real risks.

The program had learned a great deal on STS-114 from all the new cameras and imagery capability added to the vehicle after the *Columbia* accident. This capability enabled NASA to identify and assess debris sources other than foam. Gerst wanted to get the risk down as low as possible as fast as possible to keep flying safely and continue to implement new fixes to new debris problems.

Gerst started his NASA career at (GRC) in 1977 and later served as the Shuttle–*Mir* Program operations manager and program manager of the International Space Station before making his way to headquarters. He was wearing a navy blue suit and tie that was tailored to his tall, thin physique and matched his well-groomed mustache. Gerst had superb technical insight, and his mild-mannered leadership style made people comfortable around him. He received a bachelor of science in aeronautical engineering from Purdue University in 1977 and a master of science degree in mechanical engineering from the University of Toledo in 1981. In 1997 he completed coursework for a doctorate in dynamics and control with an emphasis in propulsion at Purdue University. "It was a good time to find a job, and many of the companies I was interested in working for had some involvement with NASA," stated Gerst. "It was at this point I realized that if I wanted to be directly involved with research, NASA would be the place to work. Fortunately I was lucky enough to get a job with NASA Lewis and went to work for them in the wind tunnel area. I helped develop the calibration curves for the air data probes used during entry on the space shuttle. I really wanted to do testing and get some hands-on experience similar to what I did at Purdue but to a greater extent. Working for NASA enabled me to see how the theories taught in the classroom were implemented for practical applications."

John and I had a pleasant two-hour conversation with Gerst that covered how risk was calculated, an overview of each element of the PRA model, the debris performance on STS-114, and a summary of the work being done to remove the PAL ramps and correct the other debris problems. He left satisfied we were doing the right things and summed up the meeting by saying, "Fixing the issues and determining debris risk was a tough, tough problem." The debris performance on STS-114 (particularly the large size of the PAL ramp loss) was his biggest debris worry going into the job. STS-114 illustrated the difficulty of characterizing the debris environment and showed that NASA's experts did not understand all possible failure mechanisms. The program had focused on the bipod. It thought void delta P was the risk driver for foam and did not adequately consider cryopumping or other failure modes. It was not until ET-120 was dissected after STS-114 that this and other failure mechanisms were discovered. Gerst made it clear during the meeting that in terms of the Space Shuttle Program, NASA needed to figure out a plan to correct the PAL ramp failure, either by removing the PAL ramp altogether or controlling the foam loss. It was also clear NASA needed to expedite the discussion to remove the PAL ramp and assess any consequences associated with that action. "I felt we had to continue to fly and learn more about debris performance and look for other areas of possible failure—especially for large foam losses such as the LH_2 ice/frost ramp and acreage around this area," Gerst said.

Even after the PAL ramp was removed, we all were still left wondering if there were any other areas on the tank unknowingly susceptible to foam losses.

Testing Facility Selection

I brought forward various test options to John and the SICB on November 7, 2005. John tried unsuccessfully to secure an earlier test date at the AEDC facility through his USAF network and contacts at NASA headquarters. This was the preferred option, but John had already used up all his special favors for previous wind tunnel testing completed for the first RTF effort. I recommended pursuing parallel paths by selecting both options in case we had structural load issues from the subscale test at Glenn. Although this was a more expensive alternative, I argued it was reasonable insurance given the critical nature of testing. If the results at Glenn were acceptable, NASA could cancel the AEDC test with a minimal amount invested. If the Glenn results were unacceptable, we could learn more from the better full-scale test at AEDC. The SICB agreed and designated the test at Glenn as IS-23 and the AEDC test as IS-22. I also presented two options to the SICB for testing crushed foam. The test's purpose was to examine the behavior of external tank-crushed foam in a combined environment and assess the failure mechanism to determine the liberation characteristics. This test would evaluate the combined-environment effect and supplement the ET project's hot gas and thermal vacuum testing. Test results would then be incorporated into the foam debris risk assessment. It was essential to determine the failure mode of crushed foam in order to integrate it into the PRA and generate a risk index. An unknown failure mechanism would force analysts to include crushed foam as a divoting concern, and that would erroneously drive up the foam debris risk. The more expensive F-15B option at Dryden would have been an excellent combined-environment test, but it was limited to Mach two—only a small fraction of the ascent profile. The Aerospace Corporation wind tunnel test was a viable combined-environment test that could produce combined pressure and temperature environmental results in less than three months, and it had the versatility to adjust the testing parameters more easily than the F-15B test. The wind tunnel testing could also be expanded to include acoustics and vibration.

Although I knew John favored the F-15B, I explained to the SICB it was like going to the prom and having to choose between two possible dates. The prettier prom date (the F-15B) wanted to take a limousine. The other date (the wind tunnel) was happy taking the Ford truck. I recommended the Ford, and the SICB concurred with a good laugh. John reluctantly acknowledged the merits of the Aerospace Corporation option over the F-15B but still looked disappointed at the outcome. I was pleased the Aerospace Corporation option was chosen because this gave me the most flexibility to change parameters as results became available.

Wind Tunnel Testing at Glenn Research Center

The first test team meeting at GRC was held November 29, 2005, on a cold and windy day. Being from Ohio I shuddered at the thought of running the wind tunnel test in the middle of winter where the snows from Lake Erie could bury the city in an instant. Getting trapped in Cleveland for days on end due to a blizzard was a real possibility given the high likelihood of testing delays. Other tests were going on that needed my attention as well at AEDC, MSFC, and the Aerospace Corporation. Gary Williamson, a wind tunnel test expert, and Joe Panek, one of the best instrumentation and data acquisition specialists at Glenn, led the test team at Glenn. They eagerly welcomed the test team from Houston and were committed to making the test a success. I briefed the team about debris risk in general and the high risk associated with the PAL ramp loss on STS-114. Since the program had decided to remove the PAL ramp as a prerequisite for the next shuttle launch, timely completion of this test was paramount. The biggest challenge to meeting a February test start was the lead time needed to procure the tiny instrumentation required to measure the sound pressure levels and buffet loads on the critical hardware. We managed to scrounge up a sufficient number of sensors that had been used on previous RTF tests from different NASA centers. We did so with the promise of replenishing the ones we used with a new order. Everyone recognized the urgency of the test and was happy to help. Langley assisted with the test article manufacturing and provided expertise on aerodynamic loading. I enlisted wind tunnel, load, and data acquisition experts from Ames and Goddard. John assured me I had every resource within the agency at my disposal, and I took full advantage of this to build my test team. In two days we had begun building a test plan and schedule that would get us into the tunnel the first week in February 2006.

Wind Tunnel Testing at Arnold Engineering Development Center (AEDC)

The next week the same core test team met in Tullahoma, Tennessee, for the first AEDC test meeting. Reggie Riddle, an easygoing good old boy from Tennessee, was the AEDC test lead. He had been a key factor in the success of the IA-700 RTF testing conducted two years before. However, his team was not very enthusiastic about NASA barging in and busting their test schedule to accommodate another wind tunnel test. During the lunch break, Reggie escorted me to Colonel Skelly's office. He was the base facility manager in charge of the AEDC wind tunnel facility, and he made it perfectly clear that NASA was there as a guest. Things were going to be done his way or else. The November 2006 test date for testing at AEDC had only been discussed as a possibility. He had not yet authorized it. His priority was USAF and DoD testing. If he could fit NASA in the tunnel in November, he would, but he made no promises. I felt more chewed out than welcomed and nearly saluted as I left his office, but I was at least clear where NASA and our test team stood. We built our test schedule around a November test time but knew that date was tentative at best. I made a short stop at MSFC to check on the ice liberation testing. John demanded this continue to get additional data and add to the sample population for the release model distributions. Numerous test article and test facility breakdowns besieged the ice testing, and it had to be stopped periodically due to higher-priority foam debris issues stemming from STS-114 and the lack of MAF personnel after Hurricane Katrina. It was imperative that testing continue in support of an ice debris summit planned for early 2006 to review all test data and reevaluate the ice risk. After the short visit to MSFC, I was off to the Aerospace Corporation facility in El Segundo, California.

Debris Testing at the Aerospace Corporation

I was pleasantly surprised with the progress the Aerospace Corporation team had made getting their facility ready. Randy Williams, the Aerospace Corporation test engineer, showed me the Aerospace Corporation test facility and focused my attention on a small-scale impact gun he had been working on. If I did not know any better, I could have easily mistaken Randy for a beach bum. He was an avid surfer and very knowledgeable and informative about the local surf conditions. Randy showed me a few of the charred foam samples that had been used to calibrate the thermal environment and went on to explain that the only delay to full-scale testing was getting foam samples from the ET project. He further suggested using the Aerospace Corporation impact gun to test special tile samples to define the impact capability (which was currently assumed to be zero). I agreed and contacted Justin Kerr to arrange for used orbiter tiles to be sent from KSC to the Aerospace Corporation. Randy also stated that his team could grow ice-balls and test them in their combined-environment chamber to determine how and under what conditions they would liberate. Developing an ice-ball release model would help provide a better representation of the ice-ball risk and allow more realistic updates to NSTS 08303. Ice-ball tables were based on the worst-case set of release conditions. This limited the allowable size and could result in a higher probability of scrubbing a launch if an ice-ball formed. The ice-ball that formed on the first tanking test during STS-114 would have led to a launch scrub had it been a real countdown. I agreed testing should start immediately so Randy and his test team could have preliminary ice-ball data to share during the upcoming January ice debris summit.

CHAPTER 17

John Muratore Is Replaced

Return to Flight Part II in Full Swing

John pushed all the debris teams hard over the Christmas holidays and gave them just a few days off to be with their families. He was pressing to get the wind tunnel testing at Glenn started and finish the dissection of ET-120. Everyone knew it would be back to the grind after the first of the year because a new launch date for STS-121 had tentatively been set for May 2006. That meant all wind tunnel testing had to be completed with desirable results, all debris issues from STS-114 had to be resolved satisfactorily, all integrated in-flight anomalies had to be closed, the integrated debris hazard report had to be updated, and another debris verification review had to be scheduled by thirty days before the launch. With so much open work and unresolved debris issues, many felt the launch schedule was too aggressive and would possibly cause engineers to overlook safety concerns to meet the launch schedule. However, John knew from experience that giving engineers an indefinite amount of time and an infinite amount of money would accomplish nothing without an end date. To keep the space shuttle operational, he wanted a finite time.

I was traveling to Cleveland, Tullahoma, Huntsville, and El Segundo for most of January, and when I returned to Houston, Bob Ess called me into his office and announced he was leaving the shuttle program for the new Constellation program. Bob was joining the Constellation Program as the flight test manager and would be turning over all the debris responsibilities to me. The news did not surprise me too much, but the additional duties felt a little overwhelming. And things were about to get worse.

Foam Dissection Data Meeting, MSFC, Huntsville, Alabama

The ET project had completed their dissection of ET-120 and were ready to share the data at MSFC on February 16, 2006. Results were astounding—especially for the liquid hydrogen ice/frost ramps (IFRs) and the adjacent acreage. The LO_2 IFRs' risk enveloped the LH_2 IFRs' risk prior to STS-114. That was 1 in 420 for void delta P failures. Until then there was no debris cloud available or physics-based model applied to the LH_2 IFRs because they were located aft of the LO_2 IFRs and were assumed to have a lower risk. However, dissection results showed an increased defect count and defect size in the LH_2 IFRs compared to the LO_2 IFRs. This would lead to an increased risk on both counts. Debris transport also showed higher LH_2 IFR risk due to high impact angles and transport to the thinnest orbiter tiles. The differences in defect size and count were attributed to the new bladder mold process used on STS-114. This replaced the baggie process used for the LO_2 IFRs. During the tank dissection one of the technicians stated the foam felt "funny" around one of the forward LH_2 IFRs when a large foam chunk fell off. The ET project was struggling to explain the foam loss and could only describe it as "unknown." This had huge risk implications because now the risk analysis had to assume the debris could liberate anytime during ascent and could release a large mass of debris. Even without performing the analysis, everyone knew the risk was going to be unacceptable. "Initially I thought we were in big trouble," recalled Jeff Pilet. "I didn't see any way that the debris community would ever get comfortable with large cracks under such a large piece of foam."

With the PAL ramp removal in process, the LH_2 IFRs and adjacent acreage were easily the highest debris risk on the program. John was flabbergasted as the dissection results further solidified his mistrust of the ET project. The ET project had been working on some preliminary LH_2 IFR design modifications to mitigate the acreage losses and reduce the void delta P losses from the IFR body. A smaller, lower-profile design (known as the bread loaf) used the baggie mold process and was the leading candidate.

Less foam also meant a higher likelihood of ice formation, and this potential ice hazard had yet to be checked. I did not want to trade a lower foam risk for a higher ice risk and still have the same (or higher) overall debris risk. During a break John instructed me to have the Aerospace Corporation run the risk numbers on the LH_2 IFRs using the new data. We could compute a conditional probability for the adjacent acreage using the largest historical mass loss. However, we would have to wait to perform the LH_2 IFR risk analysis until the ET project provided a debris cloud or until the Aerospace Corporation applied their linear elastic fracture mechanics model for the LH_2 IFR. John wanted the PRA calculations to take top priority, and he wanted to incorporate the new bread loaf ice/frost ramp into the Glenn testing that was due to start in two weeks. The dissection meeting ended with a number of action items aimed mostly at the ET project. The ET team's first priority was to look for an alternative LH_2 IFR design that had a lower risk for both foam and ice. The ET project also needed to finish the work on the LH_2 IFR debris cloud and determine the maximum size that could liberate from the adjacent acreage. This data needed to be forwarded to me and the Aerospace Corporation as soon as it was available so we could perform the risk assessment of these debris sources and determine the extent of potential trouble. All LH_2 ice/frost ramp PRA results, Glenn testing results, Aerospace Corporation's crushed foam testing, gap filler analysis, Tyvek cover redesign, and bipod wire redesign would be discussed at a debris technical exchange meeting to be held at JSC in mid-April. After the technical exchange meeting, an LH_2 IFR design review would be scheduled at MAF to evaluate the different redesign options.

Foam Debris Testing at Aerospace Corporation, El Segundo, California

The day after the dissection data meeting, I met with Brekke, Matt Eby, and Randy at the Aerospace Corporation in California for a readiness review of the crushed foam testing and to have them start the LH_2 IFR risk calculations. We had to begin fabricating the new LH_2 IFRs immediately and would organize the test matrix to integrate them after the test article was modified. It was a compressed schedule, but I assured John Muratore I could get it done. On the flight from Huntsville to Los Angeles, I had some time to reflect on where the program stood on closing the integrated IFAs from STS-114 and assessing the debris risk to the shuttle. It had been over six months since STS-114, Bob was off to Constellation, and it seemed each day brought new debris discoveries and challenges that only added to the workload. The debris problem seemed to be diverging instead of converging toward closure. All the additional imagery and debris-tracking assets used on STS-114 had uncovered far more problems than expected, and these indicated the shuttle was flying with a much higher debris risk than had been predicted. The first-generation foam PRA model had to be upgraded to account for seven failure modes instead of four and had to be expanded to cover all thirty separate foam debris sources identified on the external tank. Only five foam debris sources on STS-114 had been analyzed using the PRA model, and only the general acreage areas on the tank cleared deterministically. All other foam debris sources were enveloped and would have to be reassessed.

The Aerospace Corporation team had no prior experience with shuttle operations, and all its focus was on using the ground test data from NASA and the ET project to run its model. After STS-114 flew, it opened their eyes to the broader foam debris picture—one that was told by twenty-five years of prior in-flight imagery. Both ground and tank-mounted cameras provided a wealth of data on how much foam was lost, from where, and how frequently. The NASA anomaly team studying the STS-114 losses meticulously examined all the prior flight imagery and identified threats from many new foam debris sources.

Even though many losses had been occurring for years, they were new to the engineering community. One of the limitations associated with reviewing separation photographs of the tank was the absence of the release time. Unless the debris release was captured on video before SRB separation, the photos only recorded the absence of foam. The release timing uncertainty had to be accounted for leading into the STS-121 mission. This was especially true for the newly identified debris source at the LH_2 ice/frost ramps. Program managers now recognized large pieces of foam were regularly lost there. Many within the debris community studying these losses thought they were caused by a cryopumping mechanism. In that case the debris risk was very small—on the order of one in several thousand. This was because the debris released well outside ASTT. However, in the absence of concrete evidence that the losses occur late, the risk analysis had to consider a worst-case release time. This ultimately showed a peak risk of one in seventy. It was only logical for the shuttle program to start work on redesigning the LH_2 ice/frost ramp to eliminate or reduce those losses.

There was a large contingent from the ET project who followed me from Huntsville and others from JSC who joined the crushed foam test readiness review at the Aerospace Corporation. The warm weather and palm trees were a welcome relief from the wintry conditions in Huntsville and Cleveland. Other than a few minor exceptions that were addressed, everyone felt comfortable with the crushed foam test facility and test matrix that Brekke's team had designed. Initial test runs showed only minor popcorning, and this tended to cease once the foam glazed over with char from the 600°F temperature. It did not take long to give the go-ahead to the test team.

The ice-ball discussion in the afternoon was a different story. Randy displayed a few ice-ball samples he had grown in the Aerospace Corporation test facility and explained how he intended to use their combined-environment test facility to characterize ice-ball liberation after the crushed foam testing was done. The ET project had done all the ice testing up to this point and resisted nearly every aspect of the Aerospace Corporation approach. They were not happy with how the ice was grown, even though it was based on a recipe they had supplied. They were not happy with how the test was being conducted without vibroacoustic effects, even though those were going to be added later. One of the major discussions centered on the classification of "hard" ice and "frost balls" and where to draw the line between the two extremes. It was agreed that anything transparent in appearance was hard ice, and anything that appeared white was a frost ball of much lower density and mass. The difficulty lay in classifying ice formations that were "cloudy" or had a hard outer shell.

The age and condition of the foam was another matter of dispute. The ET project insisted the foam be aged in the sun at KSC. This was on the belief that the solar effects in Florida were different than those in Southern California. I joked one could get a tan just as easily on a California beach as on a Florida beach and noted that sending foam samples directly to El Segundo would save not only time but also logistical costs. No one laughed, and the negative response surprised me. However, I also was determined to push forward with the testing. My suggestion to treat the Aerospace Corporation iceball testing as a "development test" rather than a certification test where the results would be used directly to feed the PRA models appeased the critics to a certain extent. I was going to test every combination of ice formation, and if results from the ice-ball test uncovered something of interest, it could always be upgraded to a certification-level test. Previous analysis assumed the entire ice-ball mass liberated, and this drove down the acceptable diameter size in NSTS 08303. I hypothesized that regardless of ice morphology, multiple small pieces would flake off, and some ice would remain. If this were the case, the acceptable diameter size would nearly double. My logic seemed reasonable, but the ET project almost unilaterally rejected it. Although I was going to have the foam sample panels sent to California, it was going to take at least three months to get them due to higher priority needs. I left the meeting early to catch a plane to Cleveland for the start of the IS-23 wind tunnel test at Glenn with several members from the ET project still grumbling over the ice-ball testing.

RETURN TO FLIGHT

PAL Ramp Removal Wind Tunnel Testing, GRC, Cleveland

A blanket of deep snow covered Cleveland when I arrived, and the swirling wind was making a very good attempt at blowing me over on the icy surfaces. These were the very conditions I was worried would delay the testing, but when I arrived at Glenn, I found most of the test team working at the facility. It would take only a few hours to perform the first set of test runs once the test was ready to go. I had the team automate the data-reduction process from start to finish to allow immediate comparison of the test results with pretest conditions. In real-time the raw data from the test article was collected, corrected, scaled, and plotted against the pretest conditions for each Mach number and time increment. This was a huge effort that required testing sample data sets to verify the process, but it eliminated the need for lengthy analysis after the data were collected and testing secured. If the test results were below the pretest predictions, the cable tray and pressurization lines could handle the aerodynamic loading that would result from removal of the PAL ramp. This was what everyone was expecting. However, if the results were above the pretest conditions, then there would be doubt about whether the exposed hardware could survive the aerodynamic loading.

As with most complicated wind tunnel testing, there were delays. In this case there was a fire in the wind tunnel electrical control panel, broken diffusion pumps, and glitches in the data-acquisition system. The initial test runs also had to be performed at night when the center's electricity demand was lowest to ensure the wind tunnel power demand did not exceed the capability of the GRC's electrical distribution system. Day testing could take place only after the facility power demands were determined to be within the system's capability. Testing finally started the second week of March after the test team worked its way through all the problems. I eagerly waited for the first set of test results and slumped in my chair when the loads for most Mach numbers were well above the pretest conditions. We redid the calibration curves and reran the raw data sets but found no issues.

The second test run (a repeat of the first) produced the same results. This was not good and meant removing the PAL ramp to eliminate a high-risk debris source could lead to a structural failure of the cable tray and pressurization lines. I informed John of the news, instructed the test team to continue with the testing, and caught the first plane out the next morning. I carried with me results from the test runs. By the time I got to the office, John had already pulled together a tiger team of NASA aerodynamic load experts from throughout the agency, USA, and Boeing. Knowing the team needed a couple weeks to work the loads issue, John called Wayne and had the launch date moved from May to June.

Ice Debris Technical Interchange Meeting, Houston, Texas

Two weeks before the upcoming dissection review meeting at MSFC, I held an ice technical exchange meeting to review the progress on updating the ice risk. The orbiter team had been working hard to develop an ice-on-tile damage map that was more representative of ice impacts and would manifest a much lower ice risk. This and the Aerospace Corporation ice-ball testing would expand the allowable ice-ball size in NSTS 08303 and increase the launch probability if an ice-ball formed. Additional ice liberation testing measured the ice-to-foam bond strength, which was greater than the foam fracture toughness, and assessed the liberation characteristics for what was considered a "normal" ice day. The normal ice day analysis showed a much lower ice risk than the worst-case conditions used for STS-114. Despite all the great work done on ice-balls and bracket and bellows ice, we still relied on flight history and engineering judgment for the umbilical ice risk. I wanted some umbilical test data and proposed an umbilical ice test at Eglin Air Force Base to determine how much ice would liberate at liftoff from the vibroacoustic loads. Since the ice-to-foam bond was so strong, I felt that if most ice liberated during liftoff, then the remaining ice would not become a debris source due to the strong bond. If the test results confirmed this logic, I would propose a more rigorous flight rationale compared to relying on just flight history. However, this testing was low on the priority list and would not be ready to perform for another six months. That would put it in the fall—after the next two planned flights.

Orbiter Gap Filler Debris

There was a lot of argument about whether gap fillers (thin inserts placed between the tiles to insulate the tile gaps and act as buffers between tiles) should be expected or unexpected debris. Orbiter team members considered the gap filler unexpected debris, but SEI argued it was expected. Either way the risk had to be assessed assuming the gap filler could liberate. Unfortunately no one knew the failure mechanism leading to liberation. All the aerodynamic lift and drag characteristics were unknown, and a gap filler impact damage map did not exist. It was presumed the most likely time a gap filler would liberate was when the vibroacoustic and mechanical articulation loads were highest. That was at max Q and during peak ASTT. The orbiter team had initiated impact testing only to find a wide variation in the amount of damage. It was highly dependent on orientation and impact angle.

Ray Gomez built an aerodynamic model that was also highly variable depending on the orientation in the airflow. The first set of analysis results produced a very high risk of about one in ten. This did not match the historical performance due to the limitations of the release timing and aerodynamic and impact models. Because of the possible threat the orbiter team was required to inspect every gap filler on the vehicle to ensure it would not liberate before launch. However, due to the sheer number, SEI divided the thermal tiles into three different zones from high risk to low risk. Inspection began with the high risk zones (located toward the nose). The degree of risk decreased moving aft on the vehicle due to the shorter transport distance.

Foam Debris Meeting, JSC, Houston, Texas

Bob had closed several IIFAs related to debris before leaving the shuttle program, but nearly a dozen still remained. I could only conclude we needed more time, but John and the program managers were reluctant to grant that. On April 20 a debris technical interchange meeting was held at JSC to review the closure of all the integrated IFAs from STS-114 and the progress made on the LH_2 IFRs. The ET project had tested several new options, but none of the designs showed much promise. The leading candidate experienced significant foam loss during testing from just the air loads and had not yet accounted for the increased ice risk. Due to the increase in defects, the void delta P risk increased from one in four hundred to one in one hundred ten, and the adjacent acreage risk for the unknown failure mode peaked at one in seventy-five around eighty seconds. The adjacent acreage risk was substantially lower (one in five thousand) if cryopumping was deemed the failure mechanism. This was primarily because most releases would occur after ASTT. The ET project insisted the adjacent acreage losses were due to cryopumping, but other than the physics behind the phase change of nitrogen from a solid to a liquid, there was precious little time of release data available to support this failure mode. As the −423°F liquid hydrogen propellant was used up, the level in the tank would steadily decrease in a predictable fashion and allow the upper part of the tank to warm. Once the aluminum surface temperature reached −346°F, the solid nitrogen would undergo a rapid phase change and corresponding pressure increase. If there was a crack in the foam down to the aluminum surface, the condition for cryopumping would exist, and the foam would liberate. The cryopumping model was a little more conservative and assumed the foam would liberate when the substrate temperature rose to −330°F. The model also accounted for the nominal and maximum substrate temperatures provided by time and station location. If this model was accurate, only the first two of fifteen ice/frost ramps could liberate within the ASTT time frame. Releases from any other hydrogen IFR could be the maximum size and still not hurt the orbiter.

The physics supported the cryopumping failure mechanism, but the consequences were unacceptable if something else was causing the failures. Historically the LH_2 IFRs experienced three to four losses per flight from the body and one to two around the adjacent acreage. Over 95 percent of the historical releases, including the acreage loss at station Xt-1851 on STS-114, were aft of the first two locations. Despite the physics behind cryopumping, the historical releases concerned John and the debris community. Overall John was disappointed in the ET project's progress on the LH_2 IFR redesign and opposed to its declaration of cryopumping as the failure mechanism. This and the open structural load issues for the PAL ramp removal were further fueling John's irritation with the ET project. Although the testing at Glenn was finished, the tiger team John had organized to resolve the disparity between the results and pretest predictions was still working the problem. John insisted the LH_2 IFRs were as dangerous as the bipod release on STS-107 and PAL ramp release on STS-114. He thought they should be declared "probable catastrophic" risks, which would mandate the grounding of the shuttle fleet until the risk was lowered. Whether the ET project liked it or not, he was going to bring the issue to Wayne and the Program Control Board the following week to make it official.

LH_2 IFR Risk Summary, Program Control Board, Houston, Texas

I presented a summary of the LH_2 IFR risk at the April 27, 2006, meeting of the Program Control Board. This was the same summary I had presented the previous week at the foam debris meeting. The risk results coupled with flight history, applicable failure mechanisms, and potential defects suggested the LH_2 IFRs and acreage were still prime candidates for risk mitigation. This was my conclusion.

The absence of a LH_2 IFR risk classification would force John to handle the ensuing debate that would surely follow. Instead I used an analogy of swimming in an alligator-infested lake to wrap up my presentation. Although people had swum in this hypothetical lake for years without an attack, they knew there were alligators. The problem was trying to determine when an alligator might attack or if any alligator was big enough to kill a human. The key question for the swimmers ready to enter the water was whether they felt comfortable with taking the risk to swim in the lake. The same was true with the LH_2 IFR debris sources. The shuttle liberated foam from those locations on average three to four times every flight. We knew the alligators existed. Some of the losses, if released at a critical time during ascent, were large enough to cause catastrophic damage to the orbiter. We knew there were killer alligators. Although catastrophic damage from these debris sources had not occurred in the past, the extent of historical damage remained unknown. It was difficult to predict when a large alligator capable of killing might attack. The question remained whether the Shuttle Program Control Board was comfortable with the LH_2 IFR risk.

John was more assertive and recommended the LH_2 IFR risk be elevated to the highest level—"probable catastrophic." He wanted the program to commit to the IFR redesign. The debate became rather heated and the control board split—twelve for and twelve against elevating the risk level. The tiebreaking vote came down to Wayne, and he decided only to pursue the IFR redesign. He wanted the ET project and SEI to work on refining the cryopumping model and tabled the risk elevation decision until the STS-121 debris verification review. He also wanted John to meet him in his office right after the Program Control Board meeting adjourned.

John Announces the News

After meeting with Wayne, John called me into his office later that evening. He looked tired and dejected as he slumped into his chair. His normally high energy level was awkwardly subdued. "Effective today I will be taking a new assignment as the senior systems engineer, and Kim Doering will take over as SEI manager," declared John.

The announcement came as a shock, and all I could muster was a halfhearted congratulations. I knew NASA rarely fired employees. It merely deposed the person and placed him or her on a new assignment. Such was the case with John. Wayne and John had worked together since the early 1980s. Wayne considered John a friend and a genius, but the decision was made. "It was time to make a change," said Wayne, "and he [John] had trained Kim Doering well for the job."

Kim had served in a wide variety of leadership roles at NASA and in the aerospace community for more than two decades before joining the shuttle program as the element integration manager for John Muratore. She began her aerospace career right out of college in June 1984 as an engineer for Rockwell International working on the space shuttle. After working for two years on the program, Kim left for an opportunity in Germany. When she finished that assignment, Kim came back to Houston and worked on the International Space Station program as the deputy manager of the International Partners Office for a number of years. Then she worked as the deputy manager of the Orbital Space Plane Spacecraft Vehicle Engineering Office until it was phased out. After the *Columbia* accident, the JSC director suggested Kim look into supporting the SEI group, and she contacted John. "Our paths had crossed during the X-38 program, and he stated that he would love to have me aboard because we needed good people—specifically in element integration," Kim said. "Looking back on it, the element integration and SEI manager jobs were the best jobs I ever had." When John was removed as the SEI manager, Wayne Hale asked Kim if she would take the job. "I did not think I was well suited for the position because I had only been there a short time, and there were many other well-qualified people to take the position," stated Kim. "Wayne responded by saying that we already had a lot of really smart technical people, but what he really needed was someone to bring them together, and that person was me. He wanted me to build a team that worked across the elements and program, and given that challenge I took the job." Initially Kim was really worried about acceptance into SEI because John and her were so different in terms of management and leadership style. "I was concerned about whether I would be able to manage the team," recalled Kim.

Rod Wallace, a senior team member who had been part of the organization during STS-107, helped make Kim feel comfortable. Kim asked Rod for help because she did not understand all the technical aspects and intricacies of SEI. Rod responded by telling Kim she did not need to worry about the technical aspects. He would commit to her and the organization to support that role, and everyone else would follow his lead. "Rod went on to say that if I allowed the team to do its job, I would be accepted as the SEI manager," Kim said. Although Rod might have been better suited for the SEI manager job, he reassured Kim he would support her all the way. "After our discussion I felt confident from a technical authority point of view because people would listen to Rod," Kim stated.

John had established rigorous processes within SEI and demonstrated a strong leadership role in the shuttle program, but now she would have to rely on the team to help make technical decisions. Under this new management, more people had ownership of the decision-making process. It did not take Kim long to quit worrying about the team's technical abilities. She came to rely on us daily.

CHAPTER 18

Return to Flight Part II

Since John's new position removed him from SEI and the decision-making process, Kim turned over all the debris responsibility to me. Kim knew very little about debris going into the job and worked to learn as much as possible about the topic so she would feel comfortable in making informed decisions regarding debris risk. "Most days my head hurt listening to the debris discussions because it was a very challenging topic to understand—how all the liberation, transport, and impact parts of the debris risk model worked," declared Kim. "I decided I needed to trust what people were telling me about debris before I was comfortable with making decisions regarding debris risk. Over time I became more comfortable with the technical understanding of debris, and it became apparent to me whom to trust."

Setting the STS-121 Launch Date

Kim and I agreed to set the date of the STS-121 debris verification review for the end of May. That gave me only six weeks to prepare. All the scheduling was predicated on resolving the GRC wind tunnel test results and approving the removal of the PAL ramps. It was looking more like the scaling factors used in the load analysis at Glenn were incorrect and leading to overly conservative pretest predictions, but the loads tiger team still had more work to do. I also needed to get closure on the remaining integrated IFAs, finish all the debris PRA calculations, build the DVR presentation, and update the IDBR. There would be time after the DVR to make changes to the hazard report. That was due before the STS-121 flight readiness review two weeks after the DVR. Wayne moved the launch date from mid-June to July 1, 2006, to give Kim, me, and the debris team time to finish the work. STS-121 would launch seven astronauts in *Discovery* for a planned thirteen-day ISS assembly flight that was designated ULF1.1. Moving STS-121's launch date to July 1 meant there would be only five weeks to ready the next shuttle mission, STS-115, for its August 25 launch date. Kim was spending three to four hours discussing debris and touching seven or eight of the ten highest-priority issues at the RTF review every Monday and even longer at the SICB on Tuesday. In some cases the SICB would spend the entire day on debris, and on several occasions the meeting spilled over into Wednesday. The agenda for Wayne's Program Control Board on Thursday was no different, and debris topics routinely packed that time. Some board members joked the meeting should be renamed "Peters's Control Board" due to my monopolization of the meeting time to discuss debris risk. The only other high-priority item of interest was the LH_2 engine cutout (ECO) sensor failure that occurred during tanking tests and the initial launch attempt for STS-114. This failure was designated an integrated IFA requiring significant analysis and testing. Despite the extensive fault tree analysis, the root cause of failure could not definitively be established, and the IIFA was accepted as an "unexplained anomaly."

STS-121 Debris Review at the SICB

The week before the STS-121 debris verification review, the structural loads tiger team announced it had resolved the Glenn wind tunnel data. The team concluded the scaling correction factors had been in error, and the PAL ramp removal had negligible effects on other elements and the integrated vehicle aerodynamics. Static air load changes and thermal effects were determined acceptable. Although the buffet loading increased on the LO$_2$ cable trays as a result of the PAL ramp removal, it was within structural load limits. As it turned out, the buffet loading on the LH$_2$ cable trays did not increase because the cross-flow was generally outboard. This minimized the amount of shielding from the PAL ramp. The PAL ramp could be safely removed.

Also during this same week, Roy and the ISERP reviewed the debris hazard report, and Kim requested the SICB review the DVR presentation package. The ISERP met on June 12, 2006, to review the IDBR and "probable catastrophic" risks associated with the liquid hydrogen ice/frost ramps and adjacent acreage foam. Although the ISERP did consider these the highest risks to tile and special tile that could result in the loss of vehicle, the panel did not believe these risks would result in the loss of crew considering the availability of contingency shuttle crew support (CSCS). The ISERP, therefore, recommended by a seven to two vote to downgrade the risk ranking to "non-flight constraining infrequent catastrophic accepted." The panel also recommended a redesign of the LH$_2$ IFRs and adjacent acreage foam. The ISERP considered the redesign of the LH$_2$ IFRs and adjacent acreage foam imperative particularly if the STS-121 postflight risk assessment's results were bad. The SICB held the next day seconded the ISERP's recommendation only to be changed to "probable catastrophic" by the PRCB two days later.

Kim was now the SICB chair and occupied John's former seat at the head of the table. John was seated in the audience and listened intently during the LH$_2$ IFR discussion and risk classification, which I rated as "infrequent catastrophic." Kim polled the SICB members, and the vote was split. It left the final decision up to her. She agreed with the "infrequent catastrophic" risk classification based on her interpretation of the guidelines. Those guidelines stated that "infrequent catastrophic" (yellow) risk "could occur" in the program life, and "probable catastrophic" (red) risk "was expected" in the program life.

John quickly rose from his seat and passionately gave his reasons why the LH$_2$ IFR risk should be red. He felt the risk was in fact "bright red" and had all the indications of potential disaster. It was just like the bipod loss on STS-112, the mission before *Columbia*, and the rejected data on faulty solid rocket booster seals before *Challenger*. John was staring directly at Kim as he spoke. "The risk," he said, "always appears bright red to everyone after the accident." He added that Kim and the SICB would be irresponsible to let this happen.

There was a long silence when John finished. He remained standing, and Kim again stated her decision to keep the risk yellow. John just slowly shook his head and quietly took his seat. The guard had indeed changed, and I was impressed Kim had stood her ground. She was now clearly in charge of SEI.

From Kim's perspective, when we reached the point of decision on the LH_2 IFR, it was clear the spaceflight was not going to be risk-free. "Overall I felt good about making an informed decision, and I felt we reduced the risk enough to an acceptable level. I was ready to make the decision based on what people were telling me," she said. "As far as the SICB meeting, I remember the event as being intense and emotional, but I knew going into the meeting that the discussion would be emotional. Unlike many others in the meeting, I was not working on the shuttle program during the accident. Although I shared the sorrow and the shock of the *Columbia* accident, I was not there. Consequently my personal emotion was less than the people involved with the accident. This made it easier for me to make a decision that was driven more by technical merit than emotional feelings. Without the emotional aspect, I approached the decision like any other key decision."

Kim knew her promotion and the SICB decision to classify the risk as yellow were difficult for John. He had more history with debris and might have taken things personally. John and Kim had a discussion after the SICB meeting, and John told her that by not redesigning the LH_2 IFR, she was jeopardizing the crew's safety. Kim knew this because it was part of her responsibility as the SEI manager, but she still believed the risk was yellow. "I think John was surprised I did not accept his 'red' recommendation," Kim said. "I was more concerned with reaching the right technical decision without influence from the emotional arguments and John's historical influence. This ultimately turned out to be my toughest debris decision due to the emotion and risk level."

STS-121 Debris Verification Review, KSC, Cape Canaveral, Florida

The STS-121 debris verification review was scheduled to start on May 30 and last two days. It would take place at the new Operational Support Building in the fifth-floor conference room with a balcony view of the VAB, launch complex, and launchpads. Outside the meeting room, Brekke, Justin, Darby, and a few others were taking bets on whether I would finish the presentation. I was confident I could get the presentation done early the second day and offered to pay for the first two rounds of drinks if I lost. If I won they would cover my bar tab for the entire evening. It was a deal, and I slowly made my way to the front of the gymnasium-sized conference room. It had large, two-story video screens and wooden podiums strategically positioned to the left and right of a large conference table in the middle of the room. Rows of chairs (enough to seat more than two hundred people) surrounded the table on the sides and in the back. Board members began taking seats at the conference table, and the room quickly filled to capacity. John Muratore was buried in the audience along with Mike Griffin. Most board members were wearing suits or jackets for the occasion but not Mike. He was dressed in khaki slacks and a polo shirt and made it clear he did not prefer formal attire because he was just an "engineer." It was unusual to see Wayne at the head of the table as the chair instead of John. Although Kim Doering sat next to Wayne, her inexperience with debris was reason Wayne was chair. Other than Wayne and Kim, the rest of the board was the same as for the STS-114 DVR. I was presenting solo and decided to start the meeting with a joke about my zest for gambling—particularly blackjack. "When I gamble in Vegas," I said, "I usually play to lose because the odds are in my favor." There were a few chuckles, but everyone was ready to get down to business and talk LH_2 IFR risk. With the PAL ramp risk eliminated, the LH_2 IFRs were going to get most of the attention.

    I knew most people were focused on the one in seventy-five adjacent acreage risk and overlooked the context in terms of a conditional probability for worst-case conditions. After the joke I turned to an example about the likelihood of getting killed by lightning. On average, I said, the odds of being killed by lightning were about one in two million. However, the odds went go up substantially if outdoors in the middle of a thunderstorm holding up a metal object. These were examples of worst-case conditions placed on the lightning risk calculation. The same thing had been done for the LH_2 IFR adjacent acreage. The odds were much lower of getting struck if one ran to a safe building or vehicle when one first heard thunder, saw lightning, or observed dark threatening clouds developing overhead. These were examples of conditions needed to lower the risk.

Jeff and the ET project were comfortable with how the LH_2 IFR risk was assessed. At the time it was appropriate because the ET team had yet to determine the cause of the cracks in the foam. Consequently the PRA calculation had to assume the losses could occur as whole pieces anytime during flight. "I thought the debris risk presentation did a great job explaining the range of possibilities and the associated limitations for the integrated risk. The board had the data they needed to assess the risk and ultimately make the right decision to fly," Jeff said.

I backed Jeff and emphasized to the board the low debris potential based on the acreage foam design and that the ET project had implemented fabrication process improvements and nondestructive evaluation (NDE) tests as ways to mitigate the risk. This was like the person running for cover during a thunderstorm. Moreover the debris community had implemented other vehicle improvements designed to lower the risk. This included PAL ramp removal, bipod heater wire redesign, and modification of the Tyvek cover to name a few. Procedures to mitigate debris risk had also been developed for initial inspection of the orbiter gap filler, putty repair, and numerous liftoff debris. In order to more accurately represent the risk, the foam models had been adapted to account for the three additional foam failure modes observed on STS-114 and to assess the risk to special tile. The debris community evaluated all thirty foam debris sources compared to just eight assessed for STS-114. Several PRA model updates were also completed for the second-generation foam and ice debris models. This included replacing the ice model foam-on-tile damage map with a more representative ice-on-tile damage map. Aerospace Corporation made major improvements to its models by adding higher-fidelity geometry and CFD solutions, special tile risk, and more foam failures. These included crushed foam and adjacent acreage failure due to cracks or delamination.

Roy Glanville and the ISERP were growing increasingly uncomfortable with the risk classification of the ice/frost ramp and risk reduction of other foam debris sources based mostly on model updates. The PAL ramp loss on STS-114 greatly exceeded the predictions and reinforced the knowledge that, with respect to debris, the shuttle was not a mature system with well-demonstrated margins. The most effective risk reductions were only achieved by eliminating or substantially reducing the debris source. "Although the shuttle program accepted the debris community recommendations to reduce several foam debris risk rankings based on improved PRA, the ISERP considered this premature," stated Roy. He and others in the safety community felt the risk reduction must be demonstrated by flight prior to lowering the perceived risk ranking.

In hindsight the PAL ramp loss gave NASA insight into more failure mechanisms and defects based on the dissection data. Both the Aerospace Corporation and USA–Boeing models showed the ability to compare model predictions against flight performance, and they compared favorably with each other. Three of four test cases developed to compare the Boeing and the Aerospace Corporation models were nearly identical matches in terms of debris impact velocity and trajectory path, and there was only about a 12 percent difference in the debris impact velocity for the fourth test case. John Muratore was surprisingly quiet and made only one comment over the two days to help solidify the use of conditional probabilities. Despite the highlighting of all the improvements and positive changes since STS-114, the focus remained on whether the LH_2 IFR risk should be red or yellow.

RETURN TO FLIGHT PART II

The board was in a fierce debate until Mike Griffin spoke. He said he was comfortable enough with the risk that he would let his son fly the shuttle. Although Mike's son was not an astronaut or even a NASA employee, it was his litmus test for determining flight risk safety. Even if the board declared the risk red, Mike implied he would approve the flight. Flying STS-121 was a risk he was willing to take. The debate was over.

Mike chased me down in the hall while the board deliberated to appraise the DVR presentation and my debris work. It was a huge compliment when he said I was "only one of four people in the agency who understood this stuff." (I knew the other three were him, John, and Gerst.) Even though the decision was made to accept the STS-121 debris risk and flight rationale for the expected debris sources during liftoff and ascent, the LH_2 IFR battle was not over. Jeff and the ET project were confident in our ability to determine the debris risk even prior to STS-114 but were always "cautiously aware" of the PAL ramps and the debris potential they represented. "I sometimes think we were fortunate to have had the PAL ramp loss on the first Return to Flight mission," stated Jeff. "That debris event forced us to deal with the issue and eliminate it for the subsequent flights. The PRA was a great method to explain the integrated debris risk, but I relied on mitigations and controls more to get myself comfortable for flight. Where I found PRA most valuable was using it to make sure we optimized resources to mitigate debris from locations the PRA was showing as a high risk such as the PAL ramp."

Brekke, Darby, Justin, and others were at the building exit and waiting to collect on the bet. We all needed to decompress and decided to "run the gauntlet." This involved stopping for a drink at every bar between the back gate of KSC and the Cocoa Beach Hilton where most everyone was staying. It was about a five-mile stretch of bars, beach homes, restaurants, and hotels. Everyone pitched in dinner for the designated driver, and the winner was reimbursed his bar tab. I made it halfway before calling it quits. This left Justin and a few others to finish but it was Justin who eventually won it.

STS-121 Debris Discussion with Gerst

On June 9, 2006, I was frantically working on the DVR action items and hazard report updates for Roy's Integrated Safety and Engineering Review Panel when Gerst walked into my office. He wanted a status update on the STS-121 DVR action items and had a few of his own to add. Most DVR actions were complete and would be presented to the ISERP and PRCB on June 20 and 22 respectively. That was a few days before the flight readiness review. One key action still in process was the extension of ASTT out to 165 seconds with respect to mass loss and debris source to determine if a debris threat existed beyond SRB separation. I did not have to run the analysis to know that after 130 to 140 seconds, any size debris released from almost any location was not a threat due to the low dynamic pressure. However, I still wanted to complete the analysis. Gerst wanted me to truncate the mass distribution for the LH_2 IFR acreage to 0.084 pounds (from 0.25 pounds) and reassess the results. The ET project estimated the 0.25-pound mass represented the largest possible failure given the geometric constraints around the ice/frost ramp and pressurization lines. This mass was driving the conditional probability of one in seventy-five risk at eighty seconds. The smaller 0.084-pound mass represented the largest historical loss assumed to have released in one piece. I had already run the calculation, and it resulted in a worst-case conditional probability of one in three hundred. Gerst suggested using the 0.084-pound mass as a three-sigma value for the LH_2 IFR acreage distribution when performing the PRA. I had already planned to incorporate both calculation and risk results in the hazard report and present them to the ISERP and PRCB. Gerst was satisfied with that approach but added one more action item. He wanted me to assess the NDE data obtained from the next two external tanks and reassess the IFR body risk. I had already received the NDE data but could only determine conditional probabilities until the ET project provided a predicted time of release. That was still in process. Once I received the time of release data, the Aerospace Corporation could do the computations in a few days. With the ET project focused on

FRR preparations, I did not expect to get the additional data until after STS-121.

STS-121 Debris Risk Presentation to the Program Control Board

I presented the integrated debris hazard report (IDBR) to the ISERP and PRCB for approval. Kim Doering and the SICB recommended approval of IDBR and classification of IFR body and IFR adjacent acreage debris risks as "infrequent catastrophic." Much to my surprise, the ISERP decision was consistent with the SICB recommendation but certainly not unanimous. There were several organizations such as the Safety and Engineering Directorates that rejected the ISERP recommendation to approve the debris hazard report and offered dissenting opinions at the PRCB.

Just like at the SICB and ISERP meetings, the PRCB was contentious, and most of the risk debate focused on the ice/frost ramps and adjacent acreage. The risk level for an unknown failure mode peaked at one in seventy-five around eighty-five seconds. The risk level for the cryopumping failure mode was rather benign (less than 1 in 2,500), but we did not have the data to state conclusively the IFR acreage losses were due to cryopumping. Wayne pointed out that "it was a mistake to believe there were "votes" at the PRCB. Decision-making authority was vested solely in the chair. Everyone else advised the chair. Any review of positions and their rationales merely informed the chair. In the STS-121 case, I was acutely aware we were on the edge of what was acceptable, but there was very little we could do because we needed flight data. Modifications could be made, but the results would be problematic without gaining more flight data, and that would induce more months of delay. There was no substitute for flight data."

According to Wayne, "there was no choice but to classify the IFR acreage losses as a red risk and elevate it to the senior NASA management with my personal recommendation as program manager that we proceed." As it later turned out, lack of flight data to properly model the situation inflated the probability numbers. In retrospect stated Wayne "a proper computation placed the risk in the yellow category. However, that is irrelevant. Decisions must be made with the data in hand. Frequently that means getting more data, but sometimes it means accepting the risk. The bottom line was that we had been proven wrong a year before, and it would have been the height of folly to assert that we understood all aspects of the problem and were assuredly safe."

Aside from the "red" risk classification of the IFR acreage, the PRCB concurred with the SICB and ISERP recommendations and baselined IDBR for the next two missions. Steve Poulos, the orbiter project manager, was seated next to me during the PRCB. He said he was personally comfortable with the yellow classification. "Because of the accident," he said, "NASA had become overly conservative, in my opinion, as it related to risk acceptance in general. Do not get me wrong. Due diligence is absolutely necessary. However, there is definitely a point of diminishing returns, which I think we crossed in a number of the debris areas."

"Jim, Head to the Airport"

On the morning of June 24, 2006, I had just finished showering when the phone rang. It was 7:15 a.m. Kim Doering wanted to know if I could make the nine o'clock flight out of Houston Hobby Airport to Orlando. The LH_2 IFR risk was still the major concern at the FRR due to the high risk, and many on the review board favored grounding the fleet until the failure mode was better understood and a safer configuration flown. The FRR Board consisted of headquarters representation from the deputy administrator, Chief Safety Officer, and the Office of Chief Engineer, center director representation from JSC, KSC, and MSFC, director of the crew office, and the program managers for the space shuttle and ISS. Gerst chaired it, and Bryan O'Conner cochaired. Kim said Gerst's opening remarks described the FRR as a soccer game, and he did not want everyone rushing to the ball (i.e., focusing on the LH_2 IFR debris risk of one in seventy-five and ignoring the other shuttle

RETURN TO FLIGHT PART II

risks and subsystems). Gerst was expecting a debris special topic briefing and wanted me to give it. Kim had already booked my flight and wanted me to head to the airport immediately. I had no time to waste and raced through the house. I stuffed things into my suitcase, and I ran out the door while saying good-bye to my wife and kids.

As I sped off to the airport, I had to decide whether I should drive to the office, retrieve my computer, and risk missing my flight or head directly to the airport and generate my FRR presentation when I arrived at Kennedy. Needing my computer to access all the debris data and risk assessment information, I decided to chance it. I drove to JSC. If I missed the morning flight, my backup plan was to catch the next flight and arrive at KSC late in the afternoon. I zipped into the office, grabbed my laptop, and ran back to my car. It was 8:15 when I left the office, and I was fortunate that traffic was light. I made it to the airport with about twenty minutes to spare, cruised through security, and made it to the plane as the final boarding call was made.

Kim had stated before she hung up the phone that the FRR Board was most interested in LH_2 IFRs—especially the one in seventy-five risk number. I knew from past experience the risk numbers were often taken out of context because the assumptions behind the computations were ignored, and this was probably the case. From historical data on previous flights, we knew the LH_2 IFR release rate, mass distribution, and likelihood of release from each of the fifteen LH_2 IFR locations. We also knew the forward three locations experienced the brunt of thermal extremes and aerodynamic forces. The key missing component of assessing the risk for this debris source was determining the time of release. My challenge was to explain that the one in seventy-five risk number represented a peak conditional probability for an unknown failure mode for the largest possible release mass liberating at the worst possible time. It was difficult for people to understand conditional probabilities, and they often struggled with the peak one in seventy-five risk. It sounded bad, but in reality it had limited exposure time (less than 5 seconds) over the entire ascent phase. The fixation on the one in seventy-five risk number made it challenging to convince many that the integrated LH_2 IFR debris risk was really much lower and hence more appropriately classified as an acceptable yellow. The peak risk for the largest historical adjacent acreage release was one in three hundred.

I also had Darby and Brekke run a set of parametric time of release distributions to get a more representative risk assessment. I had just received the results the previous day from them both, and the results ranged from 1 in 560 for Boeing and 1 in 250 to 400 for the Aerospace Corporation. Even though the ice risk was mathematically higher, Darby felt the LH_2 IFR and adjacent acreage risk was more of an issue than ice. Darby communicated to me that, "The flight data were telling us that we missed something in our modeling heading into STS-114, and when you miss some of the physics, you have to stop and reevaluate."

I was comfortable with the analysis, given the favorable model comparison between Boeing and the Aerospace Corporation. However, the debris community had not reviewed it. SEI had been heavily criticized for the constant change in risk numbers and for presenting those changes without debris community review. Since no one other than the Boeing team and I had seen the results, I decided only to verbalize them during the presentation and to put the supporting analysis in backup. For those on the outside looking in, the risk numbers did appear to change frequently. This was due to the constant model updates driven by our ever-increasing understanding of the debris environment and improvements in the fidelity of the models to remove limitations. Besides addressing the LH_2 IFR risk, I wanted to stress the vehicle improvements and closure of the twenty-eight integrated IFAs regarding debris-related events observed during STS-114. I finished the presentation as the plane landed in Orlando. Off in the distance over thirty miles away, I could see the Vehicle Assembly Building standing as big as a nation's dream to beat the Russians to the moon. I would be heading that way for the flight readiness review.

CHAPTER 19

STS-121 Flight Readiness Review

Operations Support Building II, KSC, Florida

It was a little after lunch when I made it to the new Operations Support Building (OSB-II). Kim was glad to see me when I walked into the crowded fifth-floor conference room—the same location as the debris verification review three weeks prior. When she signaled Wayne, he rose from the center table and shuffled me off to a small office down the hall from the main conference room. He was eager to see me but appeared nervous about the upcoming debris presentation. He asked if I was ready. I handed him my memory stick and explained the presentation was similar to what he had seen at the PRCB the week before when the debris hazard report was approved. The only exception was the Boeing parametric risk analysis. That was similar to the ice risk assessment approach and more representative of the risk. He took a deep breath and stated that I should stick with the facts and only offer my opinion if asked.

I decided an icebreaker was needed and started the presentation by describing how strange the day had begun with Kim asking me if I wanted to come to KSC for a "soccer game." I mentioned how I had built the presentation during the two-hour flight from Houston to Orlando while sitting in the middle seat. The room broke out in laughter. That included Wayne, Gerst, and Mike Griffin. I breezed through the presentation without much challenge, and I included the recently completed Boeing and the Aerospace Corporation parametric analyses. As I presented Kim was anxious about having to defend the risk classifications against any disagreements or protests. "If the members could just listen to your presentation with minimal emotion or preconceived ideas, they would reach the same conclusion with respect to debris risk," she told me later. "I was a little surprised we reached the decision as smoothly as we did because I felt many on the Board or in the audience might not have understood all the nuances to the risk calculation. I think what won people over was the presentation style and [Jim's] confidence going toe-to-toe with Mike Griffin over technical content during the presentation. By doing this Jim demonstrated his expertise of the subject matter, and this solidified his credibility with not only the administrator but also members of the FRR Board and audience."

Based on his reputation and technical background, Mike Griffin was always viewed as the supreme authority figure in the room. This led many to jokingly call him NASA's "chief engineer." Consequently many tended to rely on Mike to make key technical decisions. He loved being the technical expert, and at the time there were only a few people on the shuttle program who would stand up to him. The program could have easily slipped into a pattern where everyone would look to him to determine every decision. "Fortunately the STS-121 FRR changed that perception," stated Kim. "I did not know it at the time, but this became my most memorable moment as the SEI manager. During the debris presentation, Jim responded confidently and forcefully to counter Mike's differing opinion. Jim and Mike technically sparred, and in the end Jim was right, and Mike was wrong. Other than the satisfaction of seeing it, what it communicated to everyone in the room was that Mike might not be right all the time. In addition, if people felt we were going down a wrong path that Mike happened to support, they needed to say so. One time Mike stated that SEI was being too conservative. I recall making a stand by waving my finger in front of him and saying that everyone at the table was responsible for the conservatism—not just SEI. I laugh now, but at the time I thought I would be searching for another job. The feedback I received from his staff after the meeting was that he wanted to know the name of the woman waving her finger at him during the meeting. Although you did not want to cross him, deep down he respected those who defended their differing opinions. The relationships with Mike were healthy and critical for those involved with making key debris risk decisions."

Wayne had never seen me present at a flight readiness review and was mostly concerned about me speculating in front of a management group. Without some type of disclaimer, the audience would assume the speculation was fact. "Jim did a great job," recalled Wayne, "and the entire discussion at the FRR went pretty much as I thought it would because I had been through all the information beforehand."

Ralph Roe was never shy about sharing his position, and he reiterated the NESC recommendation given to the PRCB just a week earlier that LH_2 IFR should be redesigned before STS-121 flew. At the FRR Ralph recalled, "I thought the discussion of the risk was really more important than which block it ended up in. I think these early Return to Flight FRRs were very good in terms of reestablishing how we should discuss and debate issues at this forum."

At the end of my presentation and the ensuing discussion, Mike Griffin spoke about how the debris and overall shuttle risks had to be factored into the greater system risk. That included ISS and partner flight operations. Mike had spent a considerable amount of personal time in the weeks leading up to the flight pursuing his own understanding of the statistics of foam loss. It was clear to him a catastrophic accident would be the result of a lengthy chain of events: the liberation of a piece of foam sufficiently massive to damage the orbiter within a particular time frame (neither too early nor too late) from a location where the flow field could cause it to impinge upon the orbiter, in a place where the orbiter was vulnerable, and with enough energy deposition to cause catastrophic damage. Mike's own assessment was that the actual likelihood of such a failure, assuming individual events of more nearly average than worst-case probability, would be close to that for foam loss due to cryopumping. I confirmed this rough assessment during the presentation and subsequent questioning from Mike, but I did not have any hard data to pin the failure on cryopumping. Consequently Mike thought we were as safe to fly as we were going to be in regard to the foam loss aspect. When polled both the Office of Safety and Mission Assurance and the chief engineer voted no go for flight, while Gerst and other key individuals voted yes. The issue then came to Mike for resolution, and he stated "we would fly." He also noted that, "should catastrophic damage occur, it would not affect the ascent capability of the orbiter—only its ability to perform reentry. He pointed out that we might lose an orbiter, but we would not lose the crew. We would eventually be able to get them home from ISS using Soyuz or the shuttle." Griffin stated that the data and debris risks were consistent with other existing shuttle risks classified as "infrequent catastrophic." The FRR Board concurred and started the launch countdown for STS-121 on July 1, 2006. Griffin received a lot of criticism in the media for this decision—especially the *Houston Chronicle* and the *New York Times*. The NASA Inspector General also objected to this decision and did so in writing. All in all STS-121 turned out to be Griffin's most memorable flight.

STS-121 DVR Decision Is Scrutinized

For Gerst the decision to move forward with STS-121 and fly a high-risk LH_2 IFR was one of his toughest debris decisions due to the opposing opinions. The ET-120 tank dissection data were nothing short of miraculous and probably the best data the ET project had ever gotten on foam debris. If there were ever a lesson learned, it was that we should have loaded the tank with fuel and dissected various foam insulation locations before flying STS-114. Ironically the engine cutout sensor failure—not the foam—was the real driver behind sending the tank back to MAF and ultimately dissecting it. The ET project might never have gotten the dissection data had it not been for the ECO sensor problem. The other concern heading into STS-121 was Hurricane Katrina and the effect it had on the ET project. Gerst and others on the program believed NASA would have flown sooner had it not been for Katrina.

When Gerst first started working for NASA on the wind tunnel at Lewis, he tested an early tank configuration without the PAL ramp. This experience made him comfortable with the PAL ramp removal. According to Gerst the STS-121 FRR presentation characterized the risk very well. He did not necessarily like using the "red" and "yellow" designations to classify risk, and he did not like just talking about the "mean" risk. He was more interested in the uncertainty of that risk around the mean. The big concern with a mean risk was that it might be placed in the yellow category, but the uncertainty might really put it in the red. "I think we do a disservice by trying to classify the risk level and saying we know more about the risk level than we actually do," asserted Gerst. "I was not worried about which box we were in but more of what we knew and did not know. The real question became what we would gain and learn by flying and how this data could be used to assess the next flight. Or could there be another way (such as a ground test) that would give us the same insight into the debris performance? What I took away from the discussion was that we could never do a full simulation or ground test that would give us the same information as we would get from flying."

Although the LH_2 IFR risk was high, it was not unusually high and not high enough to stand down the flight. The other aspect that played into the flight decision was the performance of different LH_2 IFR configurations tested. All the proposed configurations were actually more of a foam debris risk than the current configuration and were more prone to ice formation. The bottom line was there was not an easy design change to the LH_2 IFR that would lower the risk, and it might in fact increase the risk and uncertainty around that risk. "There was a lot of uncertainty surrounding this issue, and I had to weigh in my own mind if it was still acceptable to fly and whether I would learn something during this flight that would allow me to lower the risk for the next flight," stated Gerst.

With all this uncertainty, Gerst and the FRR Board were willing to take a onetime risk to go fly and see what came of it.

STS-121 Launch

Due to its planned trajectory, STS-121 was going to fly a lower dynamic pressure (low Q) ascent. Flying low Q instead of a nominal or high Q profile would change dynamic pressure versus time, velocity versus time, thrust versus time, and liquid propellant level versus time. A low Q trajectory meant lower stresses on the vehicle but the extra propellant needed to fly the trajectory reduced the payload capability. Since most ISS assembly missions would push the payload capacity of the vehicle, low Q trajectories were rarely flown. There was a lot of speculation the debris environment might be altered, which would change the risk posture. An action item that came out of the FRR was to determine whether the debris risk would increase, decrease, or remain the same and to report the results at the L minus two meeting. As it turned out, the debris risk for a low Q trajectory was lower due to the lower dynamic pressure and lower heat rate. There would be fewer foam losses because of lower aeroheating and later foam releases by about five to ten seconds due to external and internal temperature changes as a function of time. In addition the impact energy from releases during the 30 to 120 second time frame would be less for the low Q mission due to the lower dynamic pressure. Since all the risk computations were based on high Q trajectories that enveloped the low Q risk and due to the time constraints, I decided not to compute the PRA for STS-121.

Launch of *Discovery* was scrubbed twice—July 1 and 2—due to unacceptable weather conditions. During the second postdrain inspection of the external tanks, the foam coating on the inboard strut of the LO_2 feed line bracket at station Xt-1129 was found to be cracked. The cracked piece of foam subsequently fell off and landed on the mobile launch platform deck. I called the Aerospace Corporation and USA–Boeing debris teams into action to assess the foam and ice risk respectively should a similar failure occur. STS-121 was delayed until the debris risk assessment was completed and the ET project had determined the failure's cause. The debris teams and ET project worked all night to fulfill their tasks. The existing gap between the feed line bracket strut and monoball closeout provided a "thermal short" for ice formation, and ice had formed during the second launch attempt. During tanking the LO_2 feed line was cryoconditioned prior to filling the LH_2 tank. This caused the feed line to effectively shrink about 3.50 inches and rotated the feed line bracket aft. The aft rotation compressed the ice between the foam surfaces and resulted in a crack. When the launch attempt was scrubbed and the propellant off-loaded, the cracked foam liberated as the bracket rotated back to its original position.

It would take at least two to three days to run analysis and determine the risk if the foam had released during ascent. Since there was no time to perform a lengthy risk assessment, we decided to envelop the feed line foam risk using the LH_2 IFR and the ice risk using the feed line bracket at the same station. After a day's stand-down, the launch attempt resumed on July 4. It was the first Fourth of July launch in the history of the US space program.

On July 4, 2006, *Discovery* lifted off on time at 2:38 p.m. (EDT). STS-121 was the second RTF mission. It replaced critical hardware needed for future space station assembly and demonstrated techniques for inspecting and protecting the shuttle's thermal protection system. The mission also restored the ISS to a three-person crew for the first time since May 2003. European Space Agency astronaut Thomas Reiter was left aboard to join Expedition 13 as the third crew member. This was the most photographed shuttle mission in history. More than one hundred high-definition digital, video, and film cameras documented the launch and climb to orbit. Although there were numerous foam debris events between 165 and 173 seconds into flight, the external tank feed line camera detected several debris releases that appeared to originate from the same location. It took another day to downlink the tank separation images to pinpoint the release location. It turned out to be the adjacent acreage around a forward ice/frost ramp at station Xt-1269. This was the type of foam failure feared heading into the flight due to the high potential risk for an unknown failure mode. However, the foam failure occurred well after ASTT and represented no risk to the orbiter. The late release was also indicative of a cryopumping failure. That was a relief to me because this type of failure during these late release times was such a low risk. Although I did not like to see any foam failures, I was happy this occurred. It gave credibility to the timing of the cryopumping failure mode, and it also demonstrated that the foam failed in small pieces instead of one large chunk, and that meant even lower risk.

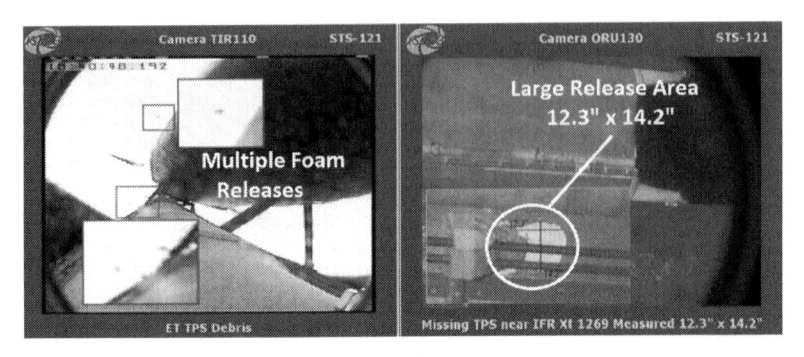

The external tank feed line camera took the picture on the left. It shows several foam debris releases between 165 and 173 seconds. An umbilical well camera took the picture on the right seconds after ET separation. The correlated images suggest the foam failure was due to cryopumping. (Courtesy of NASA.)

Another debris-related activity the crew performed on this mission was testing repairs on the reinforced carbon-carbon panels. This was accomplished on the third and final space walk. Under evaluation was a preceramic polymer sealant containing carbon–silicon carbide powder known as NOAX for use on damaged panels. Mission Specialists Piers Sellers and Michael Fossum made three gouge repairs and two crack repairs. These would be evaluated in the high-temperature arc jet facility when they returned.

STS-121 Mission Completion

On July 17, 2006, at 9:15 a.m. (EDT), *Discovery* landed on Runway 15 at Kennedy Space Center. A runway walkdown and preliminary postlanding inspection of *Discovery* were conducted at KSC that day. The orbiter sustained ninety-six total hits. Eleven had a dimension of one inch or larger. In general damage occurred evenly between the right- and left-hand sides of the vehicle. John Brekke and Matt Eby had the opportunity to participate in the postflight inspection of STS-121 in the orbiter processing facility. "My first visit to the orbiter processing facility made a lasting impression," Matt recalled. "Just a few steps past the access control desk, the room opened up, and I found myself standing underneath the orbiter."

The geometric pattern of the heat shield tiles streaked by the reentry heat was a mesmerizing sight indeed. The opportunity proved invaluable because John and Matt could see how the model prediction of damage due to foam debris compared with the real article. The typical pattern of tile dings—especially along the nose, and at the aft near the fuel feed doors—was apparent. Most nose damage was quite shallow and involved just the coating chips. The symmetric pattern on both the port and starboard sides and the smallness of the dings indicated small popcorn foam loss from the intertank had caused the damage.

Review of all STS-121 debris events indicated all ET foam losses were within risk assessments and model predictions. There were fewer foam losses than the models predicted and less impact damage than what was observed on STS-114. Two foam loss events at Xt-1270 and Xt-1399 had large masses but were not statistically unusual. The Tyvek covers performed per specification, and there were only three ascent debris IFAs: missing orbiter putty repairs, orbiter blanket repair damage, and more orbiter gap filler debris. Putty repairs were ceramic-type material applied to minor flaws in the orbiter TPS tiles in lieu of tile replacement, and these were beginning to become a debris concern due to their number and age. Somewhere between two thousand and three thousand such repairs had been made on each orbiter, and those that had been done after the first shuttle flight were now more than twenty years old. Although only two missions had flown, the orbiter was shedding unexpected debris sources that were both an ascent and reentry debris concern. The cable tray accelerometer data also showed that the PAL ramp removal did not cause significant differences in the cable tray response or any design load issues.

Hubble Space Telescope Repair Mission

On January 16, 2004, four months after the CAIB report was released, NASA administrator Sean O'Keefe officially announced the cancellation of the final scheduled servicing mission to the Hubble Space Telescope (HST). This was consistent with the CAIB recommendation that any future space shuttle mission fly in an orbit that allows it to reach the International Space Station in case of an emergency. The 28.5-degree inclination orbit required to service Hubble would prevent the shuttle from being able to get to the station, which was in a 51.6-degree inclination orbit.

The Hubble Space Telescope was launched in April 1990 from KSC, and *Discovery* carried it into orbit. Weighing approximately 25,575 pounds and measuring 43.5 feet long by 14 feet across, the HST orbits at an average altitude of 375 miles above Earth. Since its launch the HST has been gathering data from the universe with its five main remote sensing capabilities. Ever since Galileo first peered through his telescope into the night sky in 1609, atmospheric interference has limited humans' views of space. The atmospheric constituents such as carbon dioxide, nitrogen, oxygen, and particulates distort and block the various forms of light and electromagnetic energy entering from space. To mitigate these effects, early astronomers placed their telescopes on top of high mountains where the air was thinner. This helped to a certain degree, but there were still the problems of cloud cover, light pollution, and the inability of some radiation to penetrate the atmosphere. The simple solution to this problem was to place a telescope into Earth's orbit. In the late 1970s, NASA proposed the Hubble Space Telescope—named in honor of Edwin Hubble. He could be characterized as one of the most important astronomers in the twentieth century. The HST launch was originally planned to occur in 1983. Because its development and construction took longer than expected, the launch was delayed two years. In 1985 HST was declared ready for launch and was to be carried aboard a shuttle flight sometime in 1986. Unfortunately the *Challenger* disaster on January 28, 1986, derailed that launch plan.

HST sat in storage for five years before being launched on *Discovery* in April 1990. The excitement of viewing the first pictures from the most capable telescope ever developed soon turned into pure disappointment. Engineers realized there was a major problem with the optics installed on the HST. There was a flaw in the large mirror—one of the few components that could not be replaced in space. The first of a series of repair missions was launched in December 1993 on STS-61 to install new instruments and correct the optical flaw in Hubble's primary mirror. The second service mission was launched February 11, 1997, on STS-82 to greatly improve Hubble's productivity by installing new sensors that allowed NASA to probe the most distant reaches of the universe. Hubble was brought back to life after the repair of four failed gyros on the third repair mission launched in December 1999 on STS-103. The fourth repair mission, launched in March 2002 on STS-109, installed a new advanced camera that created ten times more discovery power. Eventually HST needed additional servicing and battery replacement to keep it operational. However, after the *Columbia* accident and NASA administrator O'Keefe's 2004 announcement to cancel the fifth Hubble mission, the fate of the space telescope appeared sealed. This was the case until Michael Griffin took over as NASA administrator in April 2005. During Dr. Griffin's confirmation hearing, he said he would reconsider the decision by his predecessor to cancel a planned shuttle mission to service the Hubble Space Telescope. "We should reassess the earlier decision in light of what we learn after a few Return to Flight missions," Mike said.

He stated he would review two options: sending a shuttle to refurbish the popular space telescope or mounting a simple robotic mission to deorbit Hubble and plunge it into the ocean. Griffin explained the option of sending a robotic spacecraft to refurbish Hubble with new instruments, batteries, and gyroscopes was off the table due to the complexity and expense.

Chuck Shaw, the technical assistant to the deputy shuttle program manager at Johnson Space Center, was named the Hubble mission manager. His first duty was to coordinate the decision process to assess the feasibility of the mission and to add it back into the shuttle manifest. Chuck was a former USAF officer who hired on with NASA after the *Challenger* accident and had been the flight director for thirty-two shuttle missions. While in the air force, Chuck was stationed at Vandenberg Air Force Base in California and was selected to train in Houston for shuttle operations. He wanted to fly classified missions for the Department of Defense out of a secure mission control in Colorado Springs, but he applied for and was selected as the first non-NASA flight director in 1983. Although very task focused and organized, Chuck had a polite, diplomatic way of engaging the help of those he needed to complete the job. This included me. The HST mission decision really boiled down to determining the loss of crew and loss of mission risks and whether they were both acceptable. All the major risk drivers on the shuttle program were known except the integrated debris risk and the risk of transferring crew from one shuttle to another if a rescue mission was launched. Chuck needed the integrated debris risk from me to complete his analysis. The integrated debris risk also had to account for vehicle improvements and a limited on-orbit repair capability. Chuck approached me with this request in May before the STS-121 mission and needed the data to support a decision meeting with Mike on August 2. I promised to work the analysis after the STS-121 shuttle mission but also needed to close the debris IFAs before the next mission.

Debris analysis up to this point had focused on debris source–target pairs such as LH_2 IFR foam on tile or RCC and not on the overall aggregate. I expected the debris risk for future flights to be much less than STS-114, STS-121, and STS-115 but needed to quantify the risk reduction. Future vehicle improvements included redesign of the LH_2 IFRs, reduction in IFR body losses due to the void delta P failure, reduction or elimination of the thermal crack or delamination failure, application of "new" foam to the external tanks that would be flown after the next seven tanks flew, improved foam application processes, handiwork, inspection, and controls, 100 percent inspection of gap fillers, possible ice mitigations such as coatings, improved controls for orbiter expected debris such as putty repair, shim stock, and blankets, and liftoff debris mitigations. Foam debris dominated the integrated risk computation, and it turned out to be 1 in 170. The result seemed reasonable, but I needed an independent assessment to determine if the number was valid.

Safety and Mission Assurance had developed an ascent damage assessment model (ADAM) that based the debris risk on historical impact data. ADAM was strictly a simulation of a probabilistic model that exposed the orbiter to random debris impacts based on the historical impact counts and damage size since STS-1. The integrated debris risk was one in four hundred using this technique. That was comparable to the PRA model.

Determining go-no-go flight test plan criteria for future missions to support a case that the ascent debris risk was being controlled and the LON rescue vehicle was not required was a program-level decision. The program managers had to reach a decision on how many "clean" flights were needed to validate models and demonstrate vehicle improvements as well as how many flights were needed to conclude that a modified LH_2 IFR design was acceptable. These factors and the availability in the shuttle schedule would drive the launch date. Debris performance results on STS-114 and STS-121 were not statistically significant, but they demonstrated a trend showing that added objective evidence from flight data indicated RTF design changes had been effective. The two missions also accomplished key flight test objectives related to debris. This included demonstrating the liftoff and ascent debris environments were acceptable and demonstrating debris sensor effectiveness with imagery, ground radar, wing leading edge impact sensors, and the ELVIS cameras. The effectiveness of PAL ramp removal to eliminate critical debris during ascent was demonstrated on STS-121. The good news for Hubble was that future debris risk characterization would provide a more representative risk assessment. This would include a "single" integrated foam risk number, extensive model validation from flight history, and reduction in model conservatism.

Conclusions from the previous two flights related to the RCC panels and tile repair assessments were needed for determining the probability of damage exceeding the repair capability. Assessing this probability had not been done before, and it presented its own set of challenges. Results from the STS-121 damage repair test were still being assessed, and the risk analysis results would be based on an uncertified repair capability. The damage assessment approach involved several key steps. The first step was for the orbiter team to determine the maximum damage that could be successfully repaired for both RCC and tile. They would also have to determine the mass or range of masses and conditions that could cause the maximum repairable damage. The Aerospace Corporation could filter out the risk contribution for those masses and conditions from the total risk and report the remaining risk as the probability of damage exceeding the repair capability. It was not an extensive process once the initial conditions were defined, and I expected the probability of exceeding the damage risk to be lower than the baseline. Considering repair capability, the integrated debris risk was less than half the baseline risk of 1 in 170 and a function of both the repair capability and the size of the maximum repairable damage. As the repair capability increased, the integrated debris risk would lower. The analysis also showed a minimal benefit for repair capability below four square inches (two inches by two inches), and the remaining threat to RCC and special tile zones limited the repair benefit.

Hubble Space Telescope Rescue Mission

Unlike shuttle missions to the International Space Station, which could sustain a stranded crew for up to ninety days, a space shuttle flight to the Hubble Space Telescope offered no additional supplies other than those carried aboard the orbiter. Since HST was in a different orbit than the space station, a second shuttle would have to be on standby at Launch Complex 39B to serve as a rescue vehicle in the unlikely event *Atlantis* was unable to return home safely. If a rescue mission were deemed necessary, a plan had to be developed to transfer the crew. The rescue orbiter would use the robotic manipulator system (arm) to grapple the HST orbiter and transfer the crew along this path. To complete the transfer, approximately three EVAs would be required, and three crewmembers—the pilot, commander, and robotic arm operator—would not be fully EVA trained. This type of procedure and the rendezvous of two shuttles had never been done before, and this added to the rescue mission's risk. To mitigate the transfer risk, each non-EVA crewmember would be provided very basic training.

The decision boiled down to three options: fly the Hubble mission without a LON rescue capability, fly the Hubble mission with a LON rescue capability, or do not fly the Hubble mission at all. If NASA opted to fly the Hubble mission with the LON rescue capability, the program had to decide whether to use dual- or single-pad launch operations. Single-pad operations took more time to prepare the LON vehicle for launch because of the pad preparation time needed after the first launch and the rollout and integration of the second vehicle. If a rescue mission were declared toward the end of the Hubble repair activity, the resources needed to keep the crew alive would be nearly depleted. If there were a problem with the pad or integration of the second shuttle, the rescue capability would be lost. Dual-pad operations appeared to be the best choice. The rescue shuttle would be ready to launch immediately if and when an emergency was declared. However, this option had drawbacks as well. The Constellation program was eager to get started on pad modifications for launching Constellation's first test rocket, Ares I-X. Any schedule delays would add pressure on the ground operations crew to complete the modifications in time to support the Ares I-X flight. The workload for the ground operation crews would also double due to preparing two vehicles for flight.

HST Repair Mission Decision

The first key decisional milestone was August 2, 2006. It was a complete feasibility study review and assessment of all outstanding actions, and that included the integrated debris risk. On August 22 HST battery testing would be complete to ensure there was enough longevity left to even conduct the mission. PRCB review and approval of the manifest decision package would take place sometime in September. That was, however, pending the launch schedule of the upcoming shuttle mission, STS-115. The final go-no-go decision with the NASA administrator would be October 27—right at the minimum L minus eighteen month time frame. Mike Griffin attended the August 2, 2006, meeting to hear the results of the feasibility study and learn which set of recommendations were going to be brought forward to the program managers. The analysis showed the risk for loss of crew without a rescue mission was one in sixty-six compared to one in seventy-three with rescue. Debris risk, main engine failures, and on-orbit micrometeoroid debris hits drove integrated mission risk results. They appeared flawed at first because everyone expected the risk with the rescue mission would be much lower. Although the ascent and reentry risk would be lower with the rescue mission, there was added risk contribution due to the shuttle-to-shuttle rendezvous and crew transfer. There was also a risk contribution from a possible launch scrub of the rescue mission, especially for late emergency declarations. The risk for loss of crew for rescue missions to the ISS was a little lower (one in eighty) because of the removal of the rescue mission launch scrub and transfer risk. Mike was prepared to move forward with the Hubble mission but wanted to make the final decision after STS-115. Ultimately, the HST mission would be conducted with a LON rescue mission and with dual-pad operations, even though the loss of crew risk was about the same with rescue or without. Mike stated that NASA would be doomed if there were a problem during the mission and the public found out NASA did not have a rescue mission planned.

During a meeting with agency employees at NASA's Goddard Space Flight Center on October 31, 2006, NASA administrator Michael Griffin announced there would indeed be a fifth and final servicing mission to the Hubble Space Telescope. This mission would include new gyros, batteries, and thermal blankets and two new instruments. In making the announcement, he said, "We have conducted a detailed analysis of the performance and procedures necessary to carry out a successful Hubble repair mission over the course of the last three shuttle missions. What we have learned has convinced us we are able to conduct a safe and effective servicing mission to Hubble. While there is an inherent risk in all spaceflight activities, the desire to preserve a truly international asset such as the Hubble Space Telescope makes doing this mission the right course of action."

The Hubble Space Telescope mission (designated STS-125/HST-SM4) was planned for an April 2008 launch.

CHAPTER 20

Debris Integration Group

Flight-to-Flight Debris Assessment

There were still a lot of debris questions regarding the LH_2 IFR, gap filler, and now putty repairs to be answered, and there was a tremendous amount of debris work to be done in a short period of time to prepare for the STS-115 mission. In addition to the HST risk assessments and improvement of the LH_2 IFRs, I was following the standard flight-to-flight risk assessment process I adopted after STS-114. The first step in the process was to characterize the debris releases and vehicle damage from the imagery, radar, and element postflight reports. The next step involved comparing the data in the postflight reports against flight history and risk assessment masses documented in NSTS6 0559, Volume One. If the risk assessment masses or assumptions such as release rate were exceeded, an integrated in-flight anomaly was generated. Flight observations and test results were then used to update the release, transport, and impact models to improve their fidelity and better characterize the risk. New risk results were then computed for the next flight using the updated models, and results were incorporated into the integrated debris hazard report. In addition to the risk assessment process, vehicle improvements and debris mitigations were continually being made to reduce the debris environment and lower the overall risk to the vehicle.

Establishing the Debris Integration Group

I was constantly challenging the team to find ways to decrease the debris risk and improve the risk models. However, it was difficult to keep pace with the ever-evolving debris story and the risk assessment mandate due to the frantic pace and multitude of concurrent tasks the agency and its contractors were performing. Most debris meetings and documentation were done in an ad hoc fashion, unless debris topics were addressed at the SICB and PRCB. Even then many in the debris community would only see the final products presented in a PowerPoint presentation and had little insight into the technical methodology or reasoning behind the results. The lack of debris documentation, meeting formality, and the amount of time devoted to debris topics at the SICB and PRCB were targets of constant criticism. I proposed as a solution the debris team be formally chartered. A formal charter would establish a set of team roles and responsibilities as well as operational guidelines for decision-making, documentation, and meeting times. This would alleviate the enormous workload on the SICB and PRCB and provide the technical forum needed to engage the appropriate personnel in the debris decisions. Key debris results and decisions would still be left to Kim and Wayne at their respective boards.

The idea was well received but difficult to implement due to the need to determine the correct organizational structure and chair's level of authority. There were strict requirements for establishing technical authority and chartering boards and technical panels on the shuttle program. Helen and her team at MSFC were concerned the new "debris team" would usurp her team's authority and diminish its decision-making role within the program. Although it was not stated publicly, the last thing that team wanted was to answer to another JSC board. I was most concerned about solving debris problems that made the space shuttle safer, and I argued that the new debris team would actually strengthen MSFC's role and the roles of the element projects by including them as mandatory members. After extensive discussion and compromise, the program managers settled on calling the new debris team the Debris Integration Group (DIG) with me as the chair. Kim and the SICB delegated to the DIG responsibility for integrating space shuttle debris risk–related tasks required to assess the debris risk associated with expected debris sources during liftoff, ascent, and reentry.

When Kim first took over as the SEI manager, she struggled with deciding just how far to probe and question for proof when dealing with some of the other element teams such as the ET project. "I quickly became comfortable working with folks in SEI because I knew them," stated Kim. "However, I did not have that luxury to develop those personal relationships with the external tank team like Jim did. Since I did not know them personally, I always had this internal debate to determine whether I should accept their initial presentation or probe deeper, at the expense of questioning technical expertise or judgment, and I did not want to convey distrust. Assigning Jim as the DIG chair certainly alleviated this concern."

The DIG activities emphasized analysis of debris generation and release, debris transport, impact tolerance, and the validation components of the debris assessment process consistent with the NSTS 07700 debris requirements. Specific activities included but were not limited to implementation, review, and evaluation of analyses, trade studies, test planning and execution, data analysis, postflight assessment, sensitivity studies, and the interface relationships necessary to determine the debris risk. The DIG then reflected the results in the integrated debris hazard report. Weekly meetings were planned every Wednesday with the option for special topics as required. As the chair I was responsible for setting the agenda, establishing priorities, assigning action items, generating the minutes, and reporting a DIG summary to the SICB every week. I had already established this business rhythm, so the DIG charter was just a formality to me. To keep the lines of communication open, I provided copies of the DIG agenda, action items, and meeting minutes to all DIG members and worked quickly to establish a DIG repository that served as a centralized location for all debris information. "It was an excellent forum to bring all the various elements and stakeholders together to work debris issues," stated

Justin. "The communication and collaboration were very effective at working debris issues and making progress. Everyone supporting the DIG had to talk with each other and listen to each other in order to get anything accomplished."

Jeff Pilet concurred with Justin. "The DIG served as an efficient way for the community to have deep-dive discussions on all aspects of the debris assessments, and the ET project always worked well with DIG—especially when dealing with the big picture and sometimes emotional debates."

Although the DIG was a much-needed forum to keep the lines of communication open, there were organizations (such as Helen and her team at MSFC) that had reservations. Helen explained, "I thought the DIG was to follow the same model as the technical panels and expected it to be an advisory group that would conduct analyses and make assessments independent of the program and then present results and make recommendations to SICB. As it turned out, with the SEI-led DIG, it automatically became a program entity and thus did not have the separation and independence like the tech panels. I was concerned that program pressures could thereby influence the group's positions and that too much authority was being given to the group. It appeared DIG results and recommendations made to the SICB were always rubber-stamped, so the DIG really was making program-level decisions. This seemed inappropriate to me."

Despite these reservations, the program supported DIG.

First DIG Meeting, JSC, Houston, Texas

The first DIG meeting was held on August 4, 2006, even though the PRCB did not approve the official charter until August 24. Gerst and Wayne attended, which added substantial credibility to the DIG. It was a packed conference room and agenda that started with a charter overview and introduction of the DIG members. "It was clear we needed to continue the work on debris analysis," stated Wayne. "Having a multicentered group such as the DIG anchored in the physics of the situation was very important to avoid biases."

I discussed the STS-121 IFA closure and go-forward debris plans. These included another debris verification review in November right before STS-116. Wayne wanted to open the launch window by considering night flights, which had been banned since the *Columbia* accident. Opening the launch window would increase the operational flexibility of the shuttle program and improve the launch probability. It would also enable Wayne to move the STS-116 launch day up to earlier in December instead of over the holidays. Before making the final decision, Wayne wanted the DIG to assess the lighting conditions for a night flight. This included adding a flash capability to select cameras and understanding all the debris risk such as the contribution of the LO_2 feed line brackets, putty repair, and gap filler. Top priorities for the DIG included closing the IIFAs from STS-121, addressing the highest-level debris risks, and providing Hubble Space Telescope risk assessment support, risk model updates, and validation. The LH_2 IFR bodies and adjacent acreage were the highest debris risks and received most attention—especially when it came to determining the adjacent acreage failure mechanism. Imagery during STS-121 showed the adjacent acreage loss at station Xt-1270 consisted of multiple releases between 165 and 173 seconds into flight. This time frame was well past ASTT. If internal temperature drove the loss and the loss was typical in terms LH_2 liquid level transition, then cryopumping was the prime suspect. If this were the case, then only the first two LH_2 IFRs would be a concern for internal temperature-driven foam loss during the ASTT. Based on historical data review, there was no meaningful correlation with dynamic pressure for LH_2 IFR or LH_2 acreage foam loss. This further pointed to cryopumping. I had the ET project, Aerospace Corporation, and NESC conduct a study of the possible cryopumping release times, which now included the ongoing analysis of the STS-121 station Xt-1270 loss. NESC was working an alternative release time model that assumed the observed release time as nominal within a normal distribution and forty-second width. This analysis was promising because it

DEBRIS INTEGRATION GROUP

suggested cryopumping was a failure mechanism that had a risk of less than one in ten thousand. However, the debris community was not ready to accept the new NESC model until completing further testing and assessment.

STS-115 Flight Readiness Review, KSC

There was only enough time between flights to close the STS-121 integrated IFAs and make risk updates to the LH_2 IFR body and adjacent acreage before the STS-115 flight readiness review on August 15. Prior to the review, Ray Gomez and JSC Engineering improved the fidelity of the CFD grids and corresponding CFD solutions. The orbiter team updated its foam-on-tile damage equation distribution based on additional test data. The LH_2 IFR acreage mass distribution was also updated to include the observations from STS-114 and STS-121. Consequently the LH_2 IFR body and adjacent acreage risks decreased by about 50 percent. These were the only risk number updates from the previous flight. For the putty repairs, a review of the historical data showed a total of 205 repairs lost in the last sixty flights—approximately three to four per flight. However, not all the putty repair losses before RTF were documented. This made it difficult to assess the effect of the mitigations. The installation procedure limited the putty repair size, and if the size was exceeded, the tile was replaced. After nine losses on STS-114, a complete inspection of fleet putty repairs was conducted, and poor-quality repairs were removed and tiles replaced. This action enabled the putty repair to remain as unexpected debris. However, after another five losses on STS-121, putty repairs were classified as expected debris. The orbiter team continued microinspections and tactile testing of all putty repairs as well as removal and replacement of suspect repairs. The team had been tasked to conduct limited impact testing to develop a preliminary PRA model and additional failure mode analysis to implement mitigations to counteract the liberation. However, the test results would not be available until after the flight. Although four gap fillers liberated during STS-121, the mitigations in the reworked areas proved successful with none missing or protruding from this area. Historically fifteen gap fillers were found missing or protruding per flight (five missing and ten protruding) in areas not reworked. Missing gap fillers were both an ascent debris and reentry heating concern. Protrusions could result in early transition from laminar to turbulent flow fields on the orbiter lower surface during reentry. This would result in increased heating rates and heating loads on the tile. In terms of vehicle performance, excessive heating could decrease structural subsystem margins. In a worst-case scenario, it could compromise the tile bond strength and structural integrity of the orbiter. The team continued gap filler inspection and rework toward maximum vehicle coverage by working the highest-risk areas first. Gerst and the Flight Readiness Review Board were faced with essentially the same debris risk as STS-121 but now had additional flight data that suggested the LH_2 IFR adjacent acreage losses were due to cryopumping. The cryopumping failure mode had a very low risk because any releases (except for the first two ramps) would be outside ASTT and represented no damage risk. The debris risk was accepted, and the countdown began for an August 28, 2006, launch.

Prelaunch Weather Drama, KSC

Wayne and the shuttle program managers were pleased with the debris work and rewarded my long hours and effort. They extended my wife and family an invitation to the shuttle launch. Since I would be working the mission, I invited my parents to watch my daughters and the shuttle launch from the beach. This allowed Brenda to view it from KSC in the VIP section. It was an honor and event she was looking forward to attending. As we landed in Orlando on Friday, August 25, 2006, I received an e-mail on my BlackBerry that a lightning bolt had struck pad 39B and sent one hundred thousand amps of current through the lightning protection system. When we met up with my parents in Cocoa Beach, they described the dark thunderstorm that had roared through the area just hours before. The launch was delayed a few days to study the effects of the lightning strike and ensure all the shuttle systems were functioning properly. Two days later the National Weather Service announced the formation of Hurricane Ernesto. It was moving up Florida's Atlantic Coast and would further delay the launch. People were warned to evacuate the coastal areas as the storm strengthened into a Category 1 hurricane. Not wanting to take any chances, my family and I flew back to Houston, and my parents headed back to Ohio. We were all disappointed and knew the VIP viewing would have to wait.

A rollback to the VAB was under way on August 29, 2006, when mission managers decided to cancel the rollback and return to the pad because Ernesto was losing strength. The next launch attempt on September 6, 2006, was scrubbed early in the morning (before tanking) due to a malfunction of one of the three fuel cells in the Orbiter. The problem was corrected, and two days later another launch attempt was scrubbed due to a faulty ECO sensor in the tank. This indicated the now-chronic problem had not been resolved. After some late-night troubleshooting, the launch countdown was resumed on September 9. During the tanking a Tyvek cover (which the final inspection team later found before launch) was observed missing from the F3L thruster at the front of the orbiter. Without the cover water could accumulate in the thruster nozzle, possibly form ice, and become a debris threat to the windows. The debris team deemed it unlikely for ice to form and showed that water would not be a debris threat. While working the Tyvek issue, two frost balls had formed. Fortunately, both observations were acceptable under NSTS 08303. All debris issues were cleared, and *Atlantis* finally roared off the pad to resume assembly of the space station after a four-year hiatus.

Although this was the third flight since the *Columbia* accident, the foam debris releases were occurring regularly. Fortunately most failures were occurring outside of ASTT and posed negligible risk as predicted. Because of the release timing, the debris risk from all the failure modes except void delta P seemed to be under control. However, large void delta P releases from the ice/frost ramps were still a high-risk debris concern.

In addition to the chronic foam debris releases, a disturbing trend of orbiter debris liberation was observed during STS-121 and STS-115. The debris the orbiter started shedding such as gap fillers, shim stock, putty repairs, ceramic inserts, and blankets was initially classified as unexpected. With all the debris cameras and imagery, it was easy to see orbiter debris sources once considered unexpected start releasing during ascent. Most might have considered some of the orbiter debris to be "typical," but no one ever knew for sure until additional debris cameras were added. "The orbiter debris seemed like a Whac-A-Mole game," Justin stated. "As soon as one debris risk was fixed, another would pop up. There always seemed to be another debris source that posed an additional debris risk. It was overwhelming but not unmanageable. The orbiter team had to work aggressively to determine the causes and fix the problems. When orbiter had the problem with the ceramic inserts, I remember vividly watching the inserts fall out during vibration testing. *Holy cow*, I thought. *This is a really dangerous situation.*"

There were approximately 1,500 ceramic inserts. These provided access to fasteners that attached the carrier panels to the orbiter structure. Although they did not have a failure history and none were seen lost on STS-114 or STS-121, the size, density, and location (some in areas that could reach the orbiter windows) made them dangerous.

The debris team had to constantly assess the risk of these new debris sources and the effectiveness of the mitigations. Justin and the orbiter team worked all the orbiter issues quickly and gave each the appropriate priority level. They quickly implemented visual and limited tactile inspections of ceramic inserts on the next flight and removed and replaced any loose inserts. Historical data indicated inserts on or near the base heat shield on the aft end of the orbiter were the most likely to release because of the high vibroacoustic environment. However, debris transport analysis was expected to clear those areas.

Postlanding Debris Assessment

Just before dawn on September 21, *Atlantis* touched down on Runway 33 at KSC. This concluded the STS-115 mission. The originally scheduled landing for September 20 was postponed to allow for additional inspections of the spacecraft after video from cameras aboard the orbiter showed a piece of debris in proximity to the vehicle. Inspections using the orbiter boom sensor system concluded all of *Atlantis*'s critical equipment was in good shape. The orbiter's lower surface sustained 118 total hits. Six had a dimension of one inch or larger. Tile around the orbiter windows sustained seventeen hits. One had a dimension of one inch or larger. This type of damage near the windows was becoming a growing concern when coupled with the twelve hits observed on STS-121. The big issue was determining the source of the debris hits and whether there was a threat to the orbiter windows. Overall seven integrated IFAs related to debris were declared on STS-115—six for ascent debris and one for liftoff debris. There were always two or more dozen liftoff debris items discovered during the postlaunch walkdown inspection. Rather than declaring an IFA for each individual item that fell outside of the controls, they were integrated into one. There had been a steady improvement in the reduction of liftoff debris threats over the last three flights due to the implementation of mitigations and controls. Except for feed line bracket foam loss at Xt-1377, all ET foam losses were within the current NSTS 60559 risk assessment masses. Along with the release of a Tyvek cover on the pad, another forward thrust cover released late above the maximum allowable velocity. Despite the visual and tactile inspection of putty repairs, eight more releases were observed on the orbiter's lower and upper surface areas. This was not the only debris surprise from the orbiter. For the first time since RTF, two ceramic inserts were found missing on the upper body flap and aft fuselage stub tile. Ceramic inserts were one inch in diameter and varied in thickness as a function of the tile thickness. Most inserts installed on each orbiter vehicle were located in the aft end of the vehicle around the SSMEs. The release time was unknown. The losses baffled Justin and the rest of the orbiter

RETURN TO FLIGHT

team, and they were concerned about the possible debris threat to the windows. Based on past performance, I was expecting a few LH_2 ice/frost ramp adjacent acreage losses. Sure enough there were two such failures: a 0.023-pound mass at Xt-1593 released 328 seconds into flight and a 0.016-pound mass at Xt-1787 released at 450 seconds. The late releases coupled with the NESC thermal analysis and comparison of flight data provided further evidence the adjacent acreage failure could be attributed to cryopumping.

After the solid rocket boosters was recovered, the aft booster separation motor of the starboard SRB was found damaged. It was impossible to determine the time of impact or debris source that struck the cover. Although it was probably an ice impact, other debris sources could not be ruled out. This damage would eventually be approved as an unexplained anomaly. I was hoping for improved debris performance, but with new and reoccurring orbiter losses, the workload was only going to increase.

CHAPTER 21

Night Launch Debris Surprise

When I was training to be a pilot, I was always instructed to believe my instruments and what they were reading. Like the instruments in the cockpit, the LH_2 IFR data from the previous three missions and NESC's thermal analysis indicated the adjacent acreage debris losses were primarily due to cryopumping. They were no longer attributed to void delta P or thermal cracks or delamination. If this were the case, the risk would be much less than one in one thousand and would make it much easier to classify the risk to a lower risk level. Since substrate temperatures were believed to drive cryopumping, there was considerable discussion on the time lag between substrate heating and the time of foam debris release from the IFR body and acreage. The time lag would explain the late foam loss from station Xt-1399 event at 245 seconds (well outside ASTT). This was the conclusion I communicated to Wayne in early October 2006 during an off-line meeting with the ET project and SEI.

Night Launch Decision

Wayne wanted to set a nighttime launch date for STS-116 but first wanted to know the latest debris risk and status of the onboard flash installation. I informed Wayne I was ready to lower the risk classification of several debris sources at the STS-116 debris verification review and was confident the mitigations and controls the orbiter team were implementing for gap filler, putty repair, and ceramic inserts would work. The flight history reviews conducted in the year after the first RTF mission proved invaluable in grounding our analytic and empirical modeling—particularly for the void delta P foam failure mode. It was now becoming clear this foam debris risk was overstated because the actual in-flight release rate was lower than predicted by a factor of five to ten. Since the model otherwise did well in predicting the mass rate and release timing, it seemed reasonable to scale down the risk based on the actual release rate. If there were a factor of ten fewer losses than anticipated, the risk would correspondingly decrease by a factor of ten. Not surprisingly convincing everyone to scale down risk results by such large factors was challenging. The DIG had made a number of PRA model improvements to the release, transport, and impact models and completed sensitivity analysis to assess the levels of conservatism in the foam release and impact models. All PRA analysis and risk numbers (including the LH_2 IFRs and ice debris) would be updated. I promised to update all the debris documentation and write NSTS 60559, Volume Two. This would summarize the risk assessment methodology approach and include key ground rules, assumptions, limitations, and conservatisms. In essence Volume Two would describe how the risk index numbers documented in IDBR were derived from the Monte Carlo probabilistic risk analysis. Up until this point, the only debris methodology documentation consisted of PowerPoint slides and meeting minutes. This made it very difficult to follow and understand the flight rationale basis in the hazard report. I also planned on updating all debris documents and bringing forward updated debris launch commit criteria in NSTS 08303 to include acceptable ice-ball masses and a probabilistic ice table. The debris story pleased Wayne, and he set an evening December 7, 2006, launch date.

Last of the Debris Verification Reviews, MAF, New Orleans

Once the launch date was set, I scheduled the STS-116 DVR for November 14, 2006, at MAF. Due to cancellation of the PRCB over the Thanksgiving holiday the following week, I would have to fly back to Houston immediately after the debris verification review to present the debris hazard report to Roy Glanville and the ISERP in order to be ready for the November 16 PRCB. It felt like final exam week at the Naval Academy, and I expected to pull several all-nighters. Before the DVR began, I chased down Brekke, Justin, and Darby to make another bet on whether I would finish the presentation in a day. My plane left at 6:30 p.m., and I was certain to win the bet this time. I was proud of the work accomplished by the DIG. It highlighted the numerous vehicle changes that tangibly lowered the debris risk and the PRA model updates that more accurately represented the debris risk. It was time to take credit for this work and reduce the risk classification of several debris sources—especially those classified as "infrequent catastrophic." Once this was done, the program could more effectively utilize resources addressing other high-risk areas and not just debris. For Brekke and the Aerospace Corporation team, it was rewarding to see how far the PRA model had progressed in terms of characterizing the foam debris risk. Darby and the Boeing team were also ready to "tie a bow" around the risk reduction. "The debris hazard report was mature, our methods were mature, most of the tools had been through pretty rigorous peer reviews, and there was lots of verification and validation work done to give us confidence in the results," Darby said. "This was a good exercise in risk trades because the technical community very much wanted to stand down, but program management recognized the need to continue flying and the somewhat elusive nature of a true fix. In general foam debris was all this way. It could be improved but never truly fixed, and that was one of the big differences between *Challenger* and *Columbia*. After *Challenger* the failure mode was completely eliminated by redesigning the faulty hardware. After *Columbia* foam was still required to insulate the tank, and the mitigation effort was about reducing the size of the failure. Even so foam was still going to shed."

Improvements in the model along with debris mitigations had been successful at lowering the foam debris risk—especially when considering some of the conservatisms built into the impact and release models. Risk for special tile was artificially high because there were no data on the capability, and there were no test data for hits to the tile below 0.002-pound foam debris size (the majority of releases). Any foam impact smaller than this mass size was rounded up and treated as a 0.002-pound impact, which created a bigger damage cavity. This artificially drove the foam debris risk level much higher. Brekke and I both knew there was more capability in the tile damage allowable based on this "small foam" limitation. The debris clouds were also overpredicting the release rate of foam on the order of ten to fifteen times higher than we had seen over the last three flights. Adjusting the parametric analysis for the small foam impacts and release rates consistent with flight history lowered the foam debris risks by an order of magnitude. Consequently I felt lowering the foam debris risks was justified and provided a more realistic portrayal of foam risk levels.

Significant Debris Improvements

Since the second RTF mission, several vehicle controls and mitigations had been instituted. This included inspection, repair, and replacement of gap fillers, blankets, putty repair, and ceramic inserts, numerous liftoff debris mitigations, change in the Tyvek cover bonding adhesive, and addition of an ice-ball mass allowable table to the NSTS 08303 documentation. There were also significant improvements in the third-generation foam PRA model. The CFD grid was refined using higher-fidelity geometry. This produced more-representative solutions for debris transport. The foam debris pop-off velocity and release angle for each failure mode were determined to more accurately represent the release characteristics. A number of key foam model conservatisms were identified and quantified. This enabled computation of an integrated foam debris risk and showed how the model historically overestimated the foam debris risk. Specifically the models overpredicted the release frequency as compared to observed flight history in some cases by a factor of ten. To Jeff Pilet and the ET project, seeing all their hard work start to gel and provide a realistic characterization of the debris risk was extremely gratifying.

Jeff always felt there were conservatisms relative to release rate and timing, but there were also many other limitations relative to predicting foam debris. The feed line camera was vitally instrumental in providing the timing data that showed the majority of the losses coming from the liquid hydrogen tank were cryopumping related. Consequently the foam debris risk from this source was deemed very low. "With this data in hand," Jeff recalled, "we turned the corner relative to our ability to characterize the liquid hydrogen tank adjacent acreage debris risk for this foam failure."

The third-generation PRA model was used to update all PRA analysis and risk numbers and assess the debris risk of the modified LH_2 IFRs. Sensitivity analysis was done for the first time to determine confidence levels of PRA analysis, and a more detailed PRA was performed on select enveloped foam sources—most notably the feed line. The ET project had developed additional debris tables for the LO_2 IFR and feed line fairing. Debris reconstructions from all three flights were used for model validation with respect to release, transport, and orbiter damage. Recurring liftoff debris items such as rust scale had decreased on Pad B over the last three flights. Ongoing mitigations and controls were paying off in reductions of liftoff debris threats.

Small Foam Damage Limitation

Significant limitations of the foam impact and damage model were its inability to assess foam masses between 0.002 pounds and 0.0002 pounds and its tendency to overpredict the release rates. The majority of foam debris releases were in this mass range, and the model would automatically default to the larger 0.002-pound damage profile. Consequently the model was inflating the risk contribution from small foam impacts. This meant popcorn foam dominated the foam debris risk estimates. This was inconsistent with the flight performance data, and sensitivity analysis showed the foam debris risk numbers improved by almost an order of magnitude if the contribution attributable to small foam debris was neglected. The orbiter team had the action item to create a small foam-on-tile damage map to remove the limitation and perhaps even establish a deterministic size higher than 0.0002 pounds. However, this task would take some time, and results would not be ready until January 2007.

Another significant foam model limitation was the high release rates. These were five to ten times the historical performance. Sensitivity analysis showed the foam debris risk decreased proportionally when the risk numbers were adjusted for the actual flight history release rates. The ET project was reluctant to change debris tables because of the effort involved but supported using actual flight history release rates to determine a risk range that would be documented in the debris hazard report.

Cryopumping Failure of the LH_2 Ice/Frost Ramp

Strong resistance and debate over the LH_2 IFR adjacent acreage risk and determination of the failure mode still existed. The NESC cryopumping results were consistent with losses from the last three missions. Incorporating the NESC analysis and flight observations shifted the release time to much later in the ascent. Consequently only the first two ice/frost ramps were inside ASTT with a risk contribution of 1 in 3,300. Based on the cryopumping failure mode and risk level, I recommended changing the LH_2 acreage risk from "infrequent" to "remote." MSFC's management was against the recommendation and planned to dissent at the ISERP the following day. Despite the release rate and small foam mass sensitivity analysis that showed a significant risk overprediction, the MSFC managers insisted the LH_2 IFRs be classified as "probable catastrophic." They argued the risk was not integrated for all debris failure mechanisms to all impact areas. I argued the initial high risk calculations were manifestations of the assumptions and limitations of the model, and that included our understanding of the failure mechanism. Brekke and the Aerospace Corporation team agreed. "The NESC, Aerospace Corporation, and the ET project had compiled enough data and analysis to show the liquid hydrogen IFR failures were due to cryopumping. This meant releases for all the liquid hydrogen IFRs other than the first two locations would be well outside the ASTT time frame. The contribution of risk from the first two locations was also low due to the small percentage of releases predicted inside the ASTT time frame."

Despite all the debate, the additional flight data from STS-121 and STS-115 coupled with a better understanding of the failure mechanism enabled the debris community to more accurately determine the risk and recognize it was time to focus resources on higher debris risks.

Integrated Debris Risk

There was also intense discussion over the integrated foam debris risk of 1 in 140—a byproduct of the Hubble Space Telescope risk assessment work. The number combined tile, RCC, and special tile into the computation as opposed to separating each into individual assessments. Since the release models overpredicted release rate, the flight history release rates were used to adjust the overall STS-116 PRA results. The result was consistent with what the office of Safety and Mission Assurance produced using historical debris damage from eighty-three past flights and a Monte Carlo PRA simulation of one hundred thousand iterations. There were several sidebar discussions on the eventual usefulness of such a number and whether it was worth investing more resources into perfecting it. At the very least, flight history releases could be used to calibrate the release models and determine risk ranges. This was consistent with how the ice risk assessment was tabulated.

Ice Debris Risk

There were no updates to the ice debris assessments, and the ice risk numbers for STS-116 were the same as they were for STS-115. The highest ice debris risks were on the feed line brackets at stations Xt-1623 and Xt-1871. The risks there were 1 in 150 and 1 in 220 respectively. This was based on the worst-case parametric assessment. All other ice debris risks—other than undefined umbilical ice—were well below one in one thousand. The debris community felt comfortable with the ice risk because the program could rely on the NSTS 08303 day-of-launch ice criteria as an important mitigation. Mitigations of ice debris risk were still being pursued in the area of coatings. An NESC study on ice coatings showed the leading candidate, the shuttle ice liberation coating (SILC), demonstrated a reduction of ice adhesion and appeared promising. It could potentially mitigate the ice at the feed line brackets and umbilical areas. However, NESC had some additional compatibility testing to perform before SEI would stage a full-scale umbilical ice growth and liberation demonstration test.

Orbiter Debris Risk

During STS-115 no protruding or missing gap fillers were identified in any reworked areas—a real success story. Since the PRA modeling still had significant limitations, the gap filler risk would remain classified as "infrequent catastrophic" for STS-116. Continuing the inspections and replacement of suspect gap fillers would naturally drive this risk down in time. Wayne remarked that the program was obligated to expend resources on any risk classified as "infrequent catastrophic" in an attempt to reduce the risk. Since gap fillers were an ascent as well as reentry issue, conducting an EVA for removal was a mitigation strategy if needed but also carried with it a potential mission impact. I recommended the program expend its resources on the mitigations and forgo the work needed to improve the gap filler PRA.

After STS-121 putty repairs were classified as expected debris, and STS-115 experienced eight putty repair losses from areas not reworked. After that microinspections and tactile testing of all putty repairs were completed on two of the three orbiters with more than three hundred repairs made on both vehicles. Even so putty repairs remained classified as "infrequent catastrophic" for STS-116 due to the significant uncertainty remaining with the transport, release, impact tolerance, and effectiveness of the inspections and rework. The orbiter team continued microinspections and tactile testing of all putty repairs and replacement of suspect repairs. All vehicle inspections would be complete after the next two flights. Additional inspections were performed in high-risk areas forward of the windows and RCC panels to ensure an adequate gap around the repairs. This minimized the possibility of mechanical articulation dislodging a putty repair. The orbiter team also changed the putty repair process to limit the repair mass. They conducted limited impact testing to develop a preliminary impact model as well as additional failure mode analysis. This was used to implement mitigations to counteract the liberation. A putty repair PRA was being pursued with results expected to be similar to the bellows and bracket ice risk numbers. Although no putty repair damage map existed, orbiter team members agreed that using the ice damage map was a reasonable approximation for putty repair damage. The initial putty repair impact test results showed deeper penetration than ice, and the orbiter team suggested a depth conversion factor for putty repairs to account for the deeper penetration.

The orbiter team was investigating the ceramic insert failure mode and suspected it was due to the high vibroacoustic loads. Vibroacoustic testing of the ceramic inserts was performed at the Huntington Beach test facility using a progressive wave tube (PWT) to simulate the vibroacoustic environment. The objective of the test was twofold: one, replicate ceramic insert failures to solidify the root cause of the losses, and two, characterize the risk level differences between forward and aft locations by exposing inserts to representative environments. All ceramic insert failures observed during testing required very high acoustic levels. In addition only inserts with defects failed the test environment. Therefore, there were no failures in any of the ceramic inserts that had zero defects, regardless of the vibration level. The test facility was unable to replicate or match the acoustic environment at the low frequencies (less than fifty hertz) and upper frequencies (greater than eight hundred hertz). These were documented test limitations. Although these limitations existed, the test facility was able to produce an acoustic environment with vibration levels greater than what the main engines generally produced. During main engine ignition and liftoff, the acoustic environments in the aft were much larger than anywhere else on the vehicle. This increased the possibility of liberating a ceramic insert. Insert locations forward of the OMS pod were low risk due to the limited historical failures. The acoustic testing was consistent with the flight observations in this area, and this showed resiliency even if the ceramic inserts had bond line defects. Inserts on the forward OMS pod locations had limited historical losses isolated to the forward outboard edge. The orbiter team completed a hardware modification that eliminated most forward inserts (about 190 out of 200) from the OMS pod. This minimized the likelihood of release.

Debris transport analysis was performed on aft locations where about 85 percent of the historical ceramic insert releases had occurred (including those on STS-115). The transport analysis model treated the inserts like ice based on how the ceramic shattered on impact. DTA had been provided to the orbiter, SSME, and SRB teams for assessment and was expected to be cleared in the aft areas. The SSME project conducted ceramic insert impact testing in December 2006 that established a new impact limit of 208 foot-pounds—a significant improvement over the previous 84 foot-pounds. Since there were no analytical impact conditions to the SSME above 208 foot-pounds, there was no possibility of catastrophic damage to the SSMEs. Visual inspections and limited tactile and pull tests of ceramic inserts were conducted for STS-116. Visual inspections were used to identify cracks in the bond line, tactile testing was used to identify loose inserts, and pull testing was used to identify degraded bonds. Because of the combination of visual and tactile inspections of suspect parts, the cohesive failure mode was successfully eliminated. However, not all such bonds were necessarily susceptible to the adhesive failure mode. Many inspections had already been completed, and three loose inserts on the base heat shield were found, removed, and replaced. Historical data, tile configurations, and acoustic environments had yielded indications that locations on or near the base heat shield were the most likely to experience an insert loss. Forward inserts located in areas of high-density tile had a very low historical release rate. I recommended the program managers accept the proposed flight rationale and risk assessment characterization as "remote catastrophic" for STS-116 due to the mitigations in place and the history of infrequent releases in critical areas. The ceramic inserts aft of the OMS pods would be classified as "infrequent catastrophic" until the element projects came back with their assessments. Ceramic inserts located forward of the OMS pod were still considered unexpected debris.

STS-116 DVR Summary

The ET project continued with the LH_2 IFR redesign and developed select debris tables for enveloped foam sources with high conditional probabilities. The orbiter team pursued additional testing of putty repair on tile impacts to better assess impact damage and develop a PRA model for this debris source. The team was also assigned to complete the updated foam-on-tile damage map to account for small foam debris. The DIG continued correlating the debris model with flight history to improve accuracy of the PRA models and updating mass and release rate flight history distributions after every flight. The Aerospace Corporation continued with combined-environment testing of popcorn foam, ice-balls, and small foam-on-tile damage assessments. Both the Aerospace Corporation and USA–Boeing were tasked with continuing model comparisons and placing all debris models under program configuration management control rather DIG control. NESC continued to refine the cryopumping release timing model and worked on ice mitigation coatings.

I concluded the STS-116 DVR with a few key recommendations. First, the program should begin focusing resources more toward better on-orbit debris assessment tools rather than PRA model improvements in order to minimize the possibility of unnecessary CSCS. Second, the orbiter team should accelerate replacement of tile with hardened tile in high-impact areas such as around the umbilical doors. Finally, the ET team should continue work on the debris risk reduction plan consistent with the program risk classifications. Wayne requested the DIG and ET project develop a prioritized list of debris risks consistent and defensible with the Space Shuttle Program's priorities. He also agreed the program should no longer emphasize further refining the PRA numbers but should instead take a more qualitative approach to prioritizing debris risks and corresponding mitigation strategies without overdependence on accurate probabilistic risk numbers. The DIG needed to determine which of the top risks could be mitigated sooner as opposed to later based on the relative ease of implementing the corresponding mitigation strategy in terms of performance, cost, and schedule factors. Although I won the bet by finishing the DVR presentation sooner than planned, I would have to wait until later to collect. I had a flight to catch back to Houston to prepare for the ISERP and PRCB.

STS-116 ISERP Review, JSC, Houston, Texas

As knowledge of debris performance improved along with the maturity of PRA models, I generally pushed to lower the general debris risk level. One result of the DIG's continuing efforts to improve the PRA models was that the risk numbers changed frequently. This made it challenging to keep up with all the model changes and updates to risk numbers while supporting an operational program. Roy and the ISERP believed the recommendations to reduce risk levels were suspect due to their large uncertainty bounds and limited physical data. "Although PRA sometimes gives you a relative feel for risk, it can also be a distraction," Roy asserted. "Since improved PRA models didn't change the system design or the way we deployed and operated the shuttle, I felt we needed to avoid making hazard report risk ranking simply a numbers game."

Roy and the ISERP preferred to focus on effective hazard controls achieved through test-verified designs, processes, and procedures. Roy's duty was to ensure the predicted and documented risk assessments complied with the program risk assessment and reporting requirements and allow their delivery to the program manager who would ultimately have to approve them. Slowing down the program to fully mitigate high-risk debris sources was a subjective call and something fiercely debated within the ISERP. The STS-107 accident and other debris losses since RTF clearly demonstrated the shuttle program could not always be assumed safe to fly merely by continuing to operate as in the past. The ISERP was often afforded little to no advance time to review updates to the IDBR document. This forced Roy to participate in and glean from DIG presentations the significant data to post for ISERP members' visibility. "We worked with the best data available and committed to flight with the expectation that the program would implement corrections for encountered issues during flight," explained Roy.

Confidence in our ability to determine debris risk grew over time with additional flight experience to validate processes and failure mechanism models along with higher-fidelity target capability models.

STS-116 Flight Performance

During the Program Requirements Control Board meeting, Wayne lifted the daylight-only launch restriction because of the STS-116 DVR and the improved performance of the shuttle's external tank in minimizing foam shedding during ascent. The overall debris risk was lower and better characterized. The orbiter team also had the ability to perform 100 percent inspection of the orbiter thermal protection system for damage and rely on a rescue mission as a last resort. The STS-116 debris risk assessment I presented at the flight readiness review was a summary from the DVR. It lowered all the foam debris risks from "probable catastrophic" and "infrequent catastrophic" to "remote catastrophic." I concluded by stating that the debris risk for STS-116 was the most representative to date because it was better characterized by modeling improvements resulting from flight observations and correlations. Gerst and the FRR Board agreed with the shift to night launch operations and the recommendation to continue with the LH_2 IFR redesign because this was still the highest-ranking foam debris risk due to the deficient design (thermal cracks) and multiple failure modes. Removing the night launch restriction allowed *Discovery*'s launch to take place at night for the first ten days of its launch window. This opened on December 7 and closed on December 26 and was based on solar beta angle constraints during rendezvous and docking. STS-116 was being rated one of the most challenging shuttle missions in NASA's history. During the thirteen-day mission, which was the twentieth shuttle flight to the ISS, the crew would rewire the outpost's power system and continue constructing the station by installing the P5 integrated truss segment. *Discovery*'s liftoff would be the first nighttime shuttle launch in more than four years.

From a debris perspective, I was expecting an uneventful flight. Due to limited lighting conditions, image analysis would depend mostly on film and video taken from ground cameras and SRB camera video captures. This was a fraction of the imagery assets. Products normally available from the external tank feed line-mounted camera, the umbilical camera, and handhelds would not be available for review—or so I thought. Odds were that any significant debris event would not be captured. If all went according to plan, I would not have to work over the Christmas holidays for the first time since the *Columbia* accident and would have minimal preparation work for STS-117, which was scheduled to launch in February 2007.

The first attempt on December 7 was scrubbed just minutes before the scheduled launch time due to weather. During the final inspection, the team discovered several ice formations on the oxygen vent measuring up to one-half inch in diameter. Due to their forward location on the vehicle, these ice formations were pushing the allowable limits in NSTS 08303. This sparked some debate over the size estimate accuracy. Fortunately the icing conditions did not reoccur during the next launch attempt on December 9, and there were no debris issues to work. This was a welcome first. *Discovery* roared skyward at 8:47 p.m. (EST) and illuminated the night sky in a spectacular light show.

The light debris workload I was anticipating for the flight was short-lived. During SRB separation the SRB cameras captured debris emanating from the aft booster separation motors, traveling forward, and impacting the lower orbiter surface. The bright plume from the SRBs and main engines provided enough luminosity and contrast to record the event. Surprisingly it would have been more difficult to see during the day because the bright plumes and sunlight together tended to overexpose certain images. A fraction of a second after SRB separation, a piece of nozzle booster trowelable ablative (BTA) weighing an estimated one-tenth of a pound struck the orbiter's lower surface tile. The debris team and orbiter and SRB projects worked through the night trying to determine the impact conditions and damage extent.

Later the next day, photography of the lower orbiter surface taken during on-orbit inspection operations showed several impact damage sites near the umbilical door and tile forward of the body flap. Ultimately the orbiter team concluded the damage was within the experience base of the orbiter and not a reentry heat threat. The MMT members breathed a collective sigh of relief when the orbiter team presented its conclusions, but the DIG and SRB project were now faced with the daunting task of determining the root cause of this new debris source and the corresponding debris risk. This debris event would jeopardize the launch schedule of the next mission if the risk were too high or if mitigation were required. My Christmas break would have to wait another year.

The postlanding debris inspection of *Discovery* was conducted at Kennedy Space Center on December 22, 2006, following its thirteen-day construction mission. There were a total of 172 impacts. Twenty-two had a dimension of one inch or larger. These numbers were within the shuttle program's flight experience but greater than the average of previous flights. Tile on the lower surface sustained 121 total hits. Seventeen had a dimension of one inch or larger. Approximately eighteen damage sites were concentrated in the area forward of the starboard main landing gear. Some of this damage was attributed to foam and possible ice loss during the ascent phase. Numerous damage sites around the LH_2 and LO_2 umbilical area represented the largest concentration of hits on the lower surface. The largest damage site of the vehicle measured 3.0 inches by 2.0 inches by 0.2 inches deep. It was also located in this area near the port side umbilical door. Tile damage in this region was typical for every flight, and pieces of umbilical purge barrier (baggie material) flailing in the airstream and contacting tile before being blown downstream usually caused the damage—but nothing this big.

A comparison of the STS-116 lower surface hit maps with those from STS-114, -115, and -121 missions showed no discernible trend except for hits near the umbilical doors. Despite the inability to observe foam loss during ascent due to the nighttime conditions, the program identified ten integrated debris IFAs—nine on ascent and one during liftoff. This was the case even though it was a low ice day and the Tyvek covers performed per specification. The most significant event was the BTA impact from the solid rocket booster separation motor at SRB separation. Other than the PAL ramp loss on STS-114, Wayne proclaimed this the most significant debris event to date and classified the new debris source as "infrequent catastrophic." Other notable debris events were more missing ceramic inserts, missing stiffener foam on the SRB motor, and the umbilical door debris damage IFA, which was based on a high-impact trend over the last four flights. With each progressive flight, the debris focus shifted between the elements. First it was the PAL ramp and LH_2 IFR foam issues with the external tank, and then it was the problems with the orbiter gap fillers, ceramic inserts, and blankets. Now it was the BTA and stiffener foam debris from the solid rocket booster.

Assessing the New Booster Trowelable Ablative Debris Risk

Investigation of the BTA anomaly was successful in identifying the liberated material and failure mechanism, but we lacked data to properly define the debris release conditions. A parametric study was performed using various sizes, release velocities, and angles to envelope the potential debris impact footprint and impact conditions on the orbiter. Even before STS-116 landed, the DIG and SRB project began an investigation that included testing the booster separation motors. Analysis of previous booster separation motor qualification tests provided typical sizes (and masses) of BTA debris. The initial debris transport analysis used these dimensions. Release velocity was estimated using image data collected during the booster separation motor qualification tests and from video capture of the flight observations. The image data showed debris release velocities could range from 50 feet per second to 275 feet per second, which was more than enough speed to allow forward travel. Booster separation motor qualifications tests showed failure of the thermal protection system (BTA material) was due to stress buildup near the nozzle exit. Since the testing was inconclusive in determining the maximum release angle, the DTA modeled the release angles from 90 to a maximum of 120 degrees.

Unlike all other vehicle debris sources, the debris transport analysis for the BTA was a two-step process. The position and velocity vectors computed in step one were provided as input for the step two DTA after 0.034 seconds of particle travel. This covered the time after motor ignition. Step two debris transport analysis was performed using the full booster separation motor thrust. This corresponded to 0.1 seconds after the SRB separation command. As expected the results showed BTA debris had transported to the orbiter lower surface near the umbilical door and the external tank acreage up to the intertank region. Both impact velocity and angle generally decreased as debris impacted more-forward locations, but the results of the parametric study generated unacceptable impact conditions. These unacceptable impacts provided the motivation behind the SRB project's mitigation efforts to eliminate the debris from reoccurring.

After the SRB was recovered and inspected, a substantial piece of stiffener ring foam was found missing from the port side. The stiffener rings were structural reinforcements that strengthened the high-loading areas near the nozzle. This foam loss was the only historical failure. Foam similar to that used on the external tank was applied around the circumference to minimize aerodynamic heating. The remaining foam left in the depression was charred, and that indicated a potential ascent material loss. The external tank attachment ring enveloped most of the foam closeouts on the SRB, but the maximum debris size from the stiffener ring exceeded that of the external attachment ring. This invalidated the enveloping rationale and triggered a unique debris transport analysis for the stiffener foam. The results showed there were only fifteen cases of the seventy-five thousand simulations that exceeded the tile threshold curves that indicated the onset of tile damage. Consequently the flight rationale was based on the low likelihood of debris liberation, low likelihood of transport, and low likelihood of critical damage if an impact occurred.

Determining the Cause of the Umbilical Door Impacts

The high number of impacts around the umbilical doors was a growing concern for Justin and the orbiter project due to a possible compromise of the door seals. Kapton baggies were tied around both the liquid hydrogen and liquid oxygen umbilical doors as a moisture barrier to keep ice from forming on the exposed metal surfaces. During ascent the aerodynamic forces eventually tore the baggie off as designed. The umbilical cameras often showed the baggie would not liberate completely, leaving the cord and residual Kapton material to flap around the doors. I hypothesized there was a strong correlation between the baggie performance and the impact count, but the historical data proved otherwise. In fact the umbilical door impacts did not correlate well with the baggie performance. This suggested the baggie might not have been a significant damage contributor. Historical impact data for the entire vehicle did not show a left- or right-impact bias, but aft of the external tank crossbeam, there was a distinct impact bias on the port (or liquid hydrogen) side. The region aft of the crossbeam was arguably the most complex flow field on the vehicle. Strong circulation that would actually transport debris forward characterized this area. The DIG concluded the umbilical door damage bias was due to larger, higher-density ice that formed on the colder liquid hydrogen umbilical and the general debris bias aft of the crossbeam. There was no reason to increase the risk posture in the debris hazard report, but I did recommend the orbiter team accelerate the installation of the more impact-resistant BRI-18 tile. I suggested starting with the highest-impact areas first. The damage seen thus far was consistent with damage observed during previous flights, and the worst damage would not have been catastrophic even if it had occurred directly on the door seals. Minimizing the amount of ice growth through the NESC ice inhibitor coatings was another DIG action that first required some performance testing.

Design Requirements for the LH_2 IFR

It was late December 2006 (right before the Christmas holidays), and the DIG was diligently working the STS-116 integrated IFAs and LH_2 IFR redesign. Wayne requested Kim and SEI develop debris design requirements for the LH_2 IFR redesign and provide them to the ET project in the form of a requirements letter. Since the adjacent acreage foam losses were being attributed to cryopumping and had a low risk, the debris requirements would focus on the 1 in 285 void delta P risk and the ice risk. The letter specified a threefold improvement in the void delta P (debris cloud risk) to meet the ice deterministic mass allowable and reduce the propensity for thermal cracking. Wayne approved the letter, and Jeff Pilet and the ET project embraced it without much debate. They already had a few concepts that could meet the requirements and wanted to get started on the qualification testing.

The orbiter team was still working to improve its foam-on-tile damage map by removing the small foam limitation, but it did not expect to have the work completed until the end of January. I was convinced the void delta P risk would change substantially and might even be enough to forgo the LH_2 IFR redesign, but that determination had to await the orbiter team's completion of work. The next flight (STS-117) was scheduled to launch on March 15, 2007, and the debris team was challenged to resolve the solid rocket booster BTA issue and close the ten debris integrated IFAs. STS-116's debris performance reinforced the need to diligently scrutinize all debris assumptions and risk assessment results for all elements. From that flight forward, I was always prepared to expect the unexpected.

CHAPTER 22

2007—Rough Start for NASA

NASA in the Spotlight

NASA had an ambitious shuttle manifest in 2007 with five planned construction missions to the ISS. The Space Shuttle Program was also planning on launching from the newly refurbished 39A launchpad and turning over pad 39B to the Constellation program. However, several bizarre events during the first six months of 2007 made life brutal for NASA and the human spaceflight program. The challenges NASA faced during this time were almost unimaginable. Most of NASA's leadership took everything in stride. According to Mike 2007 was a difficult year. "We had a great team in place, and we were just dealing with the problems that came at us—one by one and as professionally as we could," he stated. "As sad as some of these events were for the people involved, most had nothing to do with flying safely and professionally." That was where the NASA administrator put his attention, and the team followed suit. "The Lisa Nowak incident, the allegations of drunken astronauts, and the incredibly sad events at JSC—in the end these were just distractions, and I'm not easily distracted," Mike stated.

At the end of January, the program was preparing for STS-117, and Kim Doering announced she had been promoted to one of the four shuttle deputy program managers. Don Noah, the former ET project manager from MSFC, was going to replace her. As SEI manager, Kim had accomplished what she was asked to do—build a cohesive and confident team. People who were not as engaged before she took over became more involved and more confident with debris and other SEI tasks. "I felt we reached a healthy tension between SEI and the projects that led to the maturation of the team and development of high trust levels," stated Kim. "Although John's aggressive approach to leading SEI was necessary to establish the organization within the program, I had the luxury of seeing where the pendulum needed to swing to build the team and give team members more authority and responsibility over their work."

Kim now felt comfortable with the debris risk since the DIG had pretty much minimized it to the greatest extent possible. However, there was still some residual risk that would always be there.

In her new position as the deputy program manager, she would be exposed to other program risks and would shift her focus away from debris to other areas. The one thing Kim remembered telling Don when he took over the SEI reins was to be aware of the delicate balance between the level of autonomy and engagement between the SEI team and the element projects. SEI needed to be strong and engaged to prevent the organization from slipping back to where it was before the *Columbia* accident. Don Noah's deputy would be Don Totton, an easygoing and self-proclaimed country boy from Oklahoma. Totton had moved over to SEI from Safety and Mission Assurance just before John Muratore was dismissed. He was an instrumental figure in the safety group and had been involved with debris problem resolution ever since the *Columbia* accident. Don Totton's knowledge of shuttle safety and debris complemented Don Noah's external tank experience, and the two became known within SEI and on the shuttle program as the "Dons."

Astronaut Accused of Attempted Murder and Kidnapping

The first major ordeal in 2007 occurred on February 6 when Lisa Nowak, a forty-three-year-old US Navy captain, NASA astronaut, and mission specialist on STS-121, was charged with attempted murder and kidnapping. She had driven from Houston, Texas, to the Orlando International Airport in Florida. Allegedly she wore a diaper so she wouldn't have to stop. She wanted to confront a romantic rival, Colleen Shipman. Lisa, who was married with three kids, thought Colleen was involved with fellow astronaut William Oefelein. It was apparently a love triangle gone sour. However, Nowak claimed her relationship with Oefelein was "more than a working relationship but less than a romantic relationship."

Upon arriving at the Orlando airport, Nowak donned a disguise and tracked down Shipman. Shipman had boarded a shuttle bus to take her to her parked car, and Nowak followed her there. At first Nowak started beating on her car window, asking for a ride and a cell phone, and feigning crying. When Shipman cracked the window to tell her no, Nowak sprayed pepper spray into Shipman's car. Shipman sped off to the parking lot attendant booth, and the police were called. A search of Lisa Nowak's car at a nearby motel turned up pepper spray, a BB gun, latex gloves, copies of e-mail traffic between Colleen Shipman and William Oefelein, a knife, rubber tubing, six hundred dollars in cash, a steel hammer, and garbage bags. The story hit the national news and spread through the NASA community like wildfire. A short time after she completed her STS-121 mission, I bumped into Nowak on an elevator and introduced myself by telling her we were classmates at the Naval Academy. We had attended many of the same classes together as aerospace engineering majors. She looked at me as if I had two heads and said nothing as she stepped off the elevator. I thought it was a strange encounter but quickly dismissed the episode and went about my business.

A few days after Nowak was arrested and released on bail, an even more sensational story broke regarding the death of Anna Nicole Smith. The former Playboy model had a flamboyant made-for-reality-TV life. It featured paternity and inheritance battles, drug and weight loss struggles, and her son's mysterious death. She died February 8, 2007, in Florida at thirty-nine years of age. This tragic event relieved NASA of the national spotlight as the media quickly shifted focus to the celebrity's death. Dramatic incidents faced the shuttle program every year. According to Wayne, however, "Not all of them captured the attention of the media like the Nowak incident. Despite the numerous challenges every year, it was my responsibility as the shuttle program manager to keep the team focused on accomplishing the goal—to safely fly."

A few months after the Nowak incident, a JSC report produced by an external review team offered the following recommendations to try to determine the psychological well-being of active-duty astronauts: conduct a thirty-minute behavioral medicine assessment in conjunction with the annual medical flight physical and perform behavioral medicine flight assessments for shuttle crewmembers. Probably the most significant recommendation was incorporating enhanced aeronautical adaptability ratings (an assessment of fitness for flying duties) into the astronaut medical selection process. Ellen Ochoa, director of flight crew operations, also announced plans to have astronauts develop a formal written code of conduct to avoid such scandals in the future. As Ochoa put it, "It's an inescapable fact that human spaceflight involves humans."

STS-117 Flight Preparations

Even with all the rampant gossip about Nowak, the DIG stayed focused on STS-117 flight preparations, and by mid-February it had reduced the booster trowelable ablative risk by redesigning the BTA and eliminating the most probable release mechanism. Recent testing and debris transport analysis results confirmed the redesign and reduction of debris mass by a factor of eleven. There was no longer a transport risk to the orbiter windows, RCC panels, or lower surface tile. That included the special configuration tile, and it only had limited transport to the vertical tail and aft fuselage upper surfaces. Despite the significant reduction in risk, the DIG recommended keeping the risk classified as "infrequent catastrophic" due to the modeling limitations and uncertainties. If the first flight of the redesign was successful and the SRB team confirmed all the liberation assumptions, the risk would be lowered.

In addition to the BTA risk reduction, the orbiter team brought the DIG an updated foam-on-tile damage map that eliminated the small foam-on-tile limitation. This had significant impact on the foam-on-tile risk for the void-delta P failure. The LO_2 IFR risk dropped from one in sixty-five to one in ten thousand. The LH_2 IFR risk dropped from one in two hundred eighty-five to one in one thousand, and the intertank IFR risk dropped from one in two hundred to one in two thousand. The Aerospace Corporation also recalculated the integrated foam debris risk, and it changed from one in two hundred to one in four hundred thirty. Although I was convinced the LH_2 IFRs were a remote risk, it was going to be an uphill battle to convince Roy Glanville, safety, and engineering due to the chronic number of losses on each flight.

Another program concern was the number of debris impacts to the orbiter since RTF. After four missions the orbiter impact counts for each mission were higher than the historical average before all the modifications, process improvements, mitigations, and controls were implemented to decrease the debris risk. Based on impact counts, it appeared the debris environment was actually getting worse. However, what was statistically different was the amount of damage volume in the tile. This had decreased by a factor of two. Although it was only a small sample size, the flight data suggested the impacts were from smaller debris and/or a reduced debris environment in terms of impact conditions. The debris environment in terms of tile damage was getting better. All ten IIFAs from STS-116 had been closed, and the DIG expected the debris risk for future shuttle missions to be less than that documented in the debris hazard report for STS-117. The risk reduction was driven primarily by further inspection and replacement of faulty gap fillers, blankets, and ceramic inserts, putty repair, and future vehicle improvements such as the LH_2 IFR redesign and liftoff debris mitigations.

Freak Hailstorm Hammers the Shuttle

NASA managers had hoped to launch *Atlantis*, the first of five planned 2007 shuttle missions, on March 15. However, on February 26 (the day before the STS-117 flight readiness review) a freak hailstorm thundered over Launchpad 39A at KSC. Hail blasted the shuttle's external tank, and it suffered thousands of pits and gouges in its foam insulation. Wind gusts reached sixty-two knots, and hailstones up to two inches in diameter were found at the pad. Movable protective panels shielded *Atlantis*, and it was unscathed. The next morning at the flight readiness review, Gerst decided to press forward with the meeting while the damage was still being assessed at the pad. I enthusiastically shared the results of the SRB BTA redesign and the updated risk numbers from the new PRA runs. However, questions about the debris risk from repairs resulting from the storm constantly sidetracked me. I had already established a standard methodology for screening foam repairs. This was performed as part of the ascent foam debris PRA. Foam repair risk was never given much attention. The risk was so low, and the repairs rarely failed. Since each tank averaged several dozen repairs, and the total repair counts would always come in late, a screening methodology was developed as a way to quickly assess the repair risk. Top repair hazards were identified by screening out small repairs located further aft on the tank. This approach permitted detailed risk assessment for a manageable number of repair sites, and all others were enveloped. Repairs typically had risk levels below one in ten thousand—primarily because they never failed. No debris release meant there would be no debris risk. The hailstorm occurred before Jeff Pilet's first FRR as the ET project chief engineer. He first heard about the freak storm while having dinner at the Sandbar—an ET team favorite. "The BlackBerry phones started ringing, and when I first saw the pictures, I was again left with the thought, *How in the hell are we going to convince the community that this tank can be safely restored?* The ET project was just getting folks comfortable with understanding foam loss, and then to have thousands of impact damage sites to assess just seemed overwhelming," Jeff recalled.

MSFC's imagery team reviewed forty-seven missions and found only a single loss attributed to repair liberation on the LO_2 tank acreage. Due to imagery limitations on some of those flights, the sample population was reduced to twenty-eight covering the most recent missions. There were 605 repairs counted for the twenty-eight-mission set. This included the 254 repairs performed on another tank (ET-100) that sustained hail damage on STS-96. This corresponded to a historical release rate of approximately one in six hundred.

Later in the day I toured the launchpad, and the amount of damage to the tank's foam insulation shocked me. Most damage was on the oxygen tank in front of the orbiter—the worst possible location in terms of debris risk. There were hundreds of hits with several areas sustaining such concentrated damage it appeared single large chucks of foam were missing. It was clear the launch would be delayed, and a decision loomed about whether the tank should be repaired or replaced.

At the end of the flight readiness review, NASA managers decided to move the shuttle back to the Vehicle Assembly Building for more detailed inspections and (if possible) repairs. Much damage to the external tank (some 2,400 sites) could be resolved by flying as is, sanding and blending areas with relatively shallow crush damage, or pouring in fresh insulation that could be sanded and shaped after curing. These were considered standard repair techniques known to perform well based on past flight experience. However, up near the very top of the tank, the hail damage was so extensive (some 1,600 cataloged, often overlapping sites) that engineers concluded the best way to make repairs would be simply to remove a large section of damaged NCFI foam and replace it with hand-sprayed BX insulation.

RETURN TO FLIGHT

In the VAB technicians applied foam and molds on the external tank to hail-damaged areas. The white hole with a red circle around it (upper right) was a hole prepared for molding and material application. The red material was sealant tape that prevented the mold from leaking when the foam rose against the mold. The white or translucent square mold was an area where the foam had been applied and the foam had risen and cured against the mold surface. (Courtesy of NASA.)

Wayne addressed the foam and hail damage situation. "Engineers have got to do some work to prove first of all that foam doesn't come off in big chunks." He acknowledged this was not considered very likely. "Secondly they must prove it will stand up to the heating and not expand or degrade thermally. Impact is a major part of the testing that's ahead of us."

Engineers utilized a mock-up of the tip of the external tank to practice manually spraying on BX foam. The mock-up was then put in a test chamber to assess the aerodynamic and thermal effects. That data and a risk assessment of the work to repair the other 2,400 damage sites played into the eventual decision about which tank to use. "To ensure we were safe, we had to go through the engineering evaluation process where we built a mock-up, practiced spraying foam onto that mock-up, and put that into a test facility where we could see how it reacted to ascent heating and other things," Wayne said. "So there was that group of work going on. It came to a point where we could evaluate whether the technique was good about the end of the first week in April. That was one thing. The other thing was how quickly we could repair the over two thousand four hundred damage sites using these techniques."

If the program managers elected to have the tank repaired instead of replaced, the processing time could have been reduced, and four missions could still have flown in 2007. However, the big unknown was the debris risk of the repairs. This was a question the DIG and ET project had to answer quickly to avoid a longer launch delay and further manifest impacts. Brekke and the Aerospace Corporation proposed an alternative methodology to arrive at a prompt assessment of the repair risk. In contrast to the standard methodology already employed, the parametric analysis would conduct a probabilistic risk assessment of the entire population of hail-related repairs.

Aerospace Corporation Debris Model Peer Review, JSC, Houston, Texas

A week after the flight readiness review, I had the DIG working full force on the repair risk, and I held a planned peer review of the Aerospace Corporation debris risk model. It had been a DIG goal to place the debris models under program configuration control just like all the other analytic models the shuttle engineers used. By doing this all element projects would be alerted to changes, and that made it easier for them to assess the updates. The Dons and SICB requested I conduct the peer review before going through the rigorous review process to get the Aerospace Corporation and USA–Boeing debris models under configuration control. Up to this point, the DIG maintained configuration control over the models—much to the displeasure of many who criticized the rapid number of model changes and risk number calculations. The primary review objective was to make a determination about whether the Aerospace Corporation PRA model was ready to be placed under program configuration control as had been done with all critical models used on the shuttle. All the model methodologies, ground rules, assumptions, limitations, verifications, validations, interfaces, and documentation would be scrutinized. Inputs to the Aerospace Corporation model such as the ET project debris tables and orbiter damage maps were considered outside the review's scope.

For the Aerospace Corporation subjecting its code to a rigorous peer review process would provide better visibility into the tool's workings. This would also start an ongoing opportunity to review any changes or modifications to the tool with the shuttle community at large—independent of discussions about a particular foam debris threat. "I approached the peer review with a great deal of anticipation, and I viewed it as a significant opportunity to change the manner in which the shuttle community (including its engineers and decision-makers) viewed the foam debris risk assessment," Matt Eby said.

For the past two years, the Aerospace Corporation team had continually improved its model. For instance, the original two-dimensional forward-facing rectangles used to simulate wing leading edge impacts had been replaced by a detailed mesh that replicated the outer orbiter surface. The new mesh was capable of pinpointing the impact location to within a few inches and also returning the impact angle. This enabled the model to differentiate between glancing blows and direct strikes. While the improvements were always discussed with the shuttle community, it was always in the midst of discussing or resolving debris threats or closing IFAs to get ready for the next flight. The threat assessment for the debris source itself often overshadowed questions about the model's fidelity. The peer review provided a unique opportunity to understand and assess all the assumptions, limitations, and improvements outside the context of a specific debris threat. This provided a better level of engagement with the engineers who provided all the data that went into the model. It allowed them to see the raw code and how their inputs were used. My goal was to get the model under program configuration control and to use the peer review as a way of fostering more confidence in the model's results.

I was confident the fundamental debris model methodology was sound and logical but wanted an independent board to make an assessment. I selected an all-star review board. Bass Redd, a crafty engineering guru who had a hand in designing, building, and operating every crewed spacecraft NASA had ever built, chaired. Bass had worked as a "Graybeard" consultant for United Space Alliance ever since the *Columbia* accident and was not afraid to share his wisdom or bountiful repertoire of engineering stories. Bass knew the fundamentals and physics better than anyone. He was always suspicious of model results and tended to focus on the model code verification and validation. Verification data would demonstrate whether the code accurately represented the mathematical models, and validation would ensure the model was an accurate representation of the physics. These two aspects determined if the PRA results were legitimate. I explained to the peer review board that the verification and validation of the debris model included two phases. Phase one focused on the individual elements or components of the model while it was being constructed. Phase two covered the end-to-end model validation using comparisons with flight data. Phase one relied heavily on test data, some analysis, and flight data when available. These data sources had been reviewed over the last three years at various peer reviews, debris technical interchange meetings, debris summits, and debris verification reviews. The majority of phase one's verification and validation data were documented at the contractor and project level as well as archived after the debris reviews. Phase two utilized flight data from all the cameras, wing leading edge sensors, radar, and tile damage assessments. This phase also involved the model comparison of sample cases with the USA–Boeing model and the integrated foam debris risk numbers compared against an independent Safety and Mission Assurance calculation.

The peer review generated nearly one hundred comments. Most suggested further model refinements. The review brought some controversy when some NASA engineers voted to forgo placing the code under program configuration control. The most notable was Mark Seaford. He argued that improvements were still needed to make it a generic debris assessment tool. However, no matter what level of fidelity was achieved, there would always be opportunities for refinements. Maintaining the development process where I worked directly with Aerospace Corporation to determine what refinements to make for each and every foam debris assessment would have been a waste of my time at this point. All the flight data, debris reconstructions, and model comparisons were reviewed and documented in postflight debris reports. Bass and the board concluded that, "Since the Aerospace Corporation debris risk assessment model has been and will continue to be used for launch safety decisions, it must be placed under configuration control."

There were some required improvements to ensure accuracy and completeness, and despite some initial resistance, the board unanimously agreed to place the Aerospace Corporation PRA model under program configuration control. "I was proud that Aerospace developed a math model that NASA considered critical for developing flight rationale for shuttle flight operations," stated Matt. Brekke and the Aerospace Corporation team believed the review "overall was a necessary and healthy process." The peer review forced the Aerospace Corporation team to look closely at the model to make sure it did not miss anything. What the PRA code did at this point was incredible compared to where it had started with STS-114. I had worked with Brekke and Matt to address many uncertainties and minimize some of the conservatisms. This made the program more comfortable with the PRA results and made it easier to develop flight rationale.

External Tank Foam Repairs

Aside from the Aerospace Corporation model peer review, the large number of repairs motivated the parametric assessment for all new repairs on a per-release basis. By doing this the results could be scaled to the historical release rate for an integrated risk level. Historical data suggested release rate was one in six hundred, but most of this data were based on the performance of repairs made on STS-96. This flight also experienced hail damage and had an unusually high hit count on the orbiter. There was real concern the high hit count was from failed repairs and the one in six hundred release rate was wrong. A special DIG meeting was held on March 19 to determine whether hail damage to ET-100 and the related repairs to the ET foam were the cause of the orbiter damage on STS-96. There was compelling evidence that the significant incidence of intertank foam loss was the predominant cause of orbiter damage. The intertank was now fully vented with perforations every two inches to limit popcorning foam debris releases. Only a small portion of the intertank region was vented on STS-96. Humidity and moisture exposure also reduced the foam strength. This was evident from the "pull strength" tests for ET-100 on STS-96. JSC imagery was conclusive that no repair losses were observed from the liquid oxygen tank for assessable areas on the tank. Dr. Kaukler, an imagery expert from MSFC, concluded there was only one "suspicious" site out of seventeen areas of interest that could have been a possible repair loss. In the end the DIG decided the release rate of one in six hundred was valid.

Murder–Suicide at JSC

While working at my desk in the early afternoon of April 20, 2007, I noticed three news helicopters hovering around the center. I was not aware of any pending announcement or Public Affairs Office event and knew something unusual was happening. Just then my wife called and asked if I had seen the news about a shooting and hostage situation at JSC. The shooter, William Phillips, was an employee of Jacobs Engineering, and he worked at Building 44—a communications and tracking facility. He shot and killed one person and took another hostage. Police said Phillips was under review for poor job performance, and he feared being dismissed. The situation had begun at 1:00 p.m. when Phillips entered a conference room, pointed a .38- or .357-caliber snub-nosed revolver at one person, and ordered everybody else to leave. According to witnesses, he immediately confronted David Beverly about his job review and said, "You're the one who's going to get me fired."

The two talked for several minutes, and at approximately 1:40 p.m., gunshots were heard. Police said Beverly was initially shot twice but was still alive. Phillips left and then returned seconds later to shoot Beverly twice more. Phillips then took Fran Crenshaw (who happened to be in the area) hostage and bound her to a chair with duct tape. Phillips barricaded himself and Crenshaw inside the second floor of Building 44 for the next three hours. During this time Crenshaw attempted to calm Phillips with whom she had had a positive relationship. Later Crenshaw was able to get herself out of the tape and alert authorities to what was happening. The ordeal lasted for more than three hours and ended with Phillips finally committing suicide.

Train Carrying SRBs Derails in Alabama

Twelve days after the murder-suicide in Texas, a train carrying two solid rocket boosters from Utah to Kennedy Space Center derailed in western Alabama. It left two people injured. The accident happened around 10:00 a.m. near Pennington, Alabama—about one hundred miles west of Montgomery. Nothing was spilled because the fuel inside the segments was ammonium perchlorate. This has the consistency of a rubber eraser. The sixteen-car train went off the tracks in a forested area after a bridge collapsed. Two locomotives and one segment car tipped over on their sides. Not all the cars carrying the eight segments derailed, but as a precaution all were sent back to Utah for a manufacturer inspection. They had been intended for use on the October 20 launch of *Discovery*. NASA minimized the delay on the launch schedule resulting from the derailment by exchanging the damaged SRBs with an ample supply of other segments available at KSC.

CHAPTER 23

Loss of Space Station Attitude Control

STS-117 Flight Readiness Review, KSC, Cape Canaveral, Florida

At the STS-117 FRR on May 30, I presented the repair risk assessment results from both the standardized repair risk assessment and the parametric methodologies. There were 914 repairs made to the liquid oxygen tank and much fewer to the liquid hydrogen and intertank regions. Using the standard repair methodology, the worst-case conditional probability was one in one thousand. This finding was consistent with the parametric risk assessment results using the final repair masses and worst-case time of release distribution based on the dynamic heating. This showed a combined risk of 1 in 1,600 using the historical release rate of one loss out of every six hundred repairs. The ET project analysis and testing also showed ice was unlikely to form in repaired areas. Although there was an incremental increase in risk due to the hail damage repairs on the external tank, the risk level for the general acreage areas would remain "remote catastrophic." In typical fashion Jeff and his team did a great job developing unique inspection and acceptance requirements for the damage. Additional testing and analysis demonstrated the integrity of the foam and foam repairs and supported the flight rationale. In hindsight Jeff recalled,

"Working through the hail damage was probably the toughest set of decisions I had to deal with through the entire program."

Mike Griffin was attending the review and pointed out an apparent disparity between acreage debris risk and repair risk, which was much lower. He jokingly suggested we should consider replacing the acreage foam with repairs to lower the debris risk. I responded by saying that the repairs were similar to the healing of a broken bone. Typically the bone was much stronger, but it probably was not the best way to strengthen a human's skeletal structure. The difference, I explained, was embedded in the release rate—a function of total area. The release rate of the general acreage was much higher than the repair area and hence had a higher risk. I also announced that both the Aerospace Corporation foam debris PRA and USA–Boeing ice debris PRA models had been placed under configuration control. This included the updated foam-on-tile damage map to remove the small foam limitation. This significantly reduced all the foam debris risk numbers. In addition the foam debris risk assessment included results of a tile shear failure mode that the orbiter team uncovered during recent impact testing. Under certain mass, velocity, and impact angle conditions, the force of impact was greater than the tile bond strength, and this caused the tile to release and expose the aluminum structure. This failure mode was unusual because any type of debris could lead to a tile loss.

Although ice had not been an issue on the last four flights, the DIG was pursuing an ice coating test at Eglin Air Force Base near Fort Walton Beach, Florida. The testing would start in August and involved using an ice coating on a high-fidelity umbilical test article positioned on a vibration table that would simulate liftoff vibration loading. NASA engineers hypothesized the ice coating would inhibit ice growth and reduce the bond strength enough to allow all the ice to liberate at liftoff. The debris risk for STS-117 had been reduced because of the BTA redesign, focused gap filler and putty repair microinspections, expansion of the no-ice zone in NSTS 08303, and liftoff debris mitigations. Mike, Gerst, and the Flight Readiness Review Board were satisfied with the debris summary, and the launch date for STS-117 was set for June 8, 2007.

LOSS OF SPACE STATION ATTITUDE CONTROL

Launch Support from KSC

STS-117 was the first time I supported a flight from KSC and participated in the twenty-four-hour pad walkdown which was part of launch preparation and identification of liftoff debris. A few debris-related concerns were noted and quickly dispositioned during the walkdown.

There were several advantages to participation in launch support. This included participating in the pad walkdown, having access to the high-definition video for debris identification and review, engaging in face-to-face communications with the final inspection team and element representatives, and getting an early quick look at launch video at the KSC lab. Some perks also came with the job—beachfront accommodations and the traditional launch party after a successful liftoff. This was usually held at one of the premier hotels and featured an abundance of food and cocktails. After the pad walkdown, I met Don Noah at the Sandbar restaurant along with several others supporting the mission from MSFC and the ET and orbiter projects. The Sandbar was a dive bar located on the beach just a short walk from Ron Jon's Surf Shop. Despite the outward appearance, dozens of patrons from the shuttle program filled the place. People had been going there for years and raved about the fish tacos. Servers brought several pitchers of beer to the table and did not bother with menus unless the patrons wanted something other than tacos. I asked Don if he went there because he liked the fish tacos or because he was superstitious before launch. He said it was probably both.

The weather the next day was perfect for launch, and surprisingly there were no debris issues during tanking or launch countdown. When the countdown reached the five-minute mark, Don excused me from my console, and I raced outside the launch control complex along with dozens of others to watch the launch. All systems were go, and on June 8, 2007, *Atlantis* rocketed into a twilit night from Pad 39A on the first attempt. The ground vibrated vigorously, and the noise was deafening as the evening sky lit up like the midday sun. I could feel the radiant heat from the SRB exhaust on my face as the vehicle roared past the launch tower. This was the first flight from Pad 39A since STS-107 launched on January 16, 2003. After watching the solid rocket boosters separate from the tank and the vehicle disappear into the night, I sprinted back into mission control and the "ice castle" to review the ELVIS and launch camera replays. The ice castle was a room reserved for the final inspection and debris teams because it contained several high-definition displays and the master control console for all ground cameras. Image analysis of the flight photography and video showed foam releases from the LH_2 acreage and the intertank-to-LH_2 tank closeout flange. The most significant debris event I noted during ascent was a piece of debris impacting the left-hand side of the aft body flap hinge area 124 seconds into flight. I did not expect much tile damage due to the late time frame but did expect to see something during the on-orbit inspection. Although I had an early flight out of Orlando the following morning, I stayed with the KSC imagery team until late in the evening. We pored over the mountain of imagery data looking for debris events. Launch video footage was played one camera at a time in a large single-screen auditorium for the twenty or so imagery team members. This was a much different approach than the JSC imagery team employed. That team had fewer members and divided the work and specific cameras among those members. It was tedious work that gave me a new appreciation for their attention to detail.

Another Reentry Debris Threat, JSC Mission Control

The next day after returning to Houston, I was eager to see the orbiter inspection results and rushed over to mission control. On-orbit inspections showed several impact damage sites on the orbiter tile, but Justin's team's assessments concluded the damage was within the experience orbiter base. The lone exception was a damaged thermal blanket on the port side OMS pod. A four-inch-by-six-inch corner had peeled up during the launch and exposed the OMS pod structure.

This was a first-time occurrence and represented a reentry debris and structural failure threat to the OMS pod. It was going to take a few days to perform the debris transport analysis and determine the impact threat if the blanket liberated during reentry. In the meantime Justin and the orbiter team were preparing for the worst-case scenario, and on flight day five, they presented EVA repair options to the MMT. They were also building a thermal blanket test article that would be exposed to simulated reentry conditions in NASA's arc jet complex. The MMT postponed the EVA decision until the DTA and testing were completed the following day.

Loss of Space Station Attitude Control

I was up late in the evening reviewing the OMS pod blanket debris transport results and preparing my presentation for the flight day six MMT when I noticed some unusual e-mail traffic about the space station. NASA flight controllers and engineers had been working overtime to determine why the three computers critical to ISS operations had crashed. This raised the possibility that the already extended *Atlantis* mission would have to be lengthened further. Attitude control was automatically shifted to the Russian segment as a result of the three failed US computers. The space station guidance, navigation, and control (GNC) system was unique because it was the only system to exchange information directly between the Russian and US segments. This was because the Russian and US GNC systems had to coordinate this critical data for integrated operations. Unfortunately, when attitude control shifted, the six Russian command and control computers failed in successive order. This left the station with a complete loss of attitude control. Without attitude control the station would go into "free drift" and eventually begin to tumble out of control until its roll rate was high enough to destroy the station. The only thing that kept the ISS under control was the shuttle. It held attitude control for ten and a half hours while ground controllers and astronauts worked to recover the computer system.

The MER was abuzz as the MMT considered ISS evacuation options in the event the problem could not be resolved or the shuttle's limited propellant became exhausted. The shuttle's remaining propellant margin was nearly gone, and after the space station GNC system was partially recovered, the ISS had only limited attitude control capability when mated with the shuttle. If attitude control was lost again, the space station would have to be evacuated. The computer malfunction and loss of attitude control trivialized the problem with the OMS pod thermal blanket on *Atlantis*, but the debris transport analysis showed impact risk to the tail and rudder. This meant an EVA was needed to repair the blanket.

The next day Gerst stated at a press conference that, "The evacuation of the ISS could be a possibility if the computers that stabilize the station remain down, but the chance of that is very slim."

The majority of the MMT was focused on the loss of attitude control and the EVA for the blanket repair. Due to the urgent nature of the computer failure, my MMT presentation on the thermal blanket was deferred. It appeared the station's GNC computers and control moment gyroscopes (CMGs) went off-line due to the mass distribution changes caused by moving the new solar array truss segment from the shuttle cargo bay to the outer truss on ISS. CMGs were a set of fixed momentum wheels that rotated at a constant velocity and a gimbal to generate the desired torque needed to control momentum and maintain station attitude control. Nominal momentum levels were typically below 20 percent but peaked to near saturation when the solar array was added. Saturation occurred when the total station momentum exceeded the capabilities of the gyros. At that point station attitude control shifted to the Russian motion control system computers, and their thrusters desaturated the gyros.[12] However, the station could not perform a control moment gyroscope desaturation while the Russian computers were down, and the station managers were not confident the US gyros could take over attitude control directly from the orbiter. If the orbiter had to depart without the Russian computers working nominally, the station would likely be unable to control attitude. To make matters worse, the orbiter propellant usage needed for station attitude control was draining the tanks. That meant time was running out. While the Russian computers were down, other

---

[12] Over time, the CMGs absorb momentum to the point where they can absorb no more, which results in loss of effectiveness of the CMG array for control. Desaturation is the term used to describe the process of "unloading" the excessive momentum using the Russian thrusters.

critical life-support systems such as the Vozdukh used for carbon dioxide removal, the Elektron used for oxygen generation, and the contamination control system were down.

It appeared the electronic signals from the installation of the new solar arrays were causing the problem with the Russian computers. Electromagnetic shields or other devices could eliminate the problem. The station's director, Mike Suffredini, said the signals were comparable to the interference caused by a mobile phone near a television set. Remove the interference, and the problem would be solved. The space station attitude control problems concerned NASA enough to take energy-saving measures on *Atlantis* to extend its mission another extra day to stabilize the station. All nonessential systems aboard the shuttle were turned off.

Assuming the Russian computers could be restored long enough to switch attitude control to the gyros, the station would not be endangered so long as the gyros did not saturate. If they failed *Atlantis* could provide stabilization. This dependency led mission managers to discuss the possibility of extending *Atlantis*'s thirteen-day mission by a day or so and conserving its supplies. The mission had already been extended two days to allow the added space walk to repair the torn shuttle blanket. In a worst-case scenario—the attitude control problem lasted longer than the current shuttle mission and exceeded the control capabilities of the space station's US system—the ISS crew would have to return to Earth. "If we are in that position, we do have the option to depart," Suffredini said. He stressed that he believed the computer issue would be solved in the next few days. He added, "I'm not thinking this is something that we will not recover from."

For now the shuttle and station stack were stable, and the MMT was confident the Russian computers would be restored.

OMS Pod Blanket Repair, JSC Mission Control

Adding to the MMT's stress level on flight day seven was a heated discussion on whether to repair the torn OMS pod blanket during an EVA.

After reviewing the wind tunnel testing and DTA, the decision was made to repair the blanket. Debris transport results showed that debris emanating from the port OMS pod thermal blanket during reentry had transport mechanism to the port OMS surface, vertical tail, rudder, and main engine nozzle—albeit with an unlikely impact probability. The most likely scenario, if the blanket released at all, was that it would fly straight over the OMS pod. Furthermore, if the blanket was lost, the initial rotation would tend to send the debris away from the OMS pod, and thus it would be less likely to hit the tail.

Justin Kerr presented the arc jet test results. They showed the OMS pod structure was susceptible to reentry heating damage. The extent of damage could not be quantified, but if the structure failed it could allow high-temperature gases to compromise the OMS pod fuel tanks and lead to a possible explosion. This scenario was deemed highly unlikely but still possible. In addition to the arc jet test results, Justin presented a modified repair procedure. John Shannon, the MMT manager, was livid with the late-breaking procedural change because Justin's new recommendation was more complicated and time-consuming than one reviewed at the MMT two days earlier. It was one of the only times I saw John Shannon become riled as he chastised Justin for modifying all the EVA planning and procedures so late in the game. Conducting the repair EVA had risks in addition to stressing the attitude control system and utilizing precious resources that might be necessary if the shuttle mission needed to be extended. "I had discussed the updates the night before the presentation with all the stakeholders, and I thought everyone was on board with why we made the changes. The change was the right thing to do because the test data supported the new configuration," recalls Justin. John's response clearly surprised Justin. "The update involved changing the position of the pins used to secure the blanket because it improved the pull loads and made the repair stronger," stated Justin. "Our goal was to make sure the repair procedure matched what we had tested because any deviation would jeopardize our confidence that the repair would work during reentry."

Due to time constraints, there was differing goals between those writing the EVA procedure and the engineers doing the test. As far as engineering was concerned, it was trying to get the best design possible and did not consider the downstream effect of those design changes. The EVA team was trying to finalize the procedure such that it could be performed in the neutral buoyancy lab before it was transmitted up to the crew to perform the repair. The two teams failed to synchronize final positions—hence the confusion at the MMT. "I probably could have communicated better with John before the MMT, but the only time available for discussion was at the MMT," Justin said.

John typically listened to the orbiter project board, and that was where I had presented the repair design change in the morning. He must have missed the discussion.

The MMT vote that followed Justin's presentation proved even more interesting because Kim Doering voted against her husband, Steve (the EVA manager). He recommended conducting the EVA. Kim's final board member vote made it a split and left the final decision up to John Shannon. "I remember this flight and MMT meeting very well," proclaimed Kim. She and Steve were married right out of college, and their professional paths had crossed several times. That included when Kim was on assignment in Germany early in her career. "Steve and I shared an office with two desks that sat face-to-face, and one of our assignments was to coauthor a document together. When we started we treated each other as husband and wife, but after about two weeks and several exchanges of critical feedback, it became clear our marriage would not survive unless we treated each other as colleagues. This carried over into the MMT. I viewed Steve as just another member and did not view his professional opinion any differently than other members'. When he voted for EVA repair, I was thinking he was basing his vote on the confidence he had in his team to conduct the repair. I was thinking EVA repair was unnecessary because of the low reentry debris risk."

When talking about it later, the EVA team members expressed how they looked forward to the challenge of conducting a "contingency" EVA and the adrenaline rush of the planning, preparation, and execution. For Steve it would have been an exciting challenge.

During the same MMT, John Shannon and Steve Poulos had a fairly heated debate about what Steve considered was the real issue other than debris. Steve was seemingly unconcerned about debris but instead was focused on the orbiter not having an adequate OMS pod model to perform the analysis and determine if the structure could accommodate the additional reentry heating. The OMS pod structure was a composite, and prior to the STS-117 flight, the vehicle had spent a significant amount of time in the VAB exposed to high humidity levels. The added moisture to the orbiter composites (in particular the payload bay doors) almost became a constraint to flight. "The OMS pod had little to no impact capability and I had the moisture concern with the composite structure as well, which caused me to personally push for the repair," stated Steve.

There was a split MMT vote, and John Shannon ultimately concurred with Justin's recommendation to conduct the repair on the third EVA. However, he wanted the original repair procedure—to staple and pin the torn edges to adjacent blankets. After the MMT I joked with Justin about the procedure tirade and said that at least he was not invited for dinner that night at the Doerings'.

On flight day nine, Mission Specialist Danny Olivas spent two hours stapling and pinning down the thermal blanket on the OMS pod. Once the repair was complete, the crew was given the approval for entry. In addition four of the six Russian computers had been recovered, and the Russian motion control and navigation base had been activated. In-flight maintenance was performed to bypass current-limiting circuitry in all lanes of each Russian computer. This allowed the systems to be returned to their normal configuration. Once stable a Russian attitude solution was sent to the US GNC computers, and the space station's attitude control was transferred from the shuttle back to the station. The attitude control emergency on the ISS was over.

High Debris Impact Count on the Orbiter

On June 22, 2007, at 3:49 p.m. (EDT), *Atlantis* landed on Runway 22 at Edwards Air Force Base in California. This concluded the thirteen-day mission. The landing was diverted from Florida due to marginal weather at Kennedy, and the postlanding debris inspection of the *Atlantis* was conducted at Edwards. The orbiter sustained a total of 292 tile impacts. Seventeen had a dimension of one inch or larger. Even though no tank repairs appeared to liberate, the total number of impacts was nearly double the shuttle's historical average. The orbiter's lower surface sustained 247 total hits. Fifteen had a dimension of one inch or larger. The largest damage site measured 1.7 inches by 0.3 inches by 0.15 inches deep and was located slightly aft and starboard of the liquid oxygen umbilical door. Numerous damage sites around the liquid hydrogen and liquid oxygen umbilical area represented the largest concentration of hits on the lower surface. This had now become a chronic pattern. Some of this damage was attributed to popcorn foam releases during the ascent phase, but we were puzzled why the damage count on this flight was so much higher than normal.

After the orbiter arrived back in Florida, I flew to KSC along with John Brekke and Randy Williams to inspect the orbiter impacts and try to determine the reason behind the high impact counts. One of the technicians who escorted us through the orbiter processing facility and *Atlantis* had been working on the shuttle program since the first flight. He became teary-eyed when he talked about *Columbia* and how much he "missed her." It was almost as if he was talking about a loved one. The emotion illustrated the personal connection and pride he felt from working on the spacecraft. Randy and Brekke were really awestruck because it was the first time seeing an orbiter up close. "I noticed a lot of little dings with the concentration of hits around the umbilical doors," Brekke said.

After the tour and inspection ended, we were still collectively stumped about the increase in impact count, but the small damage sizes comforted us somewhat. I had the same feelings as Brekke when he said he did not think the damage was significant, and he suspected most of it was from small popcorn foam less. The tile damage was consistent with other flights and still much lower than before the *Columbia* accident. Since the tile damage had decreased, we agreed to conduct some additional analysis by varying factors such as heat rate to see if we could replicate the flight observations.

There were no issues with the hail damage repairs or the SRB BTA redesign, which performed as expected. The program managers accepted six integrated debris IFAs—the lowest since RTF. Five were for ascent and one for liftoff. Except for the liquid hydrogen acreage foam loss at station Xt-1160, all ET foam losses were within the current NSTS6 0559 risk assessment mass. Because the time of release for the liquid hydrogen acreage loss occurred at about 137 seconds (outside ASTT), it was most likely due to cryopumping. Reconstructions showed the Xt-1160 risk could be as high as 1 in 170 if the release occurred earlier in the cryopumping time frame. This loss event was rare and needed further investigation to ensure a release could not occur earlier. The risk level was elevated to "infrequent catastrophic" until further investigation and analysis could be performed. Based on the reoccurring OMS pod blanket issues, reentry DTA for all blanket locations would be developed to support future missions.

STS-118 Flight Preparations

STS-118 was scheduled for an August 8, 2007, launch. That left little turnaround time to close the integrated IFAs and get ready for the next mission. The Aerospace Corporation was beginning its small foam-on-tile impact testing to verify the orbiter team's damage map, increase the acceptable foam debris size above 0.0002 pounds, and determine a special tile capability. On June 28 the ET project team unveiled its modified LH_2 IFR design, and a complete review of the liquid hydrogen IFR risk assessment was given to the administrator and Space Shuttle Program management. The redesign was dimensionally the same but incorporated numerous improvements that minimized cracking, cryopumping, and void delta P losses. Although the risk level was below one in one thousand for void delta P failures and below one in five thousand for cryopumping due to the updated foam-on-tile damage map, it was still the highest foam risk. The reoccurring losses, most of which were outside ASTT, were reason enough to pursue the redesign. According to the implementation plan, the new liquid hydrogen IFR would be flown on ET-120 during the STS-120 mission scheduled for October 2007. ET-120 was the original RTF tank that was replaced after two tanking tests due to engine cutout sensor problems and dissected to further assess the foam failures. If the new design proved successful, it would be implemented on every new starting with ET-128.

Although radar had been seldom used for debris evaluation, the improvements over the last five flights helped transform radar data to debris information that was useful for debris transport analysis and orbiter impact assessment. Radar Tony had been making steady improvements since STS-114. That included focused radar processing, contact reporting during ASTT, optimization of power level sequencing during ascent, reduction of noise levels to increase debris detection, development of ballistic number curves for various debris types, and numerous other hardware and software tweaks that fine-tuned the radar to enhance performance. New postprocessing tools automated the radar contact screening process to get quicker results to aid with sensor correlation and identification of any potential debris sources imagery did not record.

There was no IDBR debris risk level or risk number updates for STS-118. This including for LH_2 tank acreage debris such as that released at station Xt-1160 on STS-117. The Aerospace Corporation used the liquid hydrogen acreage historical mass distribution and release rates to calculate the risk at the worst-case release times (between sixty and one hundred seconds) and at the cryopumping release time. The highest risk for the worst-case release time was 1 in 1,250 for the tile, and for the cryopumping release time it was 1 in 3,330 for tile shear. After the DIG reviewed the results, the risk classification was reduced from "infrequent catastrophic" back to "remote catastrophic." The middle of July was the arduous review lineup starting with Roy Glanville and the ISERP on the sixteenth. The SICB and PRCB on the seventeenth and nineteenth respectively followed. It ended with the STS-118 flight readiness review on July 25.

Intoxicated Astronauts

Right before the flight readiness review, a story about intoxicated astronauts made the headlines. America's space program once again suffered unexpected turbulence with the revelation that astronauts had been allowed to fly on the shuttle while drunk. NASA officials confirmed there were at least two occasions when crewmembers were so intoxicated before their launches they were deemed flight safety risks. The intoxication claims came in a report the space agency commissioned to investigate the behavior of its astronauts in the wake of the arrest of shuttle crewmember Lisa Nowak in February for allegedly stalking and attacking a love rival. The panel discovered "heavy use of alcohol" by unspecified astronauts in the twelve-hour period before a shuttle launch. Michael Griffin, the NASA administrator, ordered the report. NASA had since fired Nowak. She was tried in Florida on charges of assault and attempted kidnapping. "Interviews with both flight surgeons and astronauts identified some episodes of heavy use of alcohol by astronauts in the immediate preflight period, which led to flight safety concerns," the report said. "However, the individuals were still permitted to fly. The medical certification of astronauts for flight duty is not structured to detect such episodes, and any medical surveillance program by itself is unlikely to detect these episodes or change the pattern of alcohol use."

Dr. Griffin asked the panel to evaluate the medical and psychological screenings of astronauts to determine if there were problem areas.

STS-118 Flight Readiness Review, KSC

I started off the STS-118 flight readiness review with a story about how debris problems sometimes worked in NASA's favor. If the hailstorm that damaged STS-117 had not occurred, the launch would not have been delayed, and the shuttle would not have been docked to the space station when attitude control was lost. The ISS would have tumbled out of control without the shuttle when the computer failures occurred.

Of course, after the debris presentation, flaws in my logic were pointed out to me. The docked shuttle itself contributed to the high momentum and gyro saturation that initially started the problem. However, it provided some comic relief in an otherwise routine debris summary. The risk level for the solid rocket booster BTA would be lowered pending another successful flight, and additional debris mitigations were implemented on Pad 39A for liftoff debris along with inspection of all OMS pod thermal blankets. I was thinking the worst of the debris challenges were over but was proved wrong again on STS-118.

CHAPTER 24

Prepare to Launch the Rescue Mission

Significant Tile Damage to *Endeavour*

On August 8, 2007, at 6:36 p.m. (EDT), *Endeavour* lifted off Pad 39A into an early evening sky before sunset. There were no significant debris issues during the countdown, and the weather conditions produced less than typical formations of ice and frost to the liquid oxygen feed line brackets and bellows. A small foam crack was observed at station Xt-1973 on the inboard attachment fitting, but there was no offset or ice present, and the crack was cleared for launch. STS-118 carried a crew of seven. It included teacher-turned-astronaut Barbara R. Morgan. During the mission the crew delivered a new truss segment for the station's backbone, installed a spare parts platform, and swapped out a failed gyroscope used to orient the station. Morgan was the first educator mission specialist. She had served as the backup to Payload Specialist Christa McAuliffe in the Teacher in Space Project. McAuliffe and six fellow astronauts lost their lives in the 1986 *Challenger* accident. Morgan was an elementary school teacher in McCall, Idaho, before being selected as McAuliffe's backup. She returned to teaching after the *Challenger* accident. She was selected to train as a mission specialist in 1998 and was named to the STS-118 crew in 2002. She would be involved in at least one live, interactive educational in-flight event with students in Boise, Idaho. If the mission was extended, she

PREPARE TO LAUNCH THE RESCUE MISSION

would participate in two other educational events with students in Alexandria, Virginia, and Lynn, Massachusetts, to discuss her mission and the educational aspects of human spaceflight.

Ground cameras observed a late Tyvek cover release at 16.5 seconds, and the forward SRB camera observed a foam loss emanating from the outboard liquid oxygen feed line support yoke at station Xt-1623 around fifty-eight seconds. The SRB camera video showed footage of the liberated foam mass traveling down the external tank, deflecting off the +Y thrust strut, and breaking into several pieces. This resulted in multiple impacts to the orbiter aft of the starboard main landing gear door.

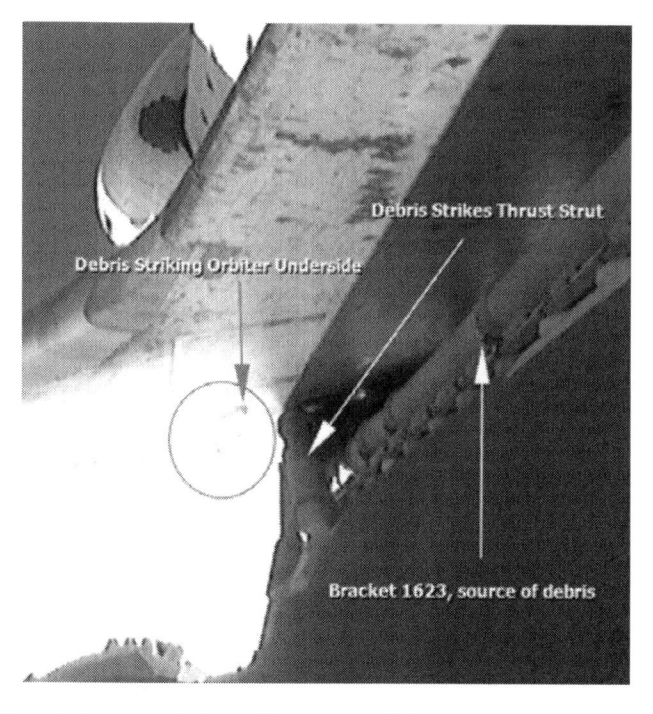

At fifty-eight seconds into flight, one of the SRB cameras recorded debris originating from the Xt-1623 feed line bracket. The debris traveled aft, impacted the thrust strut, and broke into multiple debris pieces that subsequently struck the orbiter. NASA's debris radar also detected the debris release. It classified it as something other than foam. Later it was determined to be mixed foam and ice debris. (Courtesy of NASA.)

One of the impacts resulted in a large damaged area across two lower surface tiles with the deepest part of the impact leaving an exposed strip of underlying metal. The next day (August 9), before docking with the space station, the crew used the vehicle's robotic arm and orbiter boom sensor system to take a closer look at the heat shielding on *Endeavour*'s wing leading edges. Photography showed a three-inch ding on the starboard underside, and in-depth analysis showed damage occurred through the tile to the internal framework. The extent of tile damage was confirmed the following day when Commander Scott Kelly performed the rendezvous pitch maneuver and enabled the ISS crew to take digital photos of the orbiter's tile.

Debris activities focused on evaluating the tile impact from foam liberated from the liquid oxygen feed line bracket at station Xt-1623. Steve Poulos and the orbiter team were concerned the damage depth was too severe to withstand the reentry heating and thus had to consider the repair option or rescue mission. The debris and orbiter teams formulated a risk assessment plan for the liquid oxygen feed line bracket but first needed impact data from either the DTA or imagery.

Assessing the Debris Damage, JSC Mission Control

Image analysis of the flight photography and video showed the foam loss from the liquid oxygen feed line yoke at station Xt-1623 was responsible for the large damage to the orbiter tile on the orbiter underside. I immediately called Brekke at the Aerospace Corporation and requested Matt perform a reconstruction of the observation to determine whether the damage was cause for concern. For Matt this turned out to be his most memorable debris event because we had in-flight imagery of the whole end-to-end foam loss scenario from liberation to orbiter damage. It would be a good test of the model to ascertain if it could replicate a piece of foam liberating from the LO_2 feed line, ricocheting off the aft orbiter attachment strut, and finally plowing through the entire thickness of one of the orbiter's heat shield tiles. Debris transport reconstructions were consistent with the observed impact velocity and impact angle but showed a much shallower damage depth. The damage from a relatively small piece of foam was largely unanticipated by the PRA. Either the impact mass was larger than what was observed or the impact was a "mixed" debris event such as foam and ice or foam and ablator. The debris team, imagery, and ET project concluded after intense scrutiny that the foam piece liberating from the yoke was dimensionally accurate, but the piece of foam debris alone could not have caused the amount of damage observed. This meant the debris had to be contaminated with either ice or ablator material. Unfortunately the available imagery could not conclusively indicate whether or not ablator or ice was liberated with the foam debris. None of this mattered to the orbiter team or MMT. They were trying to determine if the orbiter could survive reentry with the damaged tile. Justin had just gotten married and was on his honeymoon in Hawaii during the mission, but he did manage to stay in contact with the team as they worked the problem. He and the DAT worked night and day to assess whether there was sufficient damage to warrant recommending a rescue mission or repair or whether the tile could be left as is. The time difference (between Houston and Hawaii) did not matter because the team was working around the clock to arrive at this determination. Looking back on it later, Justin recalled, "It was a little warped I was joining daily teleconferences and reviewing analysis and PowerPoint charts while on my honeymoon. The DAT team was confident and became comfortable with the decision to fly home "as is" after reviewing the arc jet test data and analysis."

Gerst and Steve Poulos also spent a lot of time with the JSC debris team and engineers poring over analysis and test data to determine if the orbiter was safe to reenter as is. The consequences of being wrong were enormous, and because of this it was a stressful mission. Ultimately the decision whether it was safe to return home with the tile damage or to perform an on-orbit repair became a risk-trade-off discussion centered on whether the ground testing in the arc jet facility adequately showed that the tile would survive during reentry without leading to catastrophic damage. The second alternative—conducting a repair using tile filler—might itself have resulted in a rough surface that could trip the boundary layer during reentry and elevate the temperatures. Leaving the damaged tile as is (the first alternative) was better understood than performing the repair, and this was a key factor under consideration. "Our ability to replicate the tile damage and test this configuration in the arc jet during the mission ultimately gave me the confidence to fly the vehicle as is," explained Gerst.

Prepare the Rescue Vehicle

On flight day four, Dan Bell from the orbiter team, who was leading the effort in Justin's absence, compared several similar cases of damaged tile that survived entry. However, none of the cases had comparable damage depth. The MMT also acknowledged that, per the flight rules, mission termination or tile repair could only be options if the tile was considered severely damaged. Launch preparations for a rescue mission were discussed and initiated, but a decision to launch would have to wait until the orbiter team completed its reentry analysis. That took several more days to finish. On flight day six, tile repair options were presented to the MMT. An overview of the EVA4 repair procedures followed on flight day eight. The MMT had already extended the mission two days by adding EVA4 but decided to delay EVA4 to flight day eleven to give the orbiter team time to complete its reentry analysis and recommendations. The mission clock was ticking, and the crew and MMT were anxious to get the analysis results.

Mixed Debris Impact Testing
Aerospace Corporation Facility, El Segundo, California

During the mission I ordered Brekke and the Aerospace Corporation to stop work on the small foam-on-tile impact testing and instead shoot some mixed debris foam "doped" with ice and ablator. I instructed them to shoot both ablator and ice because chemical analysis of the impact site on the orbiter to determine the culprit could not be done until the vehicle returned. Brekke's team was always very responsive

PREPARE TO LAUNCH THE RESCUE MISSION

and had close interaction between Matt, who made the PRA updates, and the test team. The PRA model had neglected the possibility of foam rebounding off other parts of the shuttle stack at high velocities before impacting the orbiter. The STS-118 event had clearly demonstrated that was a distinct possibility. The Aerospace Corporation team worked all weekend trying to outline a path toward a better understanding of what had occurred on the flight. The first thing was to characterize the foam rebound so the observed trajectory could be modeled. Foam-on-foam impact tests were done to study the amount of energy dissipated in the rebound. Once Matt incorporated the preliminary rebound model, the new trajectory simulations provided impact conditions for a foam loss bouncing off the aft strut and hitting the tile. It quickly became apparent (from a combination of NASA imagery analysis and trajectory analysis) that the extreme damage depth observed could not be replicated for foam only. Either the understanding of foam impact damage was wrong, or something else had caused the observed damage. The Aerospace Corporation then started shooting mixed debris combinations of foam with either ice or ablator. (These were the only two possibilities based on the liberation location.) The results showed that an added ice layer caused slightly more damage than an equivalent volume of all foam. Also, if ablator was present in the foam, the damage could be significantly higher than a foam-only loss. The close interactions between the PRA and the Aerospace Corporation lab helped provide a better understanding of the STS-118 loss and its implications for future missions.

Although not detected by imagery, it was possible for ice to have formed on the yoke foam and for it to liberate along with the foam. It took the Aerospace Corporation only two mixed debris impact shots (one of foam doped with ablator and the other with ice) to confirm the tile damage depth. The damage reconstruction from the STS-118 impact combined with the preliminary mixed debris impact testing of foam with ablator and ice showed mixed debris sources could cause significantly more damage that could lead to possible catastrophic failure. Mixed debris sources clearly posed a higher risk than ice, ablator, or foam alone. The mixed debris combined the worst-case aerodynamic and impact effects because it flew like foam and impacted like ice.

RETURN TO FLIGHT

Mixed Debris Failure Investigation, MAF, New Orleans

In parallel with the Aerospace Corporation work, Jeff Pilet and the ET project were investigating the possible cause of failure and taking X-rays of the yokes on ET-120, which was scheduled to fly on the next mission. Due to the early release time, it was determined the most probable cause for the STS-118 liberation event was foam cracking due to mechanical interference caused by ice buildup in the gap between the yoke and the upper outboard monoball closeout followed by aeroshear. During the processing of ice characterization test articles for the titanium yoke redesign of the LO_2 feed line bracket, it was observed that applying the upper monoball closeout on the LO_2 feed line resulted in a significant load on the upper outboard foam on the yoke. The force was large enough to deflect the foam and cause a permanent bow. X-rays of the LO_2 feed line bracket yokes on ET-120 (the same design flown on STS-118) indicated the underlying ablator was cracked and considered a contributor to the yoke foam failure. Consequently the upper outboard foam on ET-120 was removed from all yokes except Xt-1973, and the crack in the ablator was confirmed. No other anomalies (such as cracks in the BX-250 foam) were found during the dissection. The bowing of the foam on the yokes was sometimes not perceptible during visual inspections. Therefore, it could not be ruled out that the yokes on the STS-118 tank did not have this type of damage to the underlying ablator. For the STS-118 mixed debris event, it appeared foam was still encapsulating the ablator material where the crack defect could have been located. However, the imagery was so blurry it was difficult to determine whether the ablator was there or not.

Debris Damage Decision, MMT, JSC Mission Control

After extensive debris transport, engineering analyses, emotional debate and tests over several days, the MMT made a decision on flight day nine. The orbiter tile was considered degraded but acceptable for reentry without repair. Even though it would have required another

PREPARE TO LAUNCH THE RESCUE MISSION

EVA, I and many others were surprised the repair option was not pursued even for the sake of flight testing and certification. MMT members believed leaving the damaged tile as is did not pose a risk to the crew during reentry. For Steve Poulos and the orbiter team, it was an easy decision. "All the work that was done prior to Return to Flight definitely paid off to help with the use-as-is decision."

The Orbiter team had all the necessary models and test data (including an arc jet test that was run during the flight) to show the damaged tile would survive reentry. This turned out to be Steve's most memorable debris event. "The fact we were able to generate the necessary data in about five days to show we were safe to reenter as is was a testament to all the hard work that was done by hundreds of people prior to and subsequent to Return to Flight."

Ralph Roe and various members of the NESC team who had been supporting the damage assessment agreed with the decision. Ralph cited the arc jet results of a similarly damaged tile. "Results for this test confirmed what the analysis was telling us," said Ralph.

Mission managers also extended the flight to fourteen days to enable a fourth space walk. However, concern over Hurricane Dean's movement toward Texas the next day caused the mission managers to end the flight one day early.

Technically the mixed debris event on STS-118 and the determination of whether the vehicle was safe for reentry was the toughest debris decision Gerst had to make. There was not a lot of public scrutiny as with the LH_2 IFR discussion during the STS-121 mission, but Gerst spent more time reaching this decision than any other. He was willing to accept the possibility that reentering as is might damage the structure and surrounding tile but was willing to take that risk based on the analysis and testing done. "I was at KSC with Mike Griffin during the landing and explained to him that we could lose a couple of tiles, and there might be some blistering, but we will be OK to land safely," recalled Gerst.

Right about that time, *Endeavour* landed at KSC on August 21, 2007, and the postlanding debris inspection immediately followed. *Endeavour* had sustained a total of 218 impacts (188 of this total on the lower surface). Thirty-one had a dimension one inch or larger. Both of these measures were above the RTF averages. The damage caused to the tile by the mixed debris release from the liquid oxygen feed line bracket yoke appeared to have grown slightly due to the loss of an adjacent putty repair aft of the damage, but overall it was relatively unchanged as predicted. Chemical analysis of the damaged area showed no traces of ablator. The tile damage was driven by the mixed debris impact and exceeded the new control limit that had been established for RTF.

Correcting the Mixed Debris Problem

The DIG was now focused on executing the LO_2 feed line bracket yoke risk analysis plan and the fix for the Tyvek covers. Determining the risk for all brackets with and without rebound as well as for mixed debris cases was the goal. Reconstruction of the STS-118 feed line foam debris event had been completed as well as the assessment from a similar tile damage event on STS-73. Debris transport analysis showed the impact from STS-73 was much different than what was observed on STS-118 and was not the result of a rebound debris hit. During STS-118 ascent, the Tyvek rain cover on the F3D thruster tore and released in two pieces. The first piece released at 4.4 seconds as designed, and the second piece released at approximately 16 seconds at 237 miles per hour. This exceeded the design requirement of 170 miles per hour. A fault tree assessment identified several possible causes for the F3D Tyvek cover tear, but the most probable cause of failure was attributed to a severe flight environment around F3D and excessive stress at the tear's origination point. A late Tyvek cover release could shatter a window. At a minimum that would lead to an aborted mission and possible loss of crew life. Several redesign efforts had been implemented since STS-114 to mitigate the anomalous performance, but failures were still occurring. The most promising solution was to add thicker material (a "doubler") as reinforcement to prevent the tearing and reduction of the adhesive bond strength in the failure area.

PREPARE TO LAUNCH THE RESCUE MISSION

On August 28 the Aerospace Corporation presented to the DIG its initial mixed debris analysis results and the test matrix for shooting additional mixed debris projectiles. The initial results showed that just small amounts of ice could cause significantly more damage than foam impacts alone. The test matrix was amended to include mixed (foam and ice) debris impacts on tile under simulated STS-118 strut impact conditions to assess the level of breakup. USA–Boeing presented a proposed mixed debris analysis approach to address the concern over debris that flew like foam and impacted like ice. The DIG recommended modeling the debris as foam for transport analyses and adjusting the debris density appropriately to account for the additional ice or ablator mass. The impacts were assessed using the appropriate ice and/or ablator damage maps to generate a risk number. The USA Graybeards (headed by Bass Redd) presented their reconstruction of the mixed debris impact to the orbiter on STS-118. Their analysis showed an orbiter impact angle of thirty-three to forty degrees at about 450 feet per second. They leaned more toward the upper end as the actual impact angle. This was consistent with the Aerospace Corporation reconstruction that showed an orbiter impact angle of about twenty-one to thirty degrees at about 575 feet per second. The orbiter impact angle used during the mission assessment was about thirty-three degrees, which was based on focused inspection results. I was satisfied with the model convergence and ability to replicate the transport and rebound conditions. It was just a matter of time before the Aerospace Corporation impact test results confirmed the use of the ablator and ice damage maps. Once this was completed, it would be possible to calculate the mixed debris PRA. The Aerospace Corporation also presented results from a focused analysis of liquid hydrogen acreage losses aft of the bipod and near the liquid oxygen feed line brackets to ensure the general liquid hydrogen acreage analysis didn't mask a potential high risk. Even though the risk was low and the releases were past ASTT, the acreage losses were still a concern due to the consistent reoccurrence.

Toward the end of September, the Aerospace Corporation provided a verbal summary of the mixed debris impact testing for ice–foam impacts on tile using a fifty-fifty split of foam and ice mass. Results for the ninety-degree impact angle case were as follows: the ice remained attached to the foam when either the foam or ice impacted a structural surface first. The tile damage depth for both cases was the same. Results for the thirty-degree impact angle case (the STS-118 impact angle) were as follows: the ice-first impact showed some breakup and release of ice but did have portions of the ice remain attached to the foam. A similar result was observed for the foam-first case with more ice remaining intact with the foam. The tile damage depth for the foam-first impact was about 20 percent deeper. This was attributed to a longer damage exposure time before the ice finally shattered away.

The ET project developed the retrofit design for the liquid oxygen feed line bracket yoke starting with ET-120. The retrofit design eliminated the ablator and reduced the possible foam mass that could be liberated. A chamfer (beveled edge) would be machined into the forward face of the yoke to aid in diverting condensation away from the yoke–monoball closeout gap. Condensation diverters and coatings that inhibited the bonding of ice to foam were also being investigated to mitigate the root cause of ice bridging on future tanks. This design mitigated the ice formation around the entire bracket by 85 percent and reduced the amount of foam applied to the entire bracket by 50 percent.

The DIG was pursuing the development of a PRA model to quantify the risk of mixed (foam and ice) debris events and for debris rebound conditions. Forward work tasks had been identified in the integrated debris hazard report to refine the mixed debris and rebound models for the liquid oxygen feed line bracket yoke. An interim disposition was given for the next three flights based on the bracket design modifications. These eliminated the possibility of cracked ablator contributing to the failure mechanism and the possibility of a mixed (foam and ablator) debris event. The condensate diverter also reduced ice formation in the gaps between the monoballs and yokes. This modification minimized ice buildup and the possibility of ice bridging, and that further mitigated the failure mode. The tanking time was also reduced an hour to minimize the possibility of ice bridging and solidification. The feed line bracket yokes were added to the debris hazard report and were given a classification of "infrequent catastrophic." The modeling uncertainties surrounding the mixed debris and rebound PRA models drove this. A new "uncharacterized debris event" risk was added to the debris hazard report to clearly identify accepted debris risk items that risk assessment did not characterize. The uncharacterized debris events included the following situations: debris impacts from rebound effects, mixed debris impacts (except for the feed line bracket yoke), secondary debris hits causing debris liberation, combined failure mode releases, and unknown or unidentified failure modes.

Ice-Ball and Umbilical Ice Debris

Despite all the work done on mixed debris events and reentry during the STS-118 mission, I held an additional technical interchange meeting to focus on ice-balls. This included evaluation of further test options and umbilical ice updates that would be used to calculate the umbilical ice risk for the first time. Up until this flight, ice debris had not been an issue, but I felt it was just a matter of time before it was. I considered postponing the meeting but had wanted to update NSTS 08303 with more realistic sizes to increase the launch probability and avoid an unnecessary launch scrub. With two more flights scheduled in the fall, it would be difficult to reschedule the meeting, so I went ahead with it. Paul Macaluso from the ET project researched the historical ice-ball events and plotted them on an external tank diagram along with color-coded risk regions. Approximately 114 formations were documented in the database. This covered a total of a hundred and two tankings across sixty-seven missions. There were seven ice-balls, twenty-nine ice/frost balls, and seventy-eight frost-balls. (These typically formed in thermal cracks and shorts that occurred during tanking.) The diagram clearly showed ice-balls formed in low-risk zones. This translated to a low launch-scrub probability based on a single case that might have led to a launch scrub. The SICB wanted a DIG recommendation on how to proceed with ice-ball testing—if at all. There were three options to consider: forgo any testing and leave NSTS

08303 as is, conduct ice-ball testing at the Aerospace Corporation, or implement the ET project ice-ball test plan that included vibroacoustic and wind tunnel tests. After reviewing the historical data, the majority opted for the quick, inexpensive Aerospace Corporation testing. Before the meeting ended, Darby Cooper presented more-realistic umbilical ice mass sizes and time of release information. These would be used to perform an umbilical PRA ice risk assessment for the first time.

Umbilical Ice Testing, Eglin Air Force Base, Florida

In addition to the umbilical ice PRA Darby and his team were working on, the testing of umbilical ice coating at Eglin AFB was about to come to an end. Preliminary results were due to the DIG toward the end of September. I had assigned Doug Drewry as the NASA test director. It was his first debris assignment after moving over from the orbiter project. Doug was an experienced, easygoing engineer who had worked with the test teams developing and testing all the new tile and RCC panel repair techniques. The Dons wanted me to train a debris backup because of the flight rate and workload and in order to groom a replacement should I eventually decide to make a career move. I had not thought about a career move because I loved the challenges of launching shuttles and fixing debris problems. Given the work's critical nature and the visibility and level of authority and autonomy involved in running the DIG, I felt as if I had one of the best jobs at NASA. In many ways it reminded me of the work I liked best as a young navy lieutenant aboard the USS *Hyman G. Rickover*.

Doug was a welcome addition to the debris team. While running the test, he discovered the best way to grow ice was by spraying a water mist onto the frigid test article that was cooled by running liquid nitrogen through the umbilical test article. This was a much better method to grow ice than the standard USAF facility procedures for spraying water in subfreezing temperatures. In general the NESC ice coating that held so much promise before the test did not appear to reduce ice formation or improve ice liberation performance on the orbiter umbilical hardware. Everyone (including me) hoped the coating would work and cause all the ice to liberate at liftoff and thus eliminate any debris threat from this source. Doug stated in one of his daily activity reports how disappointed the NESC engineers and coating developers were when the ice stuck to the test article. Many people had worked on the coating for the last three years only to have it fail during demonstration testing. Ice adhered well to the foam closeouts regardless of whether the ice coating had been applied—mostly due to the rough surface. The only positive observation was that ice did not adhere to the Kapton baggie material and liberated immediately when the vibration table was activated. I did not know it at the time, but this observation proved to be invaluable on the next flight.

STS-120 Preparations, JSC

It was another arduous task to get the debris hazard report approved through the ISERP and PRCB for STS-120. This was primarily because I recommended lowering the risk classification of another five debris sources from "infrequent catastrophic" to "remote catastrophic." These sources included the solid rocket booster BTA, ceramic inserts, the liquid hydrogen IFR adjacent acreage, IFR body, and general acreage. I also recommended the liquid oxygen feed line bracket yoke be classified as "remote catastrophic" based on the strength of the design modification to eliminate the cracked ablator failure mode and mixed (foam and ablator) debris as well as the reduction of the ice bridging and ice growth from the condensate diverter. A preliminary mixed debris and rebound methodology was developed but not to the point of maturity to produce a credible PRA. In my mind the mitigations and the low risk of foam-only debris made the absence of a PRA a moot point. Although the Tyvek cover design mitigations had been determined, they still needed to be tested. Their implementation would not be ready until STS-123—two flights way. The ISERP disagreed with my recommendations and wanted to keep the debris sources classified at a higher level. Roy Glanville and the ISERP argued the flight performance sample size was too small, and there was still a lot of uncertainty with the failure modes and modeling. He pointed out that every mission since RTF had experienced a totally unexpected debris event, and the consequences of being wrong were severe. I countered that I wanted to uncover all the debris issues, and every unexpected debris event thus far had been fixed or mitigated. We understood the physics and failure modes for each of these debris sources. Flight data and analysis from our validated debris models had confirmed each mode. If the risks were low (they were), we needed to classify them appropriately. Otherwise we risked diluting the meaning of the classification and risk level. It was not rational to spend valuable resources on low-risk, improperly classified items and ignore actual high-risk areas—not just debris. If we overestimated the debris risk, the program might declare an unnecessary rescue mission or EVA repair, and each had its own inherent risks. The tile damage from mixed debris on STS-118 was a case in point. Regardless of the risk classifications, the program was still committed to reducing the debris risk from the various debris sources. The reception was much the same at the PRCB the following day. Wayne and the PRCB recommended numerous changes, and I made them with the acknowledgment the next revision to the hazard report would have the same set of risk reduction recommendations and more flight data to justify lowering many of the debris risk classifications. I knew going into the flight readiness review that the absence of a PRA for the mixed debris event would be an issue. Prior to the STS-120 FRR, all the ablator that was causing the foam to crack was removed and thus minimized the possibility of ice formation. The preliminary mixed debris testing at the Aerospace Corporation suggested the mixed debris event was a foam and ice mixture. This could be mitigated by implementing a design modification that limited the ice buildup. Were another failure to occur, its risk would be equivalent to the foam-only risk, which was low. Even with all these mitigations, the risk still needed to be addressed.

CHAPTER 25

"Tankenstein"

History of External Tank ET-120

ET-120, which I called "Tankenstein" due to its tumultuous history, was first shipped from the Michoud Assembly Facility in New Orleans to KSC on December 31, 2004. The tank was slated to fly on the first RTF mission (STS-114) in July 2005. This was also the first tank to be modified with all the safety improvements the CAIB mandated. Back in 2005 ET-120 was fueled twice at KSC during tanking tests and was replaced before the launch due to ECO sensor failures. It was returned to MAF on October 18, 2005, where it was used as a dissection test article for the PAL ramp and liquid hydrogen IFR foam losses. Other foam applications considered risky to the overall shuttle program were also removed from ET-120 and evaluated. Once the dissections were complete, the decision was made to return the tank to a flight-ready configuration. In addition to the foam debris attention ET-120 had received, all other systems (including the structure, electrical, propulsion, and mechanical subsystems) went through verification reassessment processes. The ET-120 refurbishment work began in October 2006 to support the August 2007 launch-on-need mission for shuttle *Endeavour*'s STS-118 mission. It was then integrated as the primary tank for *Discovery*'s STS-120 mission. ET-120 also had the first four feed line bracket yokes modified with the ablator removal and incorporation of the condensate diverters to prevent another mixed debris event. On paper Tankenstein was one of the safest tanks. However, NASA did not fly paper tanks, and there were numerous concerns owing to all the modifications and refurbishments, age of foam, and any possible unseen foam damage from the shipping, handling, and processing. Many within the program were making small wagers on how the tank would perform. My money was on the typical losses from aft of the bipod, LH_2 adjacent acreage, IFR body, general acreage, and one of the modified yokes. I also wagered we would have an ice issue and see a first-time loss from a liquid oxygen IFR because it was unexpected and something we had not experienced since RTF.

Developing the STS-120 Debris Flight Rationale

After STS-118 there was little time to work the mixed debris issue and develop a PRA. Initial impact testing showed the mixed foam–ice behaved more like foam-only debris, and the ET project made this the basis of its STS-120 flight rationale. The LO_2 feed line brackets were a consistent debris threat, and the ET project's best guess was that the foam failure was due to ice formation in the gaps. Any ice that formed in a gap would transfer the shear load between the pipe and bracket. This would then crack the foam off the bracket. Although probably not explicitly stated, the risk was understood and accepted. The design had limitations that could not be mitigated at the time. However, after watching the bracket foam debris release, subsequently bounce off the thrust strut, and damage the orbiter, the program managers and ET project quickly realized immediate action was needed to mitigate or reduce the risk. During one ET project meeting, an engineer showed Jeff Pilet a test article that demonstrated how the manufacturing process used to apply the foam to the pipe was inducing significant load into the foam on the bracket side. "This led to our discovery that the ablator was actually being cracked under the foam by this process," Jeff said. "We then ordered nondestructive evaluation of the installed brackets and found the majority of the brackets had cracked ablator.

This led to the decision to remove the ablator at the bracket tops and reapply with foam only."

Although this did not mitigate the failure mechanism, it certainly reduced the debris potential by eliminating the high-density ablator. Eventually the ET project modified the new design planned for the brackets to incorporate a "zero gap" design that removed the failure mechanism and thus worked flawlessly.

Even though a mixed debris risk number was not generated, most involved such as the orbiter team felt comfortable with the design changes and flight rationale. "There was no concern at all for me," said Steve. "I was very comfortable with the work the ET project and SEI had done to prepare us for flight. The ice-foam ricochet on STS-118 was perhaps a one in ten thousand scenario. However, it was a reminder to the program that even the small risk numbers we were evaluating prior to launch could certainly happen."

The STS-118 mission exacerbated the fear of unknown debris events that had been characterized as low likelihood and accepted risk. STS-118, however, pushed the shuttle program managers to the brink of ordering either a repair or a rescue mission. Before STS-120 a mixed debris event was beyond the capability of the debris risk models to analyze due to limitations on the release and transport models and a nonexistent impact model. Because of this mixed debris events were simply identified as unlikely and were considered accepted risks.

Reviewing the feed line bracket yoke foam loss data from the five previous flights showed the releases were occurring relatively early and (in every case) before the vehicle reached max Q. Depending on the trajectory, max Q occurred around sixty-two to sixty-five seconds. Computational fluid dynamics results for each of the five feed line bracket locations showed the peak aerodynamic loading conditions occurred well before max Q. Consequently, even mixed debris released from any feed line bracket location before sixty-five seconds showed minimal risk contribution on the order of less than one in one thousand. Unfortunately, without a mixed debris impact model or representative rebound and breakup model, the debris team could not produce a mixed debris risk number. It was several more weeks before the Aerospace Corporation could get enough mixed debris impact data to make even a rudimentary impact model for risk calculation. I had ordered the Aerospace Corporation to conduct some quick mixed debris impact tests using various foam and ice combinations. The foam and ice combinations were consistent with the ET project's analysis predictions of how much ice could actually form on the feed line brackets along with the amount of mixed foam and ice combinations that could potentially be released.

During the STS-118 mission, the Aerospace Corporation was doping cylindrical pieces of foam for testing using a fifty-fifty mix of foam and ice. The flight rationale I built was predicated on the integration of all the engineering data and analysis that showed mixed debris to be a relatively low risk. Even though I did not have a risk number, Roy and the ISERP were comfortable with the flight rationale because the ET project had implemented real physical controls to reduce the risk. The bracket design modifications involved removing the ablator. This eliminated the possibility of cracked ablator contributing to the failure mechanism as well as one possible mixed debris source—foam and ablator. From Roy's perspective, with the exception of the red "probable catastrophic" risk category, which included a numerical probability in its definition, the program requirements for hazard risk ranking should have been based on: one, the effectiveness of the controls preventing the cause from manifesting itself in the first place or two, mitigating its effect when experienced. PRA risk numbers did neither. "I placed much more confidence in risk controls whose effectiveness was demonstrated by test and analysis—measured, managed, and verified to support the upcoming mission," proclaimed Roy.

STS-120 Flight Readiness Review, KSC

Although I thought the flight rationale was compelling, I knew several members of the flight readiness review audience and FFR Board were still concerned about the inability to calculate a mixed debris risk number. Without a risk number, a comparison of the relative risk of mixed debris against already accepted risks by the program was impossible. This concern was coupled with the angst the MMT suffered during the STS-118 mission. The possibility of flying a rescue mission was still fresh on everyone's mind. I knew I needed something additional and compelling to finalize the flight rationale and satisfy those who craved a risk number. I wanted to put the risk into proper perspective, so I shared a story about my infant daughter, Jordyn, and her encounter with a wasp. My daughter was taking a nap in her crib located behind a sofa in the living room. The room had a vaulted ceiling that spanned twenty-five feet across. As she slept a wasp suddenly appeared and was flying randomly around the room. It was apparently seeking an escape path. Fearing the baby would be somehow stung, my wife, Brenda, wanted to move Jordyn out of the room while I killed the wasp. I thought waking Jordyn and moving her was unnecessary because I intended to wait and kill the wasp when it landed on the opposite side of the room. I explained my plan to Brenda while rolling a newspaper into a heavy-duty swatter to pursue the threatening insect. I casually stated that the odds of Jordyn getting stung were one in a million. Brenda was not convinced and still feared Jordyn would somehow be stung. After a terse debate over the risk, she finally conceded after I boasted I would kill the wasp on the first attempt. When the wasp landed on a ceiling beam more than twenty feet from our napping daughter, I swatted it with a heavy swing. I hit my mark. However, much to my amazement, the wasp recoiled from the blast and headed straight toward the ceiling fan centered in the room. The doomed wasp bounced off the ceiling fan blade and traversed another twelve feet directly into the crib. It stung Jordyn on the ear and caused her to cry out. Brenda quickly fired an "I knew it" and an "I told you so" as she ran to console our crying daughter.

 Everyone in the flight readiness review erupted in laughter. After the audience regained its composure, I went on to explain that what we saw on STS-118 was analogous to the wasp stinging Jordyn. Both events were intuitively low risk but still possible. Although the debris team was unable to produce a risk number due to the inability of the models to handle the mixed debris, the odds of a similar occurrence were clearly very low. There were no dissenting opinions over the debris flight rationale—not even from Ralph and the NESC.

Umbilical Ice Debris

In the early morning hours on October 23, 2007, the weather conditions at KSC (starting with tanking and lasting eight hours over the entire launch countdown) were conducive to icing. The ambient temperature hovered around 79°F with relative humidity of 87 percent and winds below five knots. Light rain sprinkled the vehicle when I took to my launch station in mission control at 5:00 a.m. If the rain continued, the launch would be scrubbed due to weather, but forecasters predicted the weather would clear well before launch time. There were typical ice and frost formations on the feed line brackets and bellows with the gaps clear of ice and frost. There was a 1.5-inch diameter frost-ball located on the aft face of a cable tray to acreage interface, but it did not violate NSTS 08303. Frost-balls up to 2.20 inches in diameter were allowable in this location. Two foam cracks on the vertical attachment strut cable tray's forward face were also noted. There was no ice formation or offsets in the cracks and no launch commit violations. However, about an hour before liftoff, the light rain and heavy condensation formed a large piece of clear ice measuring 4 inches by 1.5 inches by 0.5 inches thick on the liquid hydrogen umbilical. This was a NSTS 08303 violation and considered not suitable for flight. An integrated problem report (IPR) was generated to document the violation. This was standard procedure and would jeopardize the launch unless closed. Ray Gomez and Darby's support team performed a quick DTA assuming worst-case conditions and ice liberation to envelope the potential for orbiter damage.

Darby and his team did not have much additional information to add when developing the waiver to the NSTS 08303 ice limits. "Even after six flights, the umbilical ice waiver pointed out some shortcomings in our prelaunch thinking because we hadn't covered everything," recalled Darby.

This was the very reason I was pushing the DIG for an umbilical ice PRA and to conduct the umbilical ice coating testing at Eglin AFB. Imagery and the final inspection team estimated the mass based on the observed conditions. This was used as input to the transport model. Results showed impacts to the body flap trailing edge with impact velocities less than or equal to 420 feet per second and a shallow impact angle. Analogous ice-ball PRAs for similar masses in this location had risk levels less than one in one hundred thousand. I was convinced this was not a safety of flight issue but needed more data to convince Leroy Cain, the flight director, to continue with the launch. Leroy had also been the flight director during the *Columbia* accident and would need a compelling argument to continue with the launch.

Doug Drewry, who was supporting the launch from KSC, suggested using results from the Eglin ice coating testing to close the IPR. The testing had demonstrated that ice did not bond well to Kapton and liberated immediately under a vibration load. The rest of the ice remained bonded to the foam and fire-retardant paint and was expected to remain through ascent. This was based on the MSFC ice liberation testing and strong ice-to-foam bond strength the "Rudy screwdriver" test had demonstrated. MSFC and Safety offered strong resistance because they had not been involved in the Eglin tests or seen the results. They were still preliminary and had not been reviewed at the DIG. It was also only a demonstration test and lacked the certification-level rigor needed for developing flight rationale. I argued the testing involved use of actual flight hardware, and the ice had liberated at vibration levels much lower than what would be experienced on the vehicle. The flight rationale was built mostly on the low ice-ball risk and debris transport analysis but the Eglin test data were just one supporting factor. There were only about ten minutes left in the countdown when I pressed forward with the rationale that included Eglin testing and the expectation that the ice on the Kapton purge barrier would liberate at main engine and SRB ignition. My rationale was accepted minutes before liftoff, and the umbilical ice IPR closed and allowed the launch countdown to continue.

At 11:38 a.m. *Discovery* lifted off into a partly cloudy sky and headed for the ISS on the twenty-third assembly flight. During the mission the STS-120 crew would continue the construction of the station with the installation of the *Harmony* (Node 2) module and the relocation of a power truss. At approximately –2.9 seconds MET, ground cameras observed over half of the clear ice on the umbilical liberate and fall harmlessly to the launchpad just as predicted based on the Eglin ice coating test results. At thirty-four and forty-five seconds into flight, the right SRB camera observed foam releases from stations Xt-1129 and Xt-1377 bracket yokes respectively. In both cases the debris did not appear to contact the orbiter. At 118 seconds the right lower SRB camera observed multiple pieces of foam liberating near the outboard liquid hydrogen feed line bracket at station Xt-1871. The largest piece of debris, estimated to be seven inches by six inches, struck the orbiter starboard main landing gear door before breaking into multiple pieces and falling aft. Well after ASTT several other liquid hydrogen acreage losses appeared to contact the orbiter as the tank continued to shed debris. It was time to collect on my preflight debris prediction bet and wait for the on-orbit inspections to determine the extent of the tile damage.

The RCC panel inspection was favorable and reported no signs of damage. Image analysis identified no substantial damage to the orbiter tile, and the damage assessment team concluded it was safe for reentry.

DIG Criticism

Nearly every one of my foam and ice debris predictions came to fruition. However, I did not expect to see two feed line bracket yoke releases. Even though the two yoke losses did not represent an increase in risk level or performance outside expectations, there was concern because of the dual failure. Friction was building between MSFC engineers and Safety regarding the current debris risks, and criticism surfaced that the DIG and SEI were putting schedule ahead of safety—especially when it came to the foam failures. Helen McConnaughey and people at MSFC were still unhappy with the rationale behind the umbilical ice waiver and thought the Eglin test data should have been excluded. The Dons were under pressure to address the DIG issues, and they held a special meeting in their office to discuss the options. I explained it was difficult for everyone to keep up with the flight frequency and the pace of debris work. This was especially true given the number of model changes, testing mandates, and integrated IFAs for debris from each flight. The SICB received a weekly DIG status report, and I always operated on full disclosure of all debris work and made the information readily accessible to anyone who wanted it. The problem was that the amount of debris data and information was overwhelming and difficult to comprehend unless one followed it every day. Another debris verification review was not necessary because the fleet was operational, but I agreed we needed a forum to provide an integrated overview of debris updates since the last DVR.

The Dons and I agreed to schedule a debris summit on January 16, 2008, in Houston, Texas. Although the debris summit appeased the critics, I knew there was still a growing resentment over the amount of power delegated to the DIG. The debris summit would be a good opportunity to set the stage for lowering nearly all the remaining "infrequent catastrophic" debris risks and baselining the debris hazard report through the end of the program. Debris assessments and mitigation summaries for the external tank, orbiter, solid rocket boosters, and liftoff debris would be provided along with the rationale for risk reduction and classification. I would also present the debris model validation and documentation summary.

Over the last six months, I had Justin and his orbiter team work on updating their foam and ice damage maps. The damage maps currently being used in the PRA models had taken the sixteen thousand-plus tiles and organized them into sixty-four zones based primarily on similar thickness and aerothermal performance. This was done to minimize the amount of work needed to define the damage maps, but it sacrificed overall capability. Each zone was based on the most conservative tile within that zone. I knew this conservatism was probably driving the risk but did not know by how much. The orbiter team generated new damage maps that increased the number of regions from 64 to 882.

I then had both the Aerospace Corporation and USA–Boeing recompute all the foam and ice debris risk numbers using the updated damage map. The results would not be available in time to update the debris hazard report for the next mission, STS-122. It was scheduled to launch in a few weeks on December 6, but the numbers would be available for the January debris summit. After STS-120 I became more concerned about umbilical ice and tasked the orbiter team with developing a deterministic ice-allowable table in parallel with the umbilical ice PRA. The goal was to have the deterministic results available for the next mission and eventually incorporate them into NSTS 08303 if the PRA was incomplete or produced a high risk result. I wanted to avoid the drama of developing flight rationale at the last minute to prevent a launch scrub, and I needed something more than the ice-ball PRA and Eglin AFB test results ready.

Postflight Debris Assessment

Discovery and its crew completed their fifteen-day mission and landing at KSC on November 7, 2007. The orbiter sustained a total of 311 impacts. Fifteen had a dimension one inch or larger. Although the total number of impacts was well above the historical average, the total number of hits with one dimension greater than one inch was below average. The orbiter's lower surface tile sustained 247 total hits. Twelve had a dimension of one inch or larger. In all there were five integrated IFAs relating to debris. This included four for ascent and one for liftoff debris. Two weeks after *Discovery* landed, the DIG completed all the STS-120 debris reconstructions and showed the observed losses were consistent in terms of frequency, release time, and mass with the preflight risk assessments.

STS-122 Flight Readiness Review

I presented the debris summary at the STS-122 flight readiness review on November 30. The risk for the liquid oxygen feed line bracket yoke losses, which were attributed to the light rain during tanking circumventing the condensate diverters, was negligible. Although "Tankenstein" lost a lot of foam, all other liquid hydrogen acreage losses aft of the bipod and adjacent to the IFRs also showed negligible risk at the time of release because the losses were observed after ASTT. The redesigned IFRs implemented on STS-120 performed very well with no observed losses, and the more impact resistant BRI-18 tiles the orbiter team had installed around the umbilical doors were undamaged. Liquid hydrogen acreage foam losses aft of the bipod were attributed to cryopumping from leak-path-induced damage during processing.

Additional mixed (foam and ice) debris impact testing showed a trend toward "foamlike" impacts as the mass ratio of ice to foam decreased. The maximum ice-to-foam ratio for even high-ice launch days (like that for STS-120) was at most about 30 percent. This was well below the impact test levels at 50 percent ice-to-foam ratio. This suggested the mixed ice and foam debris behaved more like foam-only debris. Preliminary analysis performed by the USA–Boeing team on the highest-risk feed line bracket using the new 882-zone damage map showed a drastic risk reduction from one in one hundred fifty to one in thirty thousand. The risk improvement was primarily due to the revision of the ice-on-tile damage model with a smaller reduction due to improved CFD fidelity. All the IFAs had been closed. It was time to go fly.

Faulty ECO Sensors Delay the STS-122 Launch

I supported the STS-122 mission from KSC and participated in the L minus one walkdown inspection on Saturday for the Sunday launch attempt. I was pleased there were no debris issues. The first and second launch attempts on December 6 and 9 were scrubbed due to an ECO sensor failure—a problem the program managers thought was resolved. Tests revealed open circuits in the external tank's electrical feedthrough connector were the most likely cause, and the launch was delayed until February 7, 2008, to correct the problem. This delay caused the planned debris summit to be postponed as well.

Over the Christmas holidays, a replacement connector with soldered contacts was installed on ET-125 to fix the ECO sensor problem. The work to restore the tank to flight configuration continued in support of the February 7 launch date. Beginning with STS-122, the engine cutout sensor system feedthrough connector would be modified on every tank to address the matter of false readings. The program managers and ET project were relieved the nagging ECO sensor problem had been isolated to the feedthrough connector and finally resolved. When Jeff Pilet had taken over as the ET chief engineer, the changes made to the sensors had supposedly solved the problem. "Everything appeared to be working fine up until the two launch scrubs on STS-122 due to ECO sensor problems. Those proved us wrong," Jeff said.

Up until that point, the troubleshooting teams had not conclusively isolated the root cause but believed the failures were most likely due to intermittent "opens" at the sensor. New information gathered from the sensor failures on STS-122 ultimately helped isolate the problem. On the initial tanking, two sensors failed simultaneously. On the second tanking, an additional sensor failed. With this data the ET project and the troubleshooting teams quickly deduced the problem was not with the sensors themselves because it was highly improbable two sensors would fail independently and simultaneously.

Jeff and the ET project then developed a failure scenario that showed how open circuits could occur at the feedthrough connector during loading and then appear normal once the tank was drained. This was the typical signature observed on previous failures. The fault was related to ice formation on the surfaces of the external connector pins that formed as the liquid level in the LH_2 tank passed the feedthrough connector location. Ice on the pin surfaces effectively became a contaminant between the pin and socket and caused the circuit to open during loading and subsequently close when the ice melted as the tank drained. The redesign worked flawlessly through the end of the program because it eliminated the potential for ice formation and open circuits during tanking.

CHAPTER 26

We Could Be Wrong

STS-122 Launch, KSC

The only debris concerns during the STS-122 countdown were a few stress cracks varying from four to eight inches on the intertank section. This had not changed from the December launch attempts. There were no offset and debris violations because the ice and frost that formed was within NSTS 08303 limits. On February 7, 2008, *Atlantis* launched from Pad 39A on the third attempt with no ECO sensor anomalies. The twenty-fourth ISS mission involved delivery and installation of the European Space Agency's *Columbus* laboratory. The *Columbus* module would be Europe's largest contribution to the station's construction. The module measured twenty-three feet long and fifteen feet in diameter. It would house experiments in life sciences, materials science, fluid physics, and other disciplines. In addition to the *Columbus* module, *Atlantis* would deliver experiments to be performed in orbit and two European Space Agency astronauts—one of whom would remain on the station to conduct those experiments.

Foam Loss at Eighty-Five Seconds

The liquid hydrogen acreage and aft bipod foam debris cryopumping losses after ASTT were becoming fairly predictable, and STS-122 was no different. Two foam losses from these locations exceeded the maximum predicted mass size, but a quick look at the ASTT curves showed "zero" risk at the time of release. The video review also showed the foam liberated in multiple small pieces rather than one large chunk. At ninety-three seconds into flight, a large (0.097-pound) piece of foam liberated from the umbilical cable tray located at the aft end of the tank. It exceeded the maximum predicted mass size and was the first loss of this type observed liberating from this location. Previous losses had been well below the mass limit and had not been considered threats to the vehicle because they had been aft of station Xt-2058—the aft debris limit. Anything aft of this station could not cause critical damage regardless of the release time, but foam pieces this size had not been considered. Two intertank flange foam losses observed by the umbilical well camera at ET separation caught my attention the most. One was located starboard at the sixty-two-degree phi position, and the other was on the port side at negative fifty-five degrees. Although the time of release for the sixty-two-degree loss was unknown, at approximately eighty-five seconds, there was a debris release observed on the port side of the vehicle that was possibly from the negative fifty-five-degree location.

Debris was seen in both the forward and aft SRB cameras at about the eighty-five-second time frame. The forward viewing camera showed multiple debris pieces, and the aft viewing camera showed a single piece of debris. It was not conclusive whether the multiple debris pieces came from the same debris source. There were no other observed foam losses on the -Y side of the tank. This made the flange loss at negative fifty-five degrees phi the "best candidate" for the event that occurred at eighty-five-seconds. Although the debris did not appear to strike the orbiter, extensive damage would have been possible had there been an impact. This was based on the size and timing within ASTT.

After extensive evaluation all the imagery analysis teams agreed on the facts and circumstantial data. They believed there was a flange loss at eighty-five seconds. However, the plot thickened because the negative fifty-five-degree release was outside the predicted

ET project debris cloud in terms of mass and release time—if the debris really liberated at eighty-five seconds. If the flange loss did occur at eighty-five seconds, then the ET project debris cloud and cryopumping and cryoingestion models were completely wrong, and if the models were wrong, then the shuttle was flying with a much-higher cryopumping debris risk. The dilemma between the imagery and the model was a huge concern because the debris event at eighty-five seconds occurred during the worst possible ASTT. It left everyone, including me, wondering if the cryopumping models and risk estimates were wrong.

Wayne Hale Steps Down as Shuttle Program Manager

The day after the STS-122 launch, Wayne Hale announced he would be leaving the Space Shuttle Program to become the deputy administrator for Strategic Partnerships. The announcement surprised me and many others—especially given the timing. It would have seemed more appropriate after the mission was over. John Shannon, who had served as Wayne's deputy since November 2005, became the new shuttle program manager. Leroy Cain, the reentry flight director during the *Columbia* disaster, would replace Shannon as the MMT chair. Although Mike Griffin forced the change, Wayne continued to provide support and frequent advice on a variety of subjects until his retirement from NASA in 2010.

During the nearly two and a half years Wayne served as the program manager, great strides were made in reducing the debris risk and calculating the PRA. The more we flew, the more knowledge we gained, and the uncertainties were whittled away. Hence the PRA calculations became more precise. In terms of confidence, Wayne confirmed using PRA to make launch decisions "was the best information we had. Even though we tried to attack the problem on every front and then found out repeatedly we were not as smart as we thought we were, in retrospect, our biggest challenge was fighting the impulse to narrow our field of view too early."

Even with all the debris improvements, Wayne was most proud of the changes in management culture that resulted in a safer system—not because of any technical change but because of a mind-set change in the program leadership. John Shannon's prior experience as a flight director, deputy program manager, and MMT chair made him a perfect person for the job. During Wayne's tenure the leadership within the program was stable, but it was still remarkable to me that, in the span of three and a half years, I had worked for three different SEI managers and three different shuttle program managers.

Flange Foam Debris Loss Assessment

Since RTF there had been five flange losses observed. In addition to the two on STS-122, there was one release on STS-117 at the negative sixty-eight-degree phi location and two releases on STS-114 at approximately the minus eighty-three- and minus ninety-degree phi locations. All these losses were within the maximum predicted mass limit for the void delta P and cryoingestion failure modes. Transport to the orbiter was limited, and risk was minimal (less than one in ten thousand) due to the small mass and phi angle location. The two losses observed on STS-122 and the others yielded a release rate since RTF of 0.5 releases per flight. That was below the 1.3 releases per flight the ET project's debris cloud predicted. Flight history before RTF could not be considered due to flange design changes implemented after STS-107. However, clearly the flange losses on STS-122 represented an increase in risk for this debris source. The debris released from the negative fifty-five-degree phi location was outside the debris cloud in terms of the mass and release time for an eighty-five-second release. Debris reconstructions for the negative fifty-five-degree phi loss at eighty-five seconds showed conditional probabilities of one in five hundred to tile and one in one hundred eighty to RCC. The sixty-two-degree phi loss showed a peak risk at about seventy seconds of one in five hundred to tile. The ET project concluded the most likely failure mechanism for both flange losses was cryoingestion, but this created an inconsistency with the release time that was predicted to occur much later than eighty-five seconds—well beyond ASTT. Since the debris cloud predicted predominantly small masses (less than 0.005 pounds), I was also concerned the negligible contribution from those small masses artificially diluted the cryoingestion risk.

Image analysis of vehicle video taken on-orbit provided an early assessment of the debris damage the orbiter sustained during ascent. Other than a piece of blanket protruding from the starboard OMS pod, no significant tile or wing leading edge damage was noted during the rendezvous pitch maneuver or robotic arm inspection. Deterministic transport analysis was performed for the blanket anomaly, and this showed no transport mechanism to the vertical tail, rudder speed brake, or SSME nozzles. Although the OMS pod was in jeopardy of impact if the blanket failed, the maximum impact velocity was not enough to cause damage. A debris reconstruction was also performed for the large (0.097-pound) aft foam release because of its size. Even though it was a large piece of foam, results showed there was still no threat for debris releases aft of station Xt-2058.

Although there were no reentry issues, if NASA was going to make the March 11 launch date for the next flight—STS-123—the DIG had to resolve the flange loss. I gave Jeff Pilet and the ET project the action to reassess their flange debris cloud and cryopumping failure model. I also had Brekke and the Aerospace Corporation calculate the flange debris risk after shifting the debris cloud twenty seconds earlier to include the eighty-five-second loss. This determined the integrated PRA for a debris cloud that would capture the eighty-five-second loss. I also instructed the Aerospace Corporaton to utilize the orbiter team's new 882-grid tile damage map. It had improved damage capability—especially for large foam. The flange PRA after shifting the debris cloud twenty seconds earlier turned out to be a low risk. It was less than one in ten thousand. Since the debris cloud released predominantly small masses (less than 0.005 pounds), I was still worried the real cryoingestion risk for larger foam releases was masked. Consequently I had the ET project and the Aerospace Corporation investigate the cryoingestion and void delta P releases separately for larger masses. I knew, however, this would take time and probably not be ready for the STS-123 flight readiness review.

STS-122 Postlanding Debris Inspection, KSC

The postlanding debris inspection of *Atlantis* was conducted at KSC immediately after the vehicle landed on February 20, 2008. During this mission *Atlantis* sustained a total of 158 impacts. Sixteen had a dimension one inch or larger. The total number of impacts was still above the shuttle flight average, but the total number of hits with one dimension greater than one inch was below the average of previous flights. There were a lot of long days ahead to close a total of nine debris integrated IFAs. Eight were for ascent debris and one for liftoff debris.

Three Different Cryopumping Models

A week after the flight, the ET project and NESC proposed new cryopumping release time models to the DIG. Flight observations for cryopumping and cryoingestion foam releases were consistently occurring much later than originally predicted due to model conservatisms. Most were outside ASTT when debris was no longer a threat. The ET project developed a semiempirical correlation model using thermal analysis predictions and historical release times. This showed good agreement with the observed cryopumping and cryoingestion losses. The model predicted late release times for cryopumping releases - especially at the forward end of the tank. The earliest-predicted cryopumping release time was 108 seconds at station Xt-1160. If the ET project model was accurate, a flange failure at eighty-five seconds was impossible.

The NESC model was a direct curve fit of the RTF data with a seventy-second dispersion around the mean. The huge variance was due to the small number of releases utilized to construct the curve fit. If the NESC model was accurate, the likelihood of a flange failure at eighty-five seconds was extremely low but possible. The DIG now had three different cryopumping models that fit the data, but the big question remained which was right. The ET project model had the most analytic rigor based on the failure physics, but it was the least conservative. There was not enough time for the DIG to baseline a new cryopumping release time model. Therefore, it deferred the model selection to the debris summit, which was now scheduled for April 16 to 17 after STS-123.

Significant Debris Issues on Every Flight

Every shuttle flight since the *Columbia* accident had had a significant debris event: STS-114: PAL ramp loss; STS-121: putty repair and gap filler losses and damaged orbiter blanket; STS-115: ceramic insert losses and feed line bracket foam loss; STS-116: SRB BTA impact and damage to the orbiter; STS-117: external tank hail damage and damaged OMS pod blanket requiring EVA repair; STS-118: mixed debris impact to the orbiter; STS-120: umbilical ice during liftoff and foam losses aft of the bipod; and STS-122: flange loss. In addition to the significant debris events, there had been a total of fifty-eight integrated IFAs in eight flights. Each had represented risk to the shuttle. Numerous redesigns, mitigations, and controls had to be implemented to address the debris issues because all IFAs had to be closed before flight operations resumed. Clearly the debris problems were not just isolated to the tank's foam insulation. The orbiter, SRB, and launchpad debris threats had also been identified as a result of improved imagery and scrutiny.

Critics pointed to these debris failures as evidence NASA was wrong about its risk assessments and ability to reduce the debris risk. Given the circumstantial evidence surrounding the possible flange loss at eighty-five seconds, many were questioning the validity of PRA models and were growing concerned we were underpredicting the risk. They wondered if the models we relied on were wrong the same way CRATER had been wrong about the damage to *Columbia* during

STS-107. Although the numerous and varied debris events did give me pause to reflect, they motivated me to push the team to solve the debris problems so every flight would be safer than the last.

The criticisms never bothered me, and I actually welcomed the feedback because sometimes it would point out weaknesses or something we had missed. In the case of NASA being wrong about debris, people made some compelling arguments. The shuttle's very design (with the orbiter mounted on the side of the SRBs and foam-covered tank) put the most vulnerable structures in the direct debris path. We could never eliminate the risk with this design. We could only reduce it to acceptable levels through debris removal or redesign, implementation of effective installation process controls to reduce liberation propensity, or hardening the orbiter to enhance its impact tolerance.

Considering it was a vehicle that possessed known catastrophic debris risks, "safe to fly" did not mean no risk. It meant the risk was mitigated enough to be considered acceptable. Nothing was ever certain except that the chance was always present that any debris source identified in the debris hazard report could cause catastrophic damage. According to Roy Glanville, the ISERP typically took issue with what were believed to be inaccurate risk rankings, understated risks, or premature claims that risk mitigations had succeeded when their impacts would not occur until later flights.

With the exception of the protuberance air load (PAL) ramp, the other debris events had been occurring since almost the first day of the Space Shuttle Program. The DIG was regularly applauded for the excellent work it did to mitigate the release of debris over time. As soon as an area of concern was identified, the DIG was off working up a solution. Beyond all the debris risk mitigation efforts, the on-orbit inspection capabilities, all the analytic tools, and the repair capabilities gave Steve and the orbiter team confidence we were safe to fly. "I thought the DIG had really learned and improved their models over time, and frankly my confidence level increased significantly from the first Return to Flight mission until I left the orbiter project in March 2008," Steve said. "I was very confident in the risk numbers, and I was always comfortable with our risk acceptance prior to flight."

For Ralph and the NESC, it was discouraging when new, previously unconsidered debris sources were found. However, Ralph felt the program took the right steps in each case to mitigate the risks. "The thing that gave me the most confidence," explained Ralph, "was how aggressively the program was working the highest-risk debris issues."

There was an obvious heightening of the level of concern for debris after the *Columbia* accident that led to taking proactive steps to resolve debris problems. Ralph credited the early Return to Flight FRRs for establishing a new way to conduct reviews that included debate on issues and expression of dissenting opinions. "I think this was a very positive change for the shuttle program and NASA. The increased discussion at the flight readiness reviews was the most important aspect of this change and provided multiple perspectives on the issues for the agency decision-makers that prior to *Columbia* might not have been heard," stated Ralph.

Mike Griffin attended almost every one of the significant in-flight technical meetings that resulted from these events. The present-day team was clearly having to make up for many years (even decades) of failure to appreciate the debris danger. As debris events occurred now, they were being pursued to their logical conclusions. Mike had great confidence in the engineering analysis teams from the start. It was completely clear to him he was dealing with teams of very smart people with both breadth and depth. "Whenever I probed I came away feeling pretty good," he recalled. "Very early on I must admit I did not have full confidence in the senior management team that was responsible for evaluating the results of those analyses. As is a matter of public record, I replaced most of the senior team. From that point on, things got better and continued to do so."

Mike and Gerst did not care so much what a particular PRA number was. They both had considerable experience with various kinds of statistical analysis and were disinclined to assign too much credibility to any given result. However, both were very interested in how debris-risk PRAs compared with the results from other assessments of potential risk and how the PRAs were changing over time as we learned more. Like many others Mike and Gerst were more interested in relative risks and trends than in the absolute numbers.

STS-123 Flight Readiness Review, KSC

The STS-123 flight readiness review was held on February 29, 2008, and I was glad to have the extra leap-year day. Most STS-122 integrated debris IFAs were closed with solid flight rationale, and the debris hazard report would remain closed through the next flight—STS-123. The only IFA being worked was the flange debris loss. That was still pending closure until the Aerospace Corporation analysis of the larger masses was complete. There was still uncertainty and a lot of debate about whether the flange loss occurred at eighty-five seconds. Even though the ET project's cryoingestion model excluded releases before 108 seconds, no imagery evidence suggested the eighty-five-second debris came from anywhere else. This was the big contradiction, and it also reflected our inability to view the entire tank at all times during ascent. The imagery limitations were even more severe during tank separation due to the viewing angle, lighting, distance from the orbiter, and crew's photographic ability. It was definitely possible foam debris could have liberated from another location and was obstructed from the cameras. No one will ever know, but to account for this uncertainty in the risk assessment, I took the most conservative route and assumed it was the flange failure at eighty-five seconds. The PRA result from the shifted debris cloud, using the Aerospace Corporation PRA model and associated updates, was one in five thousand. This risk number was a conditional probability where the total release mass was assumed. However, we knew full well the debris released in multiple pieces, which would reduce the risk even further. I recommended a one-flight disposition of the IFA to allow an updating of the debris cloud to account for cryoingestion and void delta P losses separately in the PRA. This would ensure the multitude of smaller releases was not masking the risk from larger mass releases.

Although the flange losses on STS-122 represented an increased risk for this debris source, the ET project did not identify any process deficiencies, and the release rate since RTF was two times lower than what the model predicted. I concluded the debris presentation by advertising the debris summit and reminding everyone this was the last of the old tanks left to fly. External tanks that flew after STS-123 would have newly sprayed foam and would incorporate all the new processes and redesigns. That included the new ice/frost ramps. Gerst asked if there were any questions. Brewster Shaw, a former astronaut and the vice president and general manager of Boeing's space exploration division, fired the first and only question. He was concerned about the flange risk level and asked bluntly how much attention had been given to this debris source. He wanted to know what else, if anything, we planned on doing to mitigate the risk. I explained the flange loss was our top priority but was a low-risk debris source based on conservative analysis. I also had the Aerospace Corporation performing supplemental analysis that would not be ready until the L minus two meeting, but I did not anticipate a change in risk classification. After my response an uncharacteristic silence from the audience followed. It led me to believe either I had done a great job or the failure and flight rationale mystified people. It was probably a little of both.

L Minus Two Meeting, JSC

Results from the Aerospace Corporation's first attempt to separate the flange cryoingestion and void delta P risk were completed the night before the L minus two meeting. The preliminary results confirmed my suspicion that the debris cloud and void delta P failure mode releases of predominantly small masses were masking the cryoingestion risk. The void delta P-only risk remained less than one in five thousand, but the cryoingestion-only risk was 1 in 250 for tile and 1 in 770 for RCC. Releases from the top of the tank near the bipod area drove most of the risk. The risk then tapered off rapidly as the phi angle increased. I had always questioned the value of the engineering ethics courses I had to take in college. I never thought I would be in a position that it would affect me. However, I needed to communicate this information to the program because the flange risk was a lot higher than I reported at the FRR.

It was after midnight when I called Don Noah to discuss the results. The risk was not a showstopper, but it needed to be communicated to the MMT at the L minus two meeting that afternoon. During the L minus two, I verbalized the results and explained the analysis did not take into account debris breakup, which would reduce the risk. Furthermore, we had no historical losses toward the top of the tank, and that was the area of highest risk. I did not recommend a change in risk classification but wanted the MMT to know the cryoingestion risk was much higher than reported at the flight readiness review. The DIG also needed to review the results, and more work was required to refine the inputs and conduct sensitivities on the debris breakup to understand how much it affected the risk. The MMT concurred and pressed forward with the launch countdown.

Last of the Old Tanks

After the L minus two meeting, I flew to KSC with Doug Drewry to orient him with all the debris support assets at Kennedy. I wanted him to support the mission from KSC, and I showed him the NASA debris radar facilities, Launch Control Center, KSC imagery lab in the Vehicle Assembly Building, and solid rocket booster retrieval and inspection facility. Doug and I participated in the L minus twenty-four-hour inspection at the launchpad. No debris issues were identified during this inspection, and I headed back to Houston to support the flight from JSC mission control.

The launch day conditions were very dry, and there was little condensate on the vehicle. It was rare to have no debris issues to work. On March 11, 2008, *Endeavour* lifted off the pad at 2:28 a.m. STS-123 would deliver the first pressurized component of the Japanese Kibo laboratory and a Canadian robotic device called Dextre. It would also conduct five space walks. *Endeavour*'s sixteen-day flight would be the longest shuttle mission to the ISS.

Night launches were less of a concern for debris detection because a lighting system derived from an off-the-shelf flash was added to the digital camera mounted inside the orbiter's umbilical well. It was designed to capture twenty-three illuminated photographs of the tank as it fell away from the shuttle after the main engine shut down 8.5 minutes following launch. The flash module would provide enough light to enable photography of the external tank as it separated from the shuttle amid darkness or heavy shadow. The flash was to be flown on all remaining shuttle missions and would operate day or night. Shortly after launch the vehicle passed behind a cloud and led to limited visibility for the cameras. Sources of debris liberation could not be sufficiently determined from the ground-based imagery review in order to perform reconstructions.

At SRB separation multiple pieces of debris were observed impacting the starboard wing, body flap, and port eleven. On-orbit inspection found a small area of tile damage on the starboard wing. The orbiter damage assessment team quickly cleared this. Overall it was a clean debris flight. The Tyvek covers performed as expected with all releasing before 170 miles per hour, and there were no observed flange or feed line bracket yoke losses. I was pleased there were no ascent debris-related issues to work. This gave me more time to prepare for the upcoming debris summit. *Endeavour* completed its sixteen-day mission on March 26 with a spectacular evening landing at Kennedy.

The orbiter sustained a total of ninety-eight impacts. Thirteen had a dimension of one inch or larger. For the first time since RTF, the total number of impacts was below the historical average, and so was the total number of hits with one dimension greater than one inch. There was also minimal damage around the umbilical area with just a few dings to the impact-resistant BRI-18 tile installed around the umbilical doors. There were only three debris integrated IFAs resulting from the flight—the lowest total to date. It was the first flight without a significant debris event.

Debris Summit Preparations

For the first time since working debris on the shuttle program, it felt as if the worst was over, and the debris issues were finally being mastered. There were fewer integrated IFAs and debris surprises, and the workload was starting to diminish. During STS-123 I finalized the debris summit agenda and scheduled the event for April 16 to 17, 2008, in Houston, Texas. I spent the few weeks leading up to the summit preparing my presentation materials, closing the STS-123 IIFAs, and supporting the ET team's effort to reduce the tank processing time. The ET project held a meeting at MAF on March 26 to discuss possible options. The processing time had increased by nearly ninety days as a result of all the improvements and verification steps. All the options presented appeared reasonable, but they would have minimal impact on debris risk, and it was up to the ET project to prioritize the various options and characterize the amount of time each could save.

The DIG was also working with Chuck Shaw, the Hubble flight test manager, to firm up the debris support plan for the upcoming Hubble mission. Because there would be no safe haven with the ISS or inspection of the orbiter's tile using the rendezvous pitch maneuver, it was vitally important to get debris threats identified quickly. The odds of a successful rescue mission were much better for early emergency declarations. Onboard cameras only provided a limited view of the vehicle. This meant the debris team would rely more on the radar assets to determine if there had been any impacts to the orbiter. Radar Tony and NASA had been working with the Wright Patterson AFB radar team to characterize the radar cross section and ballistic number of typical ascent debris in order to produce a quick-reference tool to be used in identifying debris. Various foam debris shapes and other debris materials such as blankets, ceramic inserts, gap fillers, putty repair, and ice molds were sent from JSC for the testing. Chris Thomas of the Wright-Patterson Radar Lab was the test director who developed the test plan and would oversee the testing scheduled to begin the last week of April.

The DIG also investigated widening the ELVIS camera view with a different lens and repositioning the camera to maximize the field of view. The ELVIS camera had been modified on STS-120 to view the first three liquid hydrogen IFRs during ascent, and the lens was changed from twelve millimeters to sixteen millimeters to expand the field of view. This onetime adjustment had improved the imagery capability, and it was just a matter of cost to implement the same changes for the Hubble mission and the remaining flights after that.

Soyuz Reentry Problem

On April 19, 2008, astronauts Yuri Malenchenko of Russia, Peggy Whitson of the United States, and Yi So-yeon of South Korea endured more than eight g's during a ballistic reentry from orbit in their Soyuz capsule. As the Soyuz entered the upper reaches of the atmosphere, they were traveling at twenty-five times the speed of sound. The aerodynamic heating from the air molecules impinging on the spacecraft heated it to 3,000°F. The rapid deceleration from a ballistic reentry caused a 200-pound man to feel as if he weighed 1,600 pounds and a 125-pound woman to feel as if she weighed half a ton. A normal reentry is a shallower entry with much lower g-loads peaking at about three g's and much lower surface temperatures. However, for the second time in a row, the Soyuz spacecraft suffered a major reentry glitch from an unknown cause. Early indicators on this entry pointed toward explosive bolts that might have failed to disengage the lower propulsion module from the central crew capsule before reentry. During this separation phase the astronauts reported an unexpected violent buffeting. The spacecraft entered the atmosphere in a dangerous orientation, and this resulted in the ballistic reentry. Fortunately the stubborn segment broke away and freed the descent module to right itself.

The trajectory reconstructions showed a high likelihood the spacecraft would have exceeded its temperature limits had the propulsion module remained attached. This would have compromised the spacecraft and probably would have killed the crew. By the time the module released, the hatch and an antenna had been severely baked. Malenchenko later reported smelling smoke inside the capsule.

The Soyuz ended up on a trajectory engineers call "ballistic" (unguided). It caused the capsule to fall more steeply to Earth at higher g-loads on the astronauts. "I saw eight point two g's on the meter, and it was pretty dramatic," said Whitson shortly after her return. "Gravity's not really my friend right now, and eight g's was especially not my friend."

The Soyuz landed on land almost three hundred miles short of its intended target, and the local townspeople helped the astronauts out of their spaceship instead of the usual recovery team. Russian space agency personnel showed up forty-five minutes later. It was hardly a smooth return. This close call with disaster was a huge concern for NASA. When the Space Shuttle Program's planned end arrived in 2010 to 2011, the Soyuz would become the only means of human transport to the ISS until NASA began flying the Orion crew exploration vehicle. That was projected to fly in 2015. Without much say in the matter, NASA was forced to defer the investigation to its Russian partners and let them examine the charred Soyuz for clues about what caused the problem. About the only good news from the landing was that Peggy Whitson, the space station's first female commander, broke the US record for time spent in space. Her 377 days in space over two space station tours edged out Mike Foale's 374 days over six spaceflight missions.

Investigators concluded the Soyuz capsule detached too late because a pyro bolt, an exploding connector that kept the module attached to the space station, failed to detonate on time. Later in the year on December 23, Space Station Commander Mike Fincke and Flight Engineer Yury Lonchakov conducted a space walk to install an electrical probe on the Russian *Pirs* air lock module designed to identify problems with the Russian Soyuz capsule. Russian scientists hoped data from the probe would help explain the malfunctions that occurred on the last two Soyuz flights that both resulted in high-g ballistic reentries.

CHAPTER 27

Debris Summit

Debris Summit, Regents Park, Houston, Texas

I was looking forward to the debris summit and was proud of all the work the Debris Integration Group had accomplished. The space shuttle was much safer and would be flying new tanks for the remaining missions. The event was being held at Regents Park—where nearly all the debris verification and major debris technical meetings took place. I started off the summit with a football analogy. NASA started deep in its end of the field after the *Columbia* accident. The debris team fought long and hard to get to this point, and we were now about to cross the goal line. We were close to finalizing the debris hazard report for the remaining shuttle flights and completing all debris testing. For the first time we had no planned vehicle modifications or mitigations remaining to be implemented. Furthermore, all the updates to the debris models had been completed, and with the exception of liftoff debris, it was time to lower the remaining "infrequent catastrophic" risks to "remote catastrophic." Several debris sources were also no longer considered catastrophic threats, even if the debris hit the orbiter at any time during flight.

The main objectives were to provide all the updated debris risk assessments since the STS-116 DVR in November 2006 and present the rationale to reduce the risk likelihood from infrequent to remote for five of the six remaining infrequent debris sources. These risks included the feed line bracket ice, feed line bellows ice, umbilical ice, orbiter putty repairs, and orbiter gap filler debris sources. It was a comprehensive debris summary and was not intended to be a peer review or a formal DVR with a decision-making board. The agenda also covered performance trends, external tank foam and ice debris, and all types of orbiter debris.

Reasons for the Increase in Orbiter Impacts

With the exception of STS-123, all the orbiter debris hit counts since RTF exceeded the historical average before the *Columbia* accident, and there was a much larger variation between the flights. Many attributed the increase in hit counts to the increase in popcorn foam debris from the vented intertank foam. Venting the foam every two to three inches limited the size of debris released and was a mitigation that performed well. Although it was assumed the main contributor was variation in popcorn foam debris from the intertank, the imagery data showed no correlation. The question remained—what was causing the large variation in hit counts if it was not the variation in the popcorn foam environment? Certainly variation in the popcorn foam environment was a contributing factor along with multiple impacts from single debris sources "skipping" across the tile, secondary debris impacts from debris breakup, and the abundance of impacts and variation around the umbilical doors due to the baggie strikes. Analysis showed there would be a much better correlation with the popcorn foam if the umbilical door impacts were not included. This suggested there was a unique phenomenon occurring due to the crossbeam (aft ET structural support) effects and recirculation impacts in this area.

Although orbiter impacts increased after RTF, the real figure of merit that demonstrated an improved debris environment was the reduction in tile volume loss. That had decreased by a factor of two since RTF. Installation of the impact-resistant BRI-18 tile and improvements to the baggie closeout around the umbilical doors by elimination of the flapping appeared to mitigate the damage. Continued damage assessment in this area identified the umbilical baggies as the most probable debris source. Subsequent orbiter testing demonstrated the baggies could not cause critical damage to umbilical door seals or surrounding BRI-18 tile.

Other Key Debris Trends

Window impacts were the other trend the DIG tracked. The debris sources with the potential to impact orbiter windows included the Tyvek covers, popcorn foam, and booster separation motor plume debris at SRB separation. Although window damage was still possible, window impact testing had cleared all the catastrophic threats. The final significant trend was noting that very few visual, radar, and wing leading edge sensor observations had occurred between seventy and one hundred seconds, which was the worst-case ASTT. Nearly 80 percent of the integrated risk occurred during this time period. Foam debris risk due to the void delta P failure mode mainly drove this. This trend confirmed the conservatisms in debris cloud models that predicted foam losses during this time frame.

Foam Debris Risk Assessments

As of STS-120, all foam debris sources had been classified as "remote catastrophic" risk. The IFR redesign and liquid oxygen feed line bracket modifications were complete and scheduled to fly on the next mission—STS-124. According to Jeff Pilet, the debris summit was a very memorable event as it was the last big debris community meeting to discuss the hazards and risks. "I thought the recommendations to reduce all the foam debris risks were well-founded and supported by our understanding of the failure modes, flight performance, and the expected improvement in performance from the new tanks. My confidence in predicting foam debris risk grew significantly once we started to get imagery data showing debris release times consistent with our understanding of the failure modes. Also the performance of the new in-line tanks on STS-123 increased my confidence because it was by far the best foam debris performance we'd ever seen on the program as evidenced by the almost nonexistent damage to the orbiter routinely observed during the on-orbit inspections."

The DIG was also participating in the production enhancement team activities with the ET project. That team was tasked to improve the tank processing time while still maintaining an acceptable risk posture. All the design changes, mitigations, and process improvements had added extra time required to process a tank. This was putting a strain on the delivery schedule, and it had to be corrected because it adversely affected the shuttle manifest.

Analyses for the updated NESC, ET project, and the Aerospace Corporation cryopumping time of release models were compared. The predicted release time for each model was compared to the thermal analysis prediction for the time in which the LH_2 tank wall reached the vaporization temperature of air (–318°F). There was a much better correlation between flight history and the updated NESC, ET project, and Aerospace Corporation models than the current PRA model. Because the cryopumping risk was so low, it did not make sense to change to another model. The ET project also presented the thermal analysis results for the intertank cryoingestion time of release. Overall the thermal analysis results were conservative in predicting structural ascent temperatures. This led to the conclusion that it was unlikely for a flange foam release due to cryoingestion at eighty-five seconds or any time during ASTT.

Ice Debris Risk Assessments

The feed line bellows and bracket ice analysis updates led to significantly lower PRA results. They were much less than one in ten thousand for all locations. Improvements to the orbiter damage models and computational fluid dynamics solutions produced a more representative ice risk. Flight history and RTF performance also indicated there was a low probability of a launch scrub due to ice-balls. This was based on the fact that ice-ball formations formed in low-risk locations. Qualitative data from ground cameras during launch suggested that most umbilical ice broke off and released before shuttle liftoff. The ice coating testing at Eglin supported this. The umbilical ice PRA was completed, and the highest risk of exceeding the orbiter tile damage threshold was one in seven hundred from a rectangular mass of 0.125 pounds. That was the largest mass assessed and much larger than anything observed. All other umbilical ice risks for various shapes used in the analysis were much less than one in ten thousand.

Orbiter Debris Risk Assessment

The orbiter debris mitigations for gap fillers, putty repairs, and ceramic inserts had been performing well. Darby said he and the Boeing team were "comfortable with reducing the orbiter debris risks to 'remote catastrophic.'

Gap filler protrusions and losses were due to process deficiencies such as improper installation and bonding. During entry gap filler protrusions could generate early boundary layer transition from laminar to turbulent flow on the vehicle's belly and cause excessive heating and decreased structural margins. It could compromise tile bonds or even compromise the vehicle's structural integrity. The effort to remove and replace suspect gap fillers began with the STS-121 mission when the problem was first discovered. Orbiter team technicians steadily worked their way aft and removed faulty gap fillers according to the prioritized threat zones. Since no location had more propensity to protrude than another, the inspections started with the highest-risk areas first. Because the PRA modeling effort had significant limitations, the gap filler risk in areas not inspected would remain classified as "infrequent catastrophic" with the expectation that the continued inspections would eventually drive down the remaining risk. The inspection and removal mitigation had been very effective over the last six flights. There had been no missing or protruding gap fillers from reworked areas. Because the mitigation was proving effective, the inspected areas were now considered unexpected debris, and the risk level was reduced to "remote catastrophic." The aft areas of the orbiter that had not been inspected were still considered expected debris.

The orbiter team had made great strides in lowering the putty repair risk by performing visual and tactile inspections and removing and fixing faulty repairs. The PRA was not ready for the debris summit, but I expected the orbiter debris risk to be even lower than the ice risk due to the smaller size and shallower impact angles compared to ice. There was no transport to the windows and limited transport to the RCC panels. I recommended reducing the risk level from infrequent to remote based on the mitigations, limitations placed on repair size, and elimination of the minimum gap conditions between the repair and adjacent tiles. The community seemed somewhat divided on whether to go forward with the putty repair risk reduction given the work that remained on determining a useful PRA. I agreed to keep the same risk level until the PRA was complete.

Approximately 1,500 ceramic inserts were installed on each orbiter to provide access to fasteners that attach the thermal protection system carrier panels to the structure. Testing that had been ongoing since the STS-116 DVR suggested the high vibroacoustic environment on the base heat shield and body flap area drove the failure mechanism at those locations. There had been no ceramic insert losses since STS-116, and all the inserts forward of the OMS pods were in the process of being removed after the orbiter team determined they were not needed. The DTA for OMS pod losses showed impacts to the main engine nozzles at 175 foot-pounds, but ceramic insert impact tests performed by the SSME project established a new limit of 208 foot-pounds. This new limit enabled the main engine to clear ceramic insert impacts deterministically. Ceramic inserts on the aft OMS pods and base heat shield were considered expected debris, but inserts and plugs everywhere else on the orbiter were considered unexpected debris. Based on the strength of the controls in place and flight performance since RTF, the risk classification for ceramic inserts was lowered to "remote catastrophic."

Liftoff Debris Assessment

Every flight since RTF had experienced multiple liftoff debris sources that exceeded a limit. This was the case despite extensive mitigation efforts, foreign object debris (FOD) prevention, debris elimination procedures, and pad walkdowns. Liftoff was essentially a controlled explosion, and the extremely high vibroacoustic levels coupled with 7 million pounds of propulsive force imparted on a forty-year-old launchpad exposed to saltwater mist from the Atlantic Ocean made it difficult (if not impossible) to eliminate liftoff debris. Characterization and classification of liftoff debris came from historically known sources, ground-based video and film imagery, final inspection team reports, and postlaunch walkdowns. Plume-driven debris during the first few seconds of liftoff was the most significant risk driver and was difficult to identify using pad walkdowns or imagery. The plume would quickly obscure cameras used to observe liftoff debris and verify existing controls. The use of infrared cameras for liftoff debris was being investigated to augment the visible light imagery and improve understanding of the liftoff environment during the time frame that plume-driven debris was a threat. Liftoff debris computational fluid dynamics modeling and debris transport had been under development by MSFC since February 2006, but not much progress had been made due to the modeling complexities involved. I was not optimistic about the model's ability to calculate a liftoff debris PRA and was inclined to stop work altogether to focus all resources on mitigation efforts. No matter how much modeling was done, the only real liftoff debris risk reduction would come through mitigation. MSFC had not made a case to lower the risk level. This made it likely the "infrequent catastrophic" risk classification would remain for the rest of the shuttle program.

Integrated Debris Hazard Report (IDBR)

The DIG's goal was to close the integrated debris hazard report for the remaining missions. This would eliminate the routine of automatically opening the report after each flight and closing it before the flight readiness review of the next mission. Of course, if a debris event exceeded a risk assessment, the IDBR would be opened until the issue was resolved. The debris community expressed some concern over designating all debris risks (except liftoff debris) as "remote catastrophic" because it could create complacency. A number of people felt strongly the shuttle program should remain "hungry" during the remaining flights. I responded by stating that the program would stay committed to debris risk reduction, further modeling refinements, testing, and maintaining the level of expertise required to respond to day-of-launch and on-orbit MMT needs for analysis and assessment regardless of the risk classification levels.

I concluded the summit by praising the debris team for their dedicated work and accomplishments. However, I also told the team the game was not over until the last shuttle flight landed safely. If there was one thing we could count on with debris, it was to expect the unexpected. Fortunately the team had been able to deal with all the unexpected events and make the shuttle a safer vehicle to fly. The debris team needed to continue its vigilance because there were still unquantified risks such as secondary debris strikes, rebound impacts, combined failure mechanisms, and possibly even unknown failure mechanisms. Even though the debris risks were low, we could still have a bad day. I reminded the audience that every day someone wins a lottery or gets struck by lightning despite the low probability.

I finished up by sharing a story about friends of a family who were planning to have children. They decided two kids were all they wanted, regardless of the children's sexes. After the birth of their first child, a girl, they quickly set off to have another. Much to their surprise, they found out they were having twins—approximately a one in one hundred possibility. Although three children were more than they planned for, it was not the end of the world, and they enthusiastically welcomed the twins. After the birth of the twins, the father elected to get a vasectomy to eliminate any further procreation. After all, three kids were more than they had planned. A little over a month after the vasectomy, the couple was shocked to discover she was pregnant again with another set of twins. The odds of pregnancy after a vasectomy were greater than one in ten thousand, and the odds of a producing two sets of twins were well over one in one thousand. Despite the long odds, they would have five kids in diapers for a long time.

The roar of laughter filled the room as I went on to explain how the story related to the debris and shuttle program. The team gave the vehicle a metaphoric vasectomy by removing, mitigating, or redesigning high-risk debris sources to the point of a very low ascent debris risk. As we continued to fly there was always a possibility a catastrophic debris event could take place. In other words another set of twins was still possible even after a vasectomy.

My Final Flight Readiness Review, STS-124

The debris special topic at the STS-124 flight readiness review held on May 13, 2008, was a summary from the debris summit, results from the completed mixed debris testing at the Aerospace Corporation, and the final results of the flange PRA. All the mixed debris impact testing was completed. This included the test conditions designed to replicate the STS-118 impact conditions in terms of mass, impact velocity, impact angle, and percentage of mixed debris. Results showed the foam-only damage model was a good estimator for the ice and foam mixed debris. However, the mixed ablator and foam debris penetrated the tile much deeper than the foam-only damage model predicted. This was more consistent with what was observed on STS-118 and suggested the damage was due to a mixed ablator–foam combination instead of ice-foam. It was a moot point, though, because the ablator for the remaining flights was removed. PRA for the flange produced a tile risk of 1 in 3,330, an RCC risk of 1 in 2,500, and tile-shear risk less than 1 in 10,000. The results were based on an updated ET project debris cloud and cryoingestion release timing model. This included the eighty-five-second loss. The release rate was also artificially increased to replicate the RTF release rate and reproduce the release rate of the larger masses.

The tremendous amount of debris assessment work since the STS-116 debris verification review had increased the understanding of the debris environment. The numerous vehicle debris design modifications and mitigations had significantly reduced the debris risk. Despite the model updates, uncharacterized debris events such as mixed debris impacts, rebound impacts, secondary debris impacts, combined failure modes, and unknown failure modes were not modeled and considered accepted risk by the program. I expected the debris hazard report to remain closed for the remainder of the shuttle program with liftoff debris as the lone remaining "infrequent catastrophic" risk. No other debris issues or open work were identified for STS-124, and there were no questions during the entire FRR presentation.

The NESC provided an assessment at the end of the review. It was consistent with Ralph's observation over the various missions. According to Ralph, "The methodology and tools developed for the probabilistic approach to debris were excellent. I think the change in philosophy we had from just fixing the proximate cause to fixing the highest-risk debris sources was the fundamental change needed to get the program to aggressively and proactively work the debris issues. I think the PRA numbers provided a good relative comparison of the risks, but I did not believe they represented the absolute values of the risks." Ralph was comfortable with the debris hazard report so long as the program continued to mitigate the highest-risk debris sources. "Although the tools were good, what changed over time was our understanding of the failure modes. That greatly improved our ability to quantify the risks."

Flight of the New External Tanks

ET-128 was assigned for the STS-124 mission. It was the first "new" tank to fly. Modifications for this tank included all-new foam applied using improved application processes, inspections, and controls. The bipod was redesigned to eliminate the large foam ramps as a debris source, and electric heaters to prevent ice formation supplemented the structure. An enhanced closeout procedure was incorporated into the flange to decrease the possibility of cryoingestion failures. The liquid oxygen feed line bellows were reshaped to include a "drip lip" that allowed moisture to run off. This prevented ice formation. A strip heater was added on the forward bellows to further reduce the amount of ice or frost formed. The PAL ramp was removed after STS-114 to eliminate that large debris source. Design changes were incorporated at all seventeen ice/frost ramp locations on the liquid hydrogen tank to reduce foam loss. Although the redesigned ramps appeared identical to the previous design, several process changes were made. Manual spray foam was applied in the ramp's base cut-out to reduce debonding and cracking. The pressurization line and cable tray bracket feet corners were rounded to reduce stresses, and the shear pinholes were sealed to reduce leak paths and foam cracking. In addition titanium thermal isolators were primed to promote adhesion, and the corners were rounded to help reduce the foam stresses.

ET-128 was also the first tank to fly with redesigned liquid oxygen feed line brackets. Titanium brackets replaced all the aluminum brackets because titanium's thermal conductivity is less than aluminum's. This modification minimized ice formation in underinsulated areas and reduced the amount of foam required to cover the brackets. Teflon was added to the upper outboard monoball attachment on the feed line bracket to eliminate ice adhesion and bridging that could lead to foam cracks. Additional foam was added to the liquid oxygen feed line to further minimize ice formation along its length. The engine cutoff sensor feedthrough connector on the liquid hydrogen tank was modified with pins and sockets soldered together to prevent an open circuit and erroneous readings. It was designed to be the safest tank ever flown.

Supporting My Last Shuttle Flight, KSC

The strange events that occur before launch have always amazed me, and STS-124 was no exception. Four days before the scheduled Saturday launch, the main toilet on the Russian segment of the ISS broke. That forced the three astronauts aboard to use the facility on the Soyuz capsule moored at the orbiting station. The main toilet worked for solid waste disposal but had an intermittent problem handling liquid waste. After three flushes the toilet required ten minutes of maintenance work by two crewmembers. "It was very inconvenient at that time because it required a lot of manual intervention," said Kirk Shireman, the deputy manager of the International Space Station program.

The shuttle program worked with the Russians and ISS officials. Everyone moved quickly to get the spare parts and a pump together in time to be sent up on *Discovery*.

During the countdown no debris issues emerged, and *Discovery* was launched on the first attempt from Pad 39A on May 31, 2008, in the early evening. The shuttle and station crews would install the thirty-seven-foot, thirty-two thousand-pound Japanese pressurized module science lab named Kibo. It would be installed on the left side of the *Harmony* connecting node and opposite the European *Columbus* science lab that had been installed in February. As *Discovery* sped off into the distance, I was sad this would be my last flight before accepting a new assignment at NASA headquarters in Washington, DC. In another month I would be taking a job as the deputy chief engineer for the Exploration Systems Mission Directorate. This position had technical oversight of the Constellation program and future lunar missions.

In 2004 President Bush announced the Vision for Space Exploration, which centered on the Constellation program. It was seen as a response to the *Columbia* disaster and a way to regain public enthusiasm for space exploration. The Vision sought to implement a sustained and affordable human and robotic program to explore the solar system and beyond and extend a human presence across the solar system. It would start with a human return to the moon by 2020. The lunar mission would be preparation for human exploration of Mars and other destinations. In pursuit of these goals, the Vision called for the space program to complete the ISS by 2010, retire the space shuttle by 2010, develop a new crew exploration vehicle (later renamed Orion) and conduct its first human spaceflight mission by 2014, explore the moon with robotic spacecraft missions by 2018, and launch crewed missions there by 2020. Lunar exploration would provide opportunities to develop and test new approaches and technologies useful for supporting sustained exploration of Mars and beyond. Watching *Discovery* head off over the horizon made me think back to how I felt after my last wrestling match, graduation day at the Naval Academy, and my final day in the military. I shrugged off my reminiscence and headed back into the Launch Control Center for the postlaunch tradition of beans and corn bread. There was already a large, bubbly crowd in the vestibule congratulating each other while serving up helpings of hot food. Mike Griffin and his wife, who attended the launch, congratulated me on the flight and all my debris work. I was looking forward to my headquarters assignment but would miss launching shuttles. However, before I shipped off to Washington, DC, I still had some unexpected debris work to finish.

Liftoff Debris Issue

During *Discovery*'s liftoff the east wall of Pad 39A's north solid rocket booster flame trench suffered significant damage. A large section of the wall (approximately eighteen feet high by seventy-three feet in length) disintegrated and liberated thousands of refractory brick pieces. These bricks were scattered from the flame trench out to beyond the pad perimeter fence about a half mile away.

RETURN TO FLIGHT

The picture shows the large area of the flame trench wall that was damaged. Thousands of bricks were scattered about the launch pad. (Courtesy of NASA.)

All the previous RTF launches had liftoff debris issues but nothing even close to this. It was, I guess, my parting debris gift from the shuttle program. Brick debris evidence was gathered in the form of imagery reports, postlaunch walkdown inspections, and reviews of all the ground camera imagery. That included infrared and radar. The goal was to determine whether any of the flame trench debris traveled from below the mobile launchpad, where the shuttle was launched, up toward the vehicle. After intensive review, there were no documented or observed instances of flame trench debris reaching the vehicle or causing damage. On-orbit inspections confirmed there was no significant tile or wing leading edge damage. Some of the imagery would eventually be used to correlate flow field features and qualitatively validate the liftoff debris CFD models. Various CFD models were applied conservatively to postulated transport mechanisms, and debris transport analysis was performed on the brick fragments liberated from all locations along the flame trench. The root cause of the refractory brick liberation was determined to be operational degradation after decades of service launching the space shuttle and Apollo rockets. No service life had ever been established for the flame trench due to inadequate or nonexistent shuttle flame trench design requirements. After the Saturn rocket and shuttle launches, the wall strength had deteriorated over the years to the point of failure.

Postflight Debris Assessment, KSC

Discovery completed its fourteen-day space station assembly mission and landed at KSC on June 14, 2008. The orbiter sustained a total of 147 impacts. Seventeen had a dimension of one inch or larger. Even with the new tank, the total number of impacts was still above the shuttle flight experience before the *Columbia* accident, but the total number of hits with one dimension greater than one inch was less than the average of previous flights. I and others in the program and ET project were relieved with the performance of the new tanks. "Not so sure it was relief, but I was definitely anxious about the new tank performance," declared Jeff Pilet. "We had a lot at stake, and I did not want to have to go back to the program at the next flight readiness review and explain any bad or unexpected performance."

All the new tank modifications and designs performed well and as expected. There were no observed losses from the liquid hydrogen IFR body, adjacent acreage, feed line bracket, or feed line bracket yoke. The tile damage was the lowest to date, and there was minimal damage observed on the BRI-18 tile around the umbilical doors. Only four integrated IFAs were opened during STS-124—two for ascent debris and two for liftoff debris.

Flame Trench Debris

I was determined to close all the STS-124 IIFAs before turning over the debris responsibilities to Doug Drewry. The next mission, STS-125, was the Hubble Space Telescope repair mission. It was planned for an October 14, 2008, launch. Due to the unique nature of the mission and lack of a safe haven, I told the Dons and John Shannon I would support the mission from headquarters and would travel to KSC or back to Houston if needed.

The two ascent integrated IFAs were minor and quickly closed, but the main focus of the DIG shifted toward addressing the flame trench debris. All the flame trench debris transport assessments showed no threat to the vehicle for debris below the mobile launch platform. This was consistent with the STS-124 video evidence. It showed that of the more than five thousand-plus pieces of flame trench debris released, none made it to the vehicle or mobile launch platform. Personnel at Kennedy inspected the mobile launch platform and found no evidence of impacts. The MSFC liftoff debris team and the USA Graybeards came to the same conclusion that there was no threat to the vehicle from brick debris traveling through the solid rocket booster flame holes in the mobile launch platform.

Preparations to repair the flame trench were already under way and would not affect the HST launch schedule. The repairs would be extensive, and they needed to protect the underlying pad concrete structure from the plume environment and preclude large losses of refractory bricks. Repairs included replacing the area where refractory brick was lost with Fondu Fyre and securing weak control joints with anchor plates. Once completed Pad 39A flame trench repairs would prevent large losses of refractory bricks or Fondu Fyre. It would remain classified as unexpected debris above the mobile launch platform deck. The risk assessment, which applied to both Pads 39A and 39B, suggested that catastrophic damage to the vehicle due to flame trench debris was "improbable." In addition the flame trench walls, main flame deflector, and bottom sections of the mobile launch platform would be instrumented with pressure transducers, vibration sensors, and temperature sensors to monitor the structural integrity of the launchpad and determine if there was a problem. The liftoff debris IFA for the flame trench brick was closed based on the strength of the debris transport analysis.

Leading My Last DIG Meeting

On July 23, 2008, I led my last DIG meeting before heading to headquarters. Darby and the Boeing team completed the putty repair PRA for release locations chosen to envelop putty repair losses since RTF. This included putty repairs on the nose and forward sidewall locations. The PRA results showed a very low risk to RCC (one in twenty-five thousand) and to tile (one in three hundred thousand). Although there was transport to the orbiter windows, there had been no putty repair releases from the areas that threatened windows. The DIG recommended lowering the putty repair risk to "remote catastrophic" based on the PRA and orbiter mitigations. The orbiter team would continue with their inspections and replace any suspect items with new repairs. If it was not feasible to replace a putty repair, the entire tile would be replaced, even though there was a possible schedule impact. Ray Gomez had updated all the ASTT curves, and they would be used for quick assessments during a mission. This would minimize the time needed to determine whether a debris release was a threat to the vehicle. This was critical for the upcoming Hubble mission. The ET project presented final producibility enhancement recommendations to the DIG. These reduced tank processing time from ninety days to less than thirty. All the producibility changes were targeted at reducing the processing time with minimal debris risk impact. The DIG agreed with the producibility recommendations.

It was time to turn over the debris reins to Doug and head for Washington.

CHAPTER 28

The Perfect Debris Storm

Hubble Repair Mission Overview

For *Atlantis* the STS-125 mission would be the final shuttle flight to NASA's Hubble Space Telescope. Although it was technically the fifth servicing flight to the telescope, the eleven-day mission was referred to as Servicing Mission 4 (SM-4). Its seven-member crew would enhance the observatory and thereby ensure the cutting-edge astronomy that had lasted nearly two decades would continue. The mission would integrate advanced technology designed to improve the telescope's discovery power by ten to seventy times. Five space walks were planned to install new instruments and thermal blankets, repair two existing instruments, refurbish subsystems, and replace gyroscopes, batteries, and a unit that stored and transmitted science data to Earth. If all went well, Hubble would have six working, complementary instruments with new capabilities and an extended operational life span through at least 2014. *Endeavour* would be prepared as the backup vehicle for *Atlantis* in the unlikely event it needed a rescue flight. *Endeavour* would remain on Launchpad 39B while *Atlantis* executed its mission. Once the shuttle was cleared to return to Earth, *Endeavour* would be moved to Pad 39A for its next flight, STS-127, to the space station. In the history of the shuttle program, it would be one of the few times two shuttles were simultaneously positioned on the pads and ready for launch. It would also be the last.

Hubble Mission Postponed

Atlantis and its twenty-two thousand pounds of Hubble cargo were in the final days of launch preparations on September 27, 2008, when HST suffered a permanent electronic system failure. The A side of the Science Instrument Command and Data Handling System (SI CDH) failed. It provided all the electronics to command Hubble's science instruments from the ground and transmit science and engineering data back to Hubble's control center. Because this system was such a critical part of Hubble's science capability, the mission was postponed to allow engineers enough time to prepare a spare SI CDH for inclusion on the servicing mission. Meanwhile, in order to restore science operations to the orbiting telescope, flight controllers on the ground successfully switched to the B side of the SI CDH electronics along with several additional spacecraft data system modules. The last-minute failure caused a seven-month postponement that moved STS-125 up to the 126th space shuttle flight, the thirtieth flight for *Atlantis*, and the second flight in 2009.

Griffin Is Replaced
NASA HQ, Washington, DC, January 20, 2009

While I was on assignment at NASA headquarters, the shuttle program continued to fly. With the Hubble mission postponed, a five-flight manifest was scheduled for 2009. This flight tempo was the most demanding since RTF. The Constellation program was also in full swing and entering into the preliminary design review phase for each element. Not long after Barack Obama was elected president, many in the agency predicted there would be a change of leadership starting with Mike Griffin. I supported NASA's Obama transition team with a series of studies that involved extending the shuttle beyond 2010 and accelerating the Constellation program. With the space shuttle scheduled to retire in 2010 and the first Constellation flight scheduled for 2015, there would be a five-year period where NASA would have to rely

RETURN TO FLIGHT

on its Russian partners to fly our astronauts to the space station at 40 to 50 million dollars per seat—a price that would continue to escalate—on the Soyuz. The five-year time frame became known as the "human spaceflight gap," which was unacceptable within the agency. The studies were focused on reducing the gap by extending the shuttle out to 2012 and accelerating the first Constellation flight to 2013, but reducing the gap would take money—lots of it. Some estimates to achieve this objective were as high as an additional 3 to 4 billion dollars per year for the next two to three years. With the nation struggling with arguably the worst financial crisis in its history, a substantial budget increase for NASA seemed highly unlikely. This was especially true considering the lower-cost alternative of flying American astronauts on the Soyuz. Eventually President Obama decided to replace Griffin with Chris Scolese, NASA's associate administrator and the agency's highest-ranking civil servant. It became effective January 20, 2009. Chris served as the acting administrator until July 2009 when Charlie Bolden replaced him. Charlie was a former astronaut and retired US Marine Corps major general whom the president nominated and the US Senate confirmed.

I sat in the HQ audience when Mike gave his farewell speech. He graciously thanked his colleagues and restated his passion for the space program. As he was finishing his speech, I wondered who would replace him and how long it would take. For Mike, there were really no tough decisions associated with debris. It always seemed to him that the proper course of action was obvious. "Fixing the foam involved a lot of engineering blocking and tackling that was made more difficult—mostly because it had been neglected for so long," stated Mike. "The shuttle program was always trying to make up for a lot of things all at once that, had they been addressed as they occurred early in the program, would have caused much less concern."

The debris work was a very challenging and interesting effort and one Mike followed quite closely. However, he did not consider that foam or other debris issues posed any "tough" decisions. "Frankly," said Mike, "I thought the ECO sensors were a lot more annoying than the foam. It took us so long to get them to malfunction and then stay broken so the root cause could be understood."

Ed Mango, NASA's recovery director after the *Columbia* accident and orbiter deputy program manager, and his team did a wonderful job determining the root cause once they finally had a "body" upon which to perform an "autopsy." The potential weakening of the RCC panels on the wing leading edges surfaced late in Mike's tenure. It was mostly as a result of better inspection methods, and it caused him greater concern than anything else on the shuttle program. As for Mike's legacy at NASA, it was for others to determine whether or not he had any long-term impact on the agency that he loved and the enterprises it carried out. I, like many others, was sad to see him go and began to worry about the direction of NASA's human spaceflight program.

The Augustine Committee

Even before a new NASA administrator was named, the Obama administration announced on May 7, 2009, the launch of an independent review of planned US human spaceflight activities with the goal of ensuring the nation was on a rigorous and sustainable path to achieving its boldest aspirations in space. A blue-ribbon panel of experts led by Norman Augustine would conduct the review. Norman was the former CEO of Lockheed Martin, and he had served on the President's Council of Advisors on Science and Technology under Democratic and Republican presidents. He led the 1990 Advisory Committee on the Future of the US Space Program. The "Review of United States Human Spaceflight Plans" was to examine ongoing and planned NASA development activities and present options for advancing a safe, innovative, affordable, and sustainable human spaceflight program in the years following the retirement of the Space Shuttle Program. The panel would work closely with NASA and seek input from Congress, the White House, the public, industry, and international partners as it developed its options. The committee was tasked to present its results in time to support an administration decision by August 2009.

Launch of *Atlantis* and the Hubble Mission

Under the leadership of Chris Scolese, the agency pressed forward with the next scheduled shuttle launch. Just hours prior to the Hubble mission launch on May 11, 2009, the final inspection team observed several ice formations on the outside edge of the liquid hydrogen umbilical. At the time of the inspection, the temperature, relative humidity, and light wind were conducive to ice formation. All the ice formations (except one that was 12 inches by 0.5 inches by 0.5 inches) were within the NSTS 08303 limits. This was the largest piece of ice the FIT had ever seen around the umbilical and was big enough to warrant a launch scrub. As the analysis was being gathered for flight rationale, the ice began to melt into three sections and was smaller than originally observed. Even though the completed analysis of the single large piece was conservative, the results were still borderline acceptable. Approximately an hour before the launch, the final inspection team was sent back to the pad for a more detailed look at all the ice around the umbilical. The team confirmed that most of the ice, including the large piece that had melted into thirds, was within acceptable limits. The vehicle was cleared for launch. Space shuttle *Atlantis* launched from Pad 39A on the first attempt at 2:01 p.m. Right after liftoff the remaining ice and frost around the umbilical fell harmlessly away from the orbiter as it had done during STS-120 and during the ice coating testing at Eglin. At 104.2 seconds into ascent, the ELVIS camera observed a dark-colored piece of debris traveling aft along the starboard fuselage and underneath the starboard wing. The imagery from another camera did confirm the single piece of debris impacted the orbiter, continued aft, and passed under the starboard wing. Although ascent imagery could not confirm the debris source, it was believed to be a small piece of foam from a liquid oxygen ice/frost ramp that liberated from station Xt-718. Radar confirmed both the debris source origination and orbiter

THE PERFECT DEBRIS STORM

impact. This was the only significant debris event, and it occurred just outside the peak ASTT time frame.

The MMT and debris team would have to wait two more days before getting results from the on-orbit inspections to determine the extent of debris damage. In the meantime the ASTT curves were assessed, and debris transport analysis was performed. These showed no catastrophic threat to the vehicle. During the servicing of Hubble, the astronauts conducted careful inspections of *Atlantis*'s exterior using the shuttle's fifty-foot-long orbiter boom sensor system attached to its forty-nine-foot-long robotic arm. Much to the relief of everyone involved with the mission, no significant damage from either launch or the days in space was found. Once mission managers gave the orbiter a clean bill of health, *Atlantis* was cleared for reentry, and *Endeavour* was released from standby.

Rare picture of two shuttles (*Atlantis* and *Endeavour*) ready for launch. (Courtesy of NASA.)

The HST servicing mission involved five intense space walks before the release of the refurbished telescope. By the end of the last space walk, all the mission objectives to improve Hubble's view of the universe and extend its life had been accomplished. The days before landing provided an opportunity for the crew to enjoy some off-duty time. It also gave them chances to speak to President Barack Obama, the International Space Station crew, and reporters back on Earth. They even testified before a US Senate committee—a first-time event from space. The STS-125 crew left the telescope ready to dazzle the world for years to come. More scientific discoveries and stunning images were now possible because of its improved view that stretched from our solar system to the far reaches of the universe. On May 24, 2009, the orbiter landed at Edwards AFB due to inclement weather at KSC. This flight demonstrated the capability and value of all shuttle debris assets and proved a mission could be successfully completed without the safe haven of the ISS.

Numerous Launch Scrubs for STS-127

After being released from its Hubble rescue mission duties, *Endeavour* was immediately moved from Pad 39B to 39A in preparation for the June 13 launch of STS-127. This space station assembly mission was scheduled to last sixteen days. This made it one of the longest of NASA's Space Shuttle Program. The main goal of STS-127 was completing the assembly of Japan's Kibo laboratory complex and delivering space station parts for future use. Five space walks and intricate robotics work by the shuttle and station crews were scheduled for this mission. The first launch attempt was on June 13, and it was scrubbed due to a gaseous hydrogen leak in the ground umbilical carrier plate. A second attempt four days later was also scrubbed for the same reason. The troubleshooting efforts that had been successful before did not resolve the problem this time around. After a lengthy delay to investigate and solve the gaseous hydrogen leak problem, *Endeavour* was rescheduled to launch on July 11. The last day at NASA headquarters for my rotational assignment was June 30. This meant I would be in Houston for the next launch attempt. My family and I enjoyed our time in Washington, DC, but were looking forward to our return home to Houston. The agency was still waiting for a new administrator to be named and for recommendations from the Augustine Committee, but at least the shuttle debris problems appeared solved.

Although I was about to begin my assignment to the Constellation program, I planned on watching the launch from mission control and seeing old friends. A severe thunderstorm on July 10 forced another delay to July 12 to allow teams to evaluate the effects of lightning strikes to the vehicle and launchpad. There had been eleven strikes within a 0.3-mile radius of the launch pad. That required a formal review to determine if the strikes were big enough to declare a lightning event. Of the eleven strikes, seven hit the shuttle lightning protection system catenary mast—the tallest object towering over the launchpad and vehicle. If a lightning event was declared, the shuttle would be delayed even further to allow the launch control team to evaluate all the shuttle's electrical systems. Although a lightning event was not declared, the launch attempts on July 12 and 13 were scrubbed during the T minus nine minute hold due to various weather-related flight rule violations. Fortunately the gaseous hydrogen leakage that had plagued the launch team before was no longer an issue. The post-drain walkdown inspections after each of the scrubbed launch attempts showed no launchpad anomalies and confirmed the facility was in good condition. The final L minus one day walkdown was conducted on July 14, and the vehicle was cleared for launch.

On July 15 there were weather violations throughout the entire day, but the weather cleared right before the launch window was reached. This was the first time in more than a year I had been present in mission control, but on this occasion I was relegated to watching the mission instead of working it. Sitting in mission control reminded me how much I really missed the excitement of launch day. Weather conditions at KSC during the tanking phase were conducive for icing and resulted in typical ice formations on the tank. At the time of the final inspection (two hours before launch), the ambient temperature was 83°F and relative humidity was 88 percent with light winds. Heavy rain occurred late during the final inspection. This washed away most of the ice and frost observations the ground cameras noted earlier. All that remained was a piece of ice classified as "slush" that had formed on the bottom of the liquid hydrogen umbilical. DTA cleared the umbilical "slush" formation, and space shuttle *Endeavour* finally lifted off from Pad 39A on its sixth attempt.

The External Tank Sheds a Shower of Foam Debris

For the first one hundred seconds into flight, *Endeavour* flew through the worst of the ASTT without any debris issues. It was just as it had done on the Hubble flight. At approximately 106 seconds, though, everyone snapped to attention as a large spray of foam appeared outboard of the liquid oxygen feed line and forward of the intertank flange. Three white spots appeared in the black tile after the debris spray impacted the orbiter fuselage. This indicated tile damage. The foam debris spray then traveled aft along the orbiter fuselage and out of view of the cameras. Surprisingly no additional impacts were observed as the remainder of the debris traveled along the fuselage and fell both over and under the wing. This was just the beginning of the debris shower. Everyone watched a steady stream of foam debris release every few seconds for the next minute in what became known as the "debris storm." Several debris releases during this time frame hit the orbiter. I was stunned. At 115 seconds, a single large piece of debris was observed falling aft and striking the orbiter on the starboard chine area. The largest pieces impacting the orbiter ranged in size from 18.5 inches by 4.0 inches to 9.0 inches by 2.0 inches. These were correlated to foam loss from the intertank acreage at station Xt-990. There were at least a dozen such foam losses from the intertank region during the ascent of STS-127.

At 126 seconds into flight, another single, large piece of debris was observed releasing near the port side bipod closeout. The bipod debris fell outboard and aft away from the tank and did not appear to contact the orbiter. A split second later, a piece of foam debris appeared to impact the forward liquid oxygen feed line fairing and break into several smaller pieces. Although there was no definitive visual evidence of the release location, it was believed to be missing foam from the aft section of the liquid oxygen IFR at station Xt-718. It was one of the rare times since RTF that a liquid oxygen ice/frost ramp had failed. There were also two other bipod foam releases that struck the orbiter—one at 158 seconds and the other at 166 seconds. Even though these two releases hit the orbiter, they were not a damage risk because the impacts were well past ASTT.

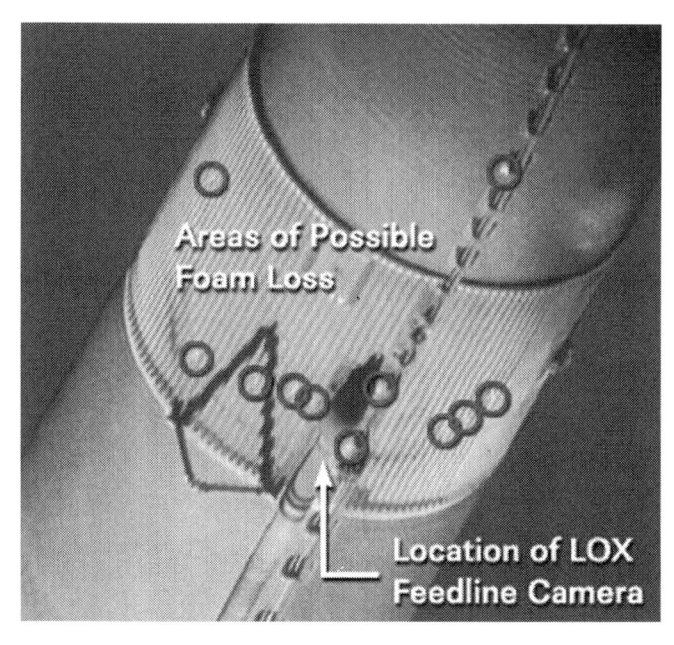

The circles indicate all the locations where large pieces of foam released from the tank on STS-127. (Courtesy of NASA.)

Although the debris storm was finally over, the shuttle team would have to wait for the on-orbit inspections to determine the extent of the damage and whether there was a reentry issue. I ran into Darby in the MCC hallway, and all he could say was, "Here we go again."

Darby and I both had some serious doubts about how well the ET project was able to control its processes and what that might mean to flying out the remaining missions. Up until that point, the debris performance of the new tanks over the last several flights had been excellent, but these failures were the worst I had ever seen. Jeff Pilet and the ET project were stunned as well. "This flight was challenging to say the least. It turned out to be my most memorable debris experience but not from a good standpoint," stated Jeff.

From their position in Firing Room Two at KSC's launch control complex, the eyes of the ET team were fixed on the ET feed line camera until tank separation. The intertank losses really showed up in the feed line camera and made for quite the visual impact. "The sighs from the other folks in the firing room still resonate with me, and watching many large pieces of foam liberate and impact the orbiter in real time is something I'll never forget," Jeff said.

There was a substantially high number of imagery observations declared as "reportable," or threats to the vehicle, emanating from the STS-127 launch. Overall there were eighty-six compared to an average of thirty-nine from the twelve previous missions. Most were related to the unusually large amount of external tank foam liberated and the resulting debris. I was part of a big crowd hovering around the imagery console in the MER as the external tank separation images were downlinked. The separation images confirmed the release locations observed during ascent. There were over a dozen foam losses of various sizes from the intertank, and both the port side bipod and liquid oxygen ice/frost ramp losses exceeded their maximum predicted sizes. Something had gone terribly wrong.

Working Shuttle Debris Again

On flight day two, the rendezvous pitch maneuver verified damage to the orbiter tile and provided an early assessment of the debris damage sustained during ascent. Before the MMT there was a lot of hallway discussion about the debris events and whether the shuttle fleet would be grounded again as it was after the PAL ramp loss on STS-121. The Dons wanted me to attend the MMT and asked if I would help the shuttle program resolve the STS-127 debris issues. I said yes but needed to clear it with my new Constellation boss, Lauri Hansen.

Surprisingly the MMT was silent as Dave Melendez from the Imagery Integration team showed pictures and videos of all the losses. Perhaps everyone was still in shock. The only thing John Shannon could muster at the end was, "Hmmm, that's interesting."

Although there was nothing official, I assumed the shuttle would be grounded. Later that evening I heard CNN report the fleet was grounded due to debris issues. Later in the day at a press briefing, John Shannon responded to questions. "We were a little bit surprised by foam loss in the intertank area," he said, "because of the structural strength and minimal deflection. It also does not experience the extreme [low] temperatures you get in the liquid hydrogen tank. So we don't typically expect to see large losses in that area. Normally the intertank experiences only popcorning, in which small air bubbles near the surface pop off from the ascent heating. What we saw here," Shannon continued, "were strips of the foam covering the intertank structure...It just kind of peeled off the primer layer of the metal, and you can actually see the metal underneath it. We don't understand why that happened. It looks like the base primer just was not holding onto the foam well."

At least ten areas of foam loss could be seen in the intertank area on the side facing the shuttle with another five possible areas on the opposite side. Even though some of the debris releases occurred during the tail end of ASTT, all the debris reconstructions showed negligible risk due to debris breakup post-ASTT.

Gerst was very concerned when he viewed the STS-127 imagery and saw all the failures. "My initial response was how fragile the vehicle was and how we should not be flying it through this type of debris environment," he recalled.

After the launch Gerst traveled from KSC to JSC. He reviewed the initial set of debris imagery and saw the debris releases, impacts to the orbiter, and other areas on the tank that debris had damaged. Until the orbiter and debris teams worked these debris events through their inspections and assessment processes, it was anyone's guess about what these failures meant in terms of vehicle risk. Only after each debris release was evaluated did Gerst finally realize the vehicle could withstand more than the models predicted. Debris evaluation took time and was tough emotionally on the managers as they anxiously waited to learn the extent of damage, whether additional inspection would be required, and if a repair or rescue mission would ultimately be needed. Gerst did not put much stock in the idea of a backup flight because he figured the same problems the first orbiter encountered might happen to the rescue vehicle. Then they would be left with two damaged vehicles and additional crewmembers in peril. "The only option at that point would be repair, and if that was the case, why jeopardize two vehicles?" Gerst said. "I had gained more confidence in the repairs based on the testing and flight demonstrations that were performed and tested in the arc jet facility."

By flight day five, I had worked with the orbiter team to assess all the tile impacts. Despite the damage MMT cleared the vehicle for entry as is. Despite what was reported in the press, Gerst and the agency never declared the shuttle program grounded. Instead he opted to work the debris issues through the normal integrated IFA process. As was the case with any IIFA, the shuttle was effectively "grounded." However, it was only until all the IIFAs were closed. There was no "official" announcement. Either way the debris issues with the bipod, liquid oxygen ice/frost ramp, and intertank foam losses had to be resolved before the next mission, STS-128, would be cleared to fly. The postlanding debris inspection of *Endeavour* was completed at KSC on July 31 right after landing. Surprisingly the orbiter sustained a total of only 107 impacts. Twenty-one had a dimension of one inch or larger. Even with all the foam failures and impacts, both damage totals were below the historical averages. This was attributed to most of the debris releases occurring outside of ASTT.

Determining the Cause of Failure, MAF, New Orleans

I spent the month of July working with the ET project and debris team to determine the root cause and corrective action for the STS-127 foam failures. This was needed to generate the flight rationale required to close the integrated debris IFAs and hold the launch date for STS-128—August 28. The ET project scheduled a two-day debris meeting at MAF starting on July 29 to present results of their investigation on the foam losses. A lot of questions needed to be answered, and several unusual circumstances needed to be reconciled with the tank on this flight. The tank had been subjected to five launch scrubs and six propellant loads, severe weather that included hail and multiple lightning strikes around the vehicle, and possible foam adhesion contamination issues due to exposure during Hurricane Katrina. During the two-day meeting, all aspects of application and process performance, failure modes and critical environments, flight and test results, and possible design, process, and inspection enhancements were covered for the liquid oxygen ice/frost ramp, bipod closeout, and intertank acreage.

Although it felt like the old DVR and debris summit days with multidiscipline participants throughout the agency and contractor community involvement, I sensed a change in attitude toward debris. The emotion and level of concern for a serious debris event had subsided. It was probably due to a combination of things affecting the program and the agency. Charlie Bolden had just been confirmed as NASA's new administrator during the STS-127 mission and did not appear to have the same level of shuttle debris interest as Mike Griffin did. Perhaps it did not seem reasonable to ground the fleet with only a few missions left before retirement. In addition, there might have been pressure to continue space shuttle operations due to the Soyuz reentry concerns. There was also fear the shuttle program would be canceled if grounded in favor of accelerating the Constellation program, and grounding the fleet might leave a bad impression with the Augustine Committee, which was still assessing NASA's human spaceflight future.

Liquid Oxygen Ice/Frost Ramps

During the meeting the ET project started with the liquid oxygen ice/frost ramp debris releases. They included a listing of the various debris failure mechanisms. There was a slight trend toward an increase in liquid oxygen IFR foam debris releases since implementation of a horizontal application process designed to reduce the processing time. The

STS-127 loss might have been the result of subsurface cracking and/or reduced strength the horizontal pour process induced. However, dissection data showed no discernible differences, and the void delta P analysis predicted the time of release after 260 seconds given the observed depth. This was inconsistent with the observed release time on STS-127, but previous test and flight experience showed foam loss was most likely to occur in multiple, smaller pieces, which gave the appearance of a larger crater. In other words, the foam could have initially failed at a shallower depth and then eroded away to a much larger cavity. Booster separation motor plumes from the SRB separation were also evaluated, but this was not identified as a contributing factor to the foam loss. Nondestructive evaluation performed on ET-132 (the next tank scheduled to fly) showed no anomalies or cracks. The ET project concluded void delta P failure was the most likely candidate and recommended all horizontally poured ice/frost ramps undergo nondestructive evaluation before flying. In addition the ET project presented a different mass distribution in the debris cloud. This provided a more realistic representation of the post-RTF history. The PRA using the new debris cloud for tile and RCC went from one in ten thousand to one in one thousand. This shifted the risk up in the debris hazard report from remote to "infrequent catastrophic"—at least until additional cycling testing and process improvements were investigated.

Bipod Foam Debris

Bipod debris had been a chronic problem since RTF started. Most losses were attributed to cryoingestion caused by a leak path along the bipod heater wires. The same was the case for STS-127. However, the release mass was larger than expected and beyond the maximum predicted. The time of release was consistent with the physics-based understanding of cryoingestion failure. The most likely cause was ingestion of liquid nitrogen into a void or subsurface crack near or around the bipod heater wire routed from the intertank. That was a known reservoir. The increase in mass was likely related to additional subsurface cracking or gaps between the wires caused by the multiple cryocycles from the launch scrubs. The crack growth was consistent with thermal vacuum testing. Nondestructive evaluation performed on the ET-132 bipod at KSC also showed no anomalies, cracks, or voids. The ET project recommended continued NDE and planned to review the heater wire harness installation process to determine if any additional enhancements were warranted. Even though the mass distribution was increased to account for cryocycling, the PRA remained unchanged because the timing observed during flight put losses generally too late to pose a significant hazard. The risk classification in the debris hazard report would remain unchanged and classified as "remote catastrophic."

Intertank Acreage Foam Debris

Other than popcorn foam debris, foam loss from the intertank acreage was unexpected. The area was vented and not susceptible to large voids or void delta P failure. The most logical explanation was an adhesive failure mode, but this raised the question why a similar failure had not been seen on other flights. The ET project utilized a fishbone failure analysis technique to evaluate the intertank foam loss cause and contributing factors. No violations to the process as written were uncovered, but the ET project suspected a combination of inadequate technician experience, poor precleaning, and the potential for primer overspray dust, could result in poor foam adhesion. The ET-131 tank, which was flown on STS-127, also experienced a record number of accumulated rain and lighting events—more than any other tank in the program's history. Contamination prior to foam application could cause a weak bond between the foam and primed substrate. On ET-131 there was a higher potential for contamination prior to foam application due to the long tank processing timeline from tank structure completion to foam application. The tank structure (built in 2003) was then moved several times around the plant until the foam was finally applied in 2007. After Hurricane Katrina the plant also lost environmental control for approximately one month. Although there was no evidence of direct weather exposure to the tank, there was the potential for salt contamination from a higher atmospheric content due to the storm. If salts were present on the tank surface from Katrina or other exposure instances, the solvent used to clean the tank did not effectively remove it.

Another significant factor contributing to the adhesion failure was inadequate cleaning prior to foam application due to the experience levels of cleaning technicians. Although all technicians were trained and certified, some were new on the job, and there was a learning curve even for the experienced employees. It had been nearly four years since a technician had applied intertank spray to a tank. ET-131 was the first since RTF. None of the severe weather exposure and environmental concerns at KSC (rain, suspected hail damage, high winds, and lightning) were identified as having any correlation to the foam adhesion failure mode. Based on physical understanding of the bond adhesion failure mode where the interface stress was dominated by vacuum conditions and ascent heating effects, the time of release for the adhesive failure was expected to occur relatively late in ASTT— past one hundred seconds. Even with the late release timing, if the adhesive failure mode was considered, the PRA risk for tile and RCC soared from one in ten thousand to one in two hundred. Because of the increased risk, modifying the cleaning process and solvent and conducting additional pull tests and inspections on all remaining flights would mitigate adhesive failure.

I struggled with the explanation of the failure mechanism being attributed to contamination, improper cleaning, and inexperienced application. A similar issue with contamination had been addressed after Hurricane Katrina and was associated with removing bug spray used to control the mosquito outbreaks. Some of the processing buildings at MAF had been compromised, and there was a concern that some of the spray landed on the exposed tank surfaces. The ET project did an informal pull test of foam sprayed on surfaces with and without bug spray. Surprisingly the pull test results were about the same, and the contamination issue had been closed. The decision to perform pull tests on the remaining flights was the only thing that made me comfortable with the risk, and I knew it would be an area of potential contention at the upcoming flight readiness review.

STS-128 Flight Readiness Review

Gerst and others at the flight readiness review were really concerned about the flight rationale heading into STS-128 because of the numerous foam losses on STS-127 and the determination of the reasons behind the failure. Jeff Pilet presented the ET project summary starting with the liquid oxygen ice/frost ramp. The IFR loss was not surprising, but when it struck the orbiter, it became a concern. Consequently the ET project performed an extensive amount of work to show the foam defect void size and frequency for the ramps were consistent with expectations and what was being used in the debris cloud for PRA. Nondestructive evaluation was expanded to include the liquid oxygen IFRs to ensure the potential for void-driven foam loss events would be minimized. The bipod loss, which was larger than the documented maximum mass limit, was due to cryopumping. The release timing primarily drove that. This was consistent with the understanding of the failure mode. Although the resulting debris mass for cryopumping failures was difficult to estimate, the maximum expected mass limit was increased based on the STS-127 tank performance. Even with the increase in mass, the bipod debris risk was shown to be acceptable based on the late expected time of release. Jeff and the ET project concluded that the observed intertank performance was most likely due to inadequate surface preparation prior to the foam application. The tank on STS-127 was the first intertank sprayed with foam after Hurricane Katrina, and a less experienced crewmember had cleaned it. On all future flights, the intertank foam would be subjected to pull tests to ensure adequate adhesion and bond strength.

Even with the pull testing, several organizations still were uncomfortable with the tank safety. This included Helen and her team at

Marshall and Roy with the ISERP. "I was very concerned about the intertank debris losses since we had never seen any losses from that area of this magnitude before," explained Helen. "We were not confident in the flight rationale that was subsequently developed, and we submitted a dissenting opinion against the 'go for launch' recommendation from Don Noah and SEI."

Ralph and the NESC team generally agreed with the flight rationale and also recommended NDE of all the ice/frost ramps prior to the STS-128 flight. "As the CAIB pointed out, our success prior to *Columbia* might have made us complacent, and we did not want to fall back into that trap," Ralph said. "Foam application is very process-sensitive, and as the foam releases from STS-127 reminded us, we needed to maintain the post-*Columbia* vigilance even after a string of very successful flights."

Roy and the ISERP agreed with elevating the risk classification of the liquid oxygen ice/frost ramps from remote to "infrequent catastrophic" and leaving the bipod risk classification as "remote catastrophic" because the impact conditions were generally benign with kinetic energy well below the tile damage thresholds and RCC capabilities. However, the foam debris losses on STS-127 from the intertank acreage indicated an unacceptable debris risk. "There was great similarity between the tank cleaning processes prior to foam application for STS-128, which clearly had limitations in controls that could result in localized areas of contamination undetectable by plug pull tests," stated Roy. "The increase in risk coupled with a release rate of at least one debris release per flight made the likelihood of a catastrophic ET intertank foam loss on STS-128 a high 'infrequent' to 'probable' [unacceptable] risk."

I sat patiently in the audience and watched as Doug Drewry for SEI presented the foam PRA results. Doug and I were expecting a lot of questions, and he was more than happy to defer any relating to the PRA to me. However, Doug sailed through the PRA presentation with only a single question and a few comments. The ET project presented after SEI and monopolized most of the meeting time. Jeff provided sound analytic data and explanation behind the liquid oxygen IFR and bipod foam losses. Then, after lengthy discussion, he attributed the STS-127 intertank foam failures to process deficiencies and contamination. However, the key to his flight rationale were the pull test results on STS-128. These showed the foam adhesion was normal. Gerst stated he would have ordered the tank stripped and the foam reapplied had the pull test showed weak adhesion. "Looking back on the flight, I was reminded that, as a manager, I should be extra skeptical because the tendency of a team is to relax after fixing or resolving a problem when another one might be looming on the horizon," Gerst said. "It's an effective technique for avoiding groupthink."

He went on to add that STS-127 served as a reminder we all needed to remain diligent as managers and keep probing and questioning the work we were doing on debris. Gerst and the FRR Board acknowledged the dissenting opinions. However, the shuttle program managers accepted the explanations, flight rationale, and risk. A few weeks later, after two launch scrubs, STS-128 was launched on August 28, 2009.

CHAPTER 29

End of an Era for Human Spaceflight

Shuttle Flight Operations

The tank performance on STS-128 was excellent. Not a single intertank foam release was recorded, and all the LH_2 and LO_2 ice/frost ramps were intact except for nominal popcorning and erosion as a result of ascent loads and temperatures. A small foam release was discovered on the LH_2 intertank flange closeout, and two occurred around the bipod area, but all were well outside ASTT. After STS-128 six remaining flights were on the manifest with the possibility of an additional flight designated STS-135. The STS-135 mission, if not needed as a rescue flight for STS-134, would carry a four-member crew, a fully loaded multipurpose logistics module, and other supplies to the ISS. Having a crew of four meant that if they became stranded on the ISS, they would stay on the station and rotate back to Earth on the Russian Soyuz. In this scenario NASA would not need another shuttle on standby for a rescue.

On August 20, 2010, shuttle program managers approved the additional flight from a safety and logistics standpoint. If Congress gave final approval for funding one more shuttle mission, *Atlantis* would be targeted to launch in early 2011. Although the shuttle was set to be retired by 2010, Congress approved the funding for flight operations into 2011 and the additional STS-135 mission. Before these flights could get off the ground, however, a series of delays due to technical problems with the external tank and (to a lesser extent) the payload affected the STS-133 mission. This caused the launch, initially scheduled for September 2010, to be pushed back to November and then finally February 2011. Consequently the remaining three flights of the Space Shuttle Program would all be flown in 2011. The last flight was scheduled for July.

Over the course of the remaining seven flights, the external tank performed flawlessly, and no other significant debris events were recorded. The new solvent, cleaning, and foam application procedures implemented on STS-128 for the intertank acreage proved effective and eliminated any further losses. After the *Columbia* accident, the program systematically implemented changes to improve the debris performance and lower the real risk, which was more important than the final risk number. Gerst and others were interested more in how the understanding of the debris performance and failure modes improved than in the PRA numbers. Although the ET project conducted more and more nondestructive evaluation of the foam, it never could provide an effective way to predict performance. In some cases the NDE would discover a foam defect, yet the foam did not fail in flight. In other cases the NDE found no foam defects, yet the foam failed. "It's an area I didn't fully understand, and that always bothered me," Gerst said. "It would always remain a concern when using nondestructive evaluation in the flight rationale."

What did improve in terms of foam loss was the understanding of the different failure modes. The PRA also did not capture the full risk to the vehicle if a debris failure were to occur. For example, one could determine the PRA for the gap filler debris risk to the orbiter, but this did not include the gap filler risk of tripping the boundary layer and exposing the vehicle to higher reentry temperatures than the tiles and structure could withstand. The shuttle needed to keep flying in order to understand the debris environment and performance and uncover any debris failures that had not been accounted for in the PRA. The key message when dealing with debris was to remain vigilant and aggressive at lowering the risk.

Conclusions from the Augustine Committee

The Augustine Committee provided a summary report to the NASA administrator on September 8, 2009. Its final report followed on October 22, 2009. In its summary, the Augustine Committee concluded the following: "The US human spaceflight program appears to be on an unsustainable trajectory. It is perpetuating the perilous practice of pursuing goals that do not match allocated resources. Space operations are among the most complex and unforgiving pursuits ever undertaken by humans. It really is rocket science. Space operations become all the more difficult when means do not match aspirations. Such is the case today." The report recommended NASA delay putting astronauts on Mars because it was too hard and costly. Instead NASA should consider going to the moon or building rockets that could zip around the inner solar system and visit asteroids or a Martian moon. NASA should keep the International Space Station going until 2020 rather than crash it into the Pacific in 2016. The report suggested it could help underwrite commercial spaceflight the same way the United States gave the airline business a boost in the 1920s with airmail. The committee determined the return on investment of ISS to both the United States and the international partners would be significantly enhanced by an extension of the station's life to 2020. It seemed unwise to deorbit the station after twenty-five years of assembly and only five years of operational life.

For the Space Shuttle Program, the plan was to retire it at the end of fiscal year (FY) 2010. Its final flight would be scheduled for the last month of that fiscal year. Although the requirement to complete the last mission before the end of FY 2010 was relaxed, Congress had to approve additional budget for continuing shuttle operations in 2011. The committee acknowledged that once the shuttle was retired, there would be a gap in America's capability to launch humans into space. That gap would extend until the next US human-rated launch system became available, which the committee estimated would be at least seven years based on NASA's current plans. This would the longest gap in US human launch capability since the US human space program began.

Constellation Program Canceled

Not long after the release of the Augustine Committee report, the cancellation of the Constellation program was announced. The cancellation announcement on February 1, 2010, officially ended aspirations to return astronauts to the moon by 2020 and shelved President Bush's Vision for Space Exploration that had been developed in the aftermath of the *Columbia* accident in 2003. In place of the moon mission, President Obama's space vision offered very little in terms of human exploration of the solar system. What the administration called a "bold new initiative" did not spell out a future destination or timetable for getting there. The Orion crew capsule would continue development under another name (multipurpose exploration vehicle), but no mission or designated launch was set. The Ares I rocket was dismissed in favor of a heavy lift vehicle operating under a program called the Space Launch System, which was also without a mission and would require years of development. For many NASA employees, a program without a mission meant a destination to nowhere with a high likelihood of future cancellation. Instead of preparing to use the Constellation's Ares I rocket and Orion crew capsule to ferry astronauts to the ISS, some of the funding would instead go to financing "space taxi services" from commercial companies. If implemented NASA tomorrow would be fundamentally different than NASA today because the space agency would no longer operate its own spacecraft. Instead it would buy tickets for its astronauts from domestic commercial providers or international partners.

The Final Shuttle Flight

I felt sad as I walked over to mission control to observe the last space shuttle flight. In a crowded room, I watched the final countdown with the rest of the SEI team as I had done for eleven missions after the *Columbia* accident. It was surprisingly quiet in the control room because there were no vehicle problems or any debris issues. It was

RETURN TO FLIGHT

a bittersweet moment for everyone—especially those who had spent their entire careers primarily supporting the shuttle program. All were alone with their thoughts. The majority of the contractors who supported the shuttle program knew layoff notices were pending immediately after the mission was over. Their futures were even more unclear and probably would involve career changes and no longer working on human spaceflight. After the debris storm on STS-127, the tank performance had been exceptional, and the vehicle had only had a few minor debris issues to resolve over the final six flights. The final inspection team found no discrepancies and cleared the vehicle for launch. On Florida's Space Coast, nearly a million spectators gathered along the beaches, rivers, and causeways to watch history in the making. Despite a gloomy prelaunch weather forecast on July 8, 2011, space shuttle *Atlantis* thundered off Launchpad 39A at NASA's Kennedy Space Center at 11:29 a.m. (EDT). The liftoff marked the last space shuttle launch after thirty years of operations. The thirteen-day mission carried more than 9,400 pounds of spare parts, spare equipment, and other supplies in the *Raffaello* multipurpose logistics module. This included 2,677 pounds of food. The supplies were delivered to sustain space station operations and six crewmembers for the next year. On the return leg of the mission, the twenty-one-foot-long, fifteen-foot-diameter *Raffaello* brought back nearly 5,700 pounds of unneeded materials from the space station. With the mission accomplished, *Atlantis* and its crew undocked from the orbiting space station and headed home. On July 21, 2011, the shuttle landed at 5:57 a.m. in Florida and brought to a close the space shuttle era.

The once-bustling halls on the seventh floor of Building One at JSC, home to the shuttle program's Systems Engineering and Integration Offices, were now quiet, and all the offices are empty.

There was a lot I missed about working on the SEI team and supporting the shuttle program. I missed John's enthusiasm, Gerst's wisdom, Mike's intellect, Wayne's patience, Roy's diligence, Justin's humor, and the dedication of the men and women who supported the DIG and worked to resolve all the debris problems encountered during flight. Similar to my service on the nuclear submarine, I missed the camaraderie, the technical challenges, and the level of responsibility as the DIG chair. There were numerous memorable challenges such as the PAL ramp loss on STS-114, the fierce debate over the IFR risk during the STS-121 flight readiness review, the freak hailstorm and on-orbit drama with STS-117, and the reentry debate concerning the mixed debris damage on STS-118. All were now part of the Space Shuttle Program's legacy. But the most memorable time for me was 2007 due to the unusual events and technical challenges. The program challenges during this time reminded me of my personal struggles in 1995 when my first marriage failed, surviving a plane crashed, retaking my PhD board and moving to Houston. Looking back on it, both the program and my personal struggles were overcome with persistence and fortitude.

Commercial Crew Program

About nine months after the Constellation program was canceled and before the last space shuttle flight, I took a position as the risk manager on the newly formed Commercial Crew Program. Ed Mango, NASA's recovery director after the *Columbia* accident and deputy manager of the orbiter project for several RTF missions, was selected as the first manager of the Commercial Crew Program. His deputy was Brent Jett, the former astronaut who directed recovery of the deceased *Columbia* crew. The US government, through NASA, was investing in the development of a US commercial crew space transportation capability with the goal of achieving safe, reliable, and cost-effective access to and return from the International Space Station. The program objective was to foster the development of a certified end-to-end crew transportation system for use in low Earth orbit. This would effectively close the human spaceflight gap and reduce the reliance on the Russians. Through this development and certification process, NASA would be helping lay the foundation for future commercial space transportation capabilities for both NASA and other customers. Once NASA certified the transportation capability, the agency could purchase transportation services to meet its ISS crew rotation and emergency return obligations. It was clear after the Constellation program was canceled in February 2010 and the release of the National Space Policy in June 2010 that President Obama and NASA were committed to building up a robust and competitive commercial space sector.

The National Space Policy in 2015 states that the United States is committed to encouraging and facilitating the growth of a US commercial space sector that support US needs, is globally competitive, and advances US leadership in the generation of new markets and innovation-driven entrepreneurship. The Commercial Crew Program would approach meeting these objectives through use of a Space Act Agreement—a set of legally enforceable promises between NASA and the other party. Space Act Agreements would not be subjected to normal federal acquisition regulations, and this would allow NASA to invest in the development of commercial crew space transportation capabilities via performance milestone-based instruments that support design, development, testing, and demonstration of commercial crew transportation systems. The commercial partner to an agreement would be responsible for providing any necessary supplemental investments for the development, and NASA would only pay if the milestone was achieved.

This approach was not without precedent. In 2005 the Commercial Cargo Program Office was established to provide an alternate commercial source to deliver cargo to the ISS. There were also several commercial companies up for the challenge. On December 8, 2011, SpaceX became the first commercial company in history to reenter a spacecraft from Earth orbit. SpaceX launched its *Dragon* spacecraft into orbit atop a Falcon 9 rocket from Launch Complex Forty at Cape Canaveral. *Dragon* orbited the Earth, reentered the atmosphere, and landed less than one mile from the center of the targeted landing zone in the Pacific Ocean. It was a monumental achievement and served as a demonstration that it could be done with a crewed mission. In May 2011 the Commercial Crew Program Office was officially established to provide a commercial source for transporting NASA crews to the ISS. It is serendipitous now that I work in Building One. The Commercial

Crew Program now occupies all the former Space Shuttle Program offices. The shuttle conference room, which is lined with all the mission patches, crew pictures, and accomplishments, and where all the debris debates took place is now used to discuss Commercial Crew Program issues.

Debris Retrospective

The process of determining, assessing, and characterizing debris risks following the *Columbia* accident in February 2003 was an enormous undertaking by NASA. Numerous improvements were made after STS-107 including but not limited to: removal of the PAL ramps, redesign of the bipod closeout, additional liftoff and debris mitigations, better understanding and characterization of all the foam failure modes, additional foam and ice testing, model refinements to remove conservatism and improve accuracy, and more validation to increase confidence in the analysis tools. The risk acceptance rationale was a careful synthesis of all the available information with the estimated risk indexes being just one part of the final rationale. Although improvements had been made, some limitations still existed when the shuttle program retired in all areas of debris release, aerodynamic transport, and impact tolerance. Although a remote possibility, the potential for exceeding the risk assessment documented in the debris hazard report was always present as was the possibility of uncovering an additional failure mode. As the Space Shuttle Program continued to monitor the debris performance of the launch vehicle of all flights after *Columbia*, high debris risk sources were mitigated based on the additional knowledge gained from flight data.

The risk from the 1.6-pound foam released on STS-107 was reduced by a factor of more than one thousand. The highest foam debris risk to the RCC panels had been reduced since RTF from approximately 1 in 250 to 1 in 2,500. The highest foam debris risk to tile was reduced from 1 in 300 to 1 in 3,330. The forward bellows heater eliminated a 1 in 250 ice risk. After the removal of the bipod ramp and protuberance air load ramp after STS-114, the largest piece of foam released since RTF was more than twenty-five times smaller than the 1.6-pound foam piece that struck *Columbia*. On average the volume of bottom-surface orbiter tile damage decreased by a factor of two since RTF. This was attributed to a reduction in the amount and size of foam debris. All "probable catastrophic" debris risks had been eliminated, and the twenty-five "infrequent catastrophic" risks were reduced to one: liftoff debris. Since RTF the overall integrated debris risk for the shuttle had been reduced from probable to "remote catastrophic."

Wayne Hale frequently gave the same speech about safety and flying the space shuttle. He iterated time and again that the shuttle was not safe to fly. It was not even remotely close to what the ordinary person would call safe. Spaceflight in general was a very risky business that operated at the extreme limit of technological capability. In terms of debris, the best solution, in theory, was always to build a spacecraft that was not susceptible to debris liberation during the launch phase. The shuttle was not safe and needed to be replaced with a safer (but fundamentally still *un*safe) spacecraft—one that would, in all likelihood, be less capable than the shuttle. The trade-off was not between safe and unsafe but between risk and gain. Was the purpose of the flight worth the risk that would be incurred? "If you want to be safe," Wayne would say, "stay in bed."

Those who were working for the human spaceflight program during the *Columbia* accident were profoundly affected. Many who had close relationships with crewmembers suffered great personal loss and were left wondering if enough had been done to try to prevent the disaster from happening in the first place. Since RTF my answer would be an unconditional yes. Future space exploration will not be without risk. Yet, the very future of humankind might well come to depend on space exploration and planetary colonization. The space environment is hazardous and extremely unforgiving. This will most likely lead to other disasters as the space program advances. There are certainly many challenges ahead, but Gus Grissom, the Apollo 1 commander who was killed in a launchpad fire, summed it up best. "If we die, we want people to accept it. We're in a risky business, and we hope that if anything happens to us, it will not delay the program. The conquest of space is worth the risk of life."

Perhaps one day in the future, humans will once again view Earth from another celestial body just as the Apollo astronauts did.

The losses of space shuttles *Challenger* and *Columbia* and their crews were stark reminders of the inherent risks and severe challenges spaceflight and space exploration pose. In preparation for future human exploration, we must advance the ability to live and work safely in space while developing the new technologies to extend humanity's reach to the moon, Mars, and beyond. These technologies also might provide applications that could be used to address problems on Earth. Like the explorers of the past and the pioneers of flight in the last century, we cannot identify all we stand to gain from space exploration. Like their efforts, the success of future US space explorers will unfold over generations. I am confident, nonetheless, the eventual return will be great. To explore space is the destiny of the human race, but the real question is whether humankind will choose to do so.

Printed in Great Britain
by Amazon